Poultry Meat Processing

Edited by
Alan R. Sams, Ph.D.
Department of Poultry Science
Texas A&M University

CRC Press
Boca Raton London New York Washington, D.C.

FIRST INDIAN REPRINT 2005

Library of Congress Cataloging-in-Publication Data

Poultry meat processing / Edited by Alan R. Sams
p.cm.
Includes bibliographical references and index.
ISBN 0-8493-0120-3.
1. Poultry—Processing. I. Title.
TS1968 .S36 2001
664'.93—dc21

00-046763
CIP

Direct all inquiries to CRC Press LLC, 2000 N.W. Corporate Blvd., Boca Raton, Florida 33431.

Visit the CRC Press Web site at www.crcpress.com

International Standard Book Number 0-8493-0120-3
Library of Congress Card Number 00-046763

Printed at Brijbasi Art Press Ltd., I-72, Sector-9, Noida, U.P., India.

Dedication

This book is dedicated to the memories of Dr. Pam Hargis and Dr. Doug Janky, two individuals who each had a profound and lasting impact on the poultry, food, and nutritional sciences, as well as the people involved in them.

Preface

This book is the product of some of the best poultry and food scientists in the world today. Its concept was born from the need for a good instructional textbook in the poultry processing and product quality courses taught by many of the contributors. The text is an instructional and not necessarily exhaustive review of the scientific literature in each of its component areas. In addition to its teaching use, this book will also be a useful reference for academic researchers, industry personnel, and extension specialists/agents seeking further knowledge.

Most of the contributors are active participants in the S-292 USDA Multi-State Research Project, and the collaborative relationships fostered by this project have made this book possible. I thank the contributors for their time and meaningful input.

I am also deeply indebted to Mrs. Elizabeth Hirschler for her excellent technical and creative assistance, without which this book would not have been possible.

Alan R. Sams, Ph.D.
Editor

Contributors

James C. Acton
Department of Food Science
Clemson University
Clemson, SC

Sacit F. Bilgili
Department of Poultry Science
Auburn University
Auburn, AL

J. Allen Byrd
Southern Plains Agricultural Research Center
College Station, TX

David J. Caldwell
Departments of Veterinary Pathobiology and Poultry Science
Texas A&M University
College Station, TX

Muhammad Chaudry
Islamic Food and Nutrition Council
Chicago, IL

Donald E. Conner
Department of Poultry Science
Auburn University
Auburn, AL

Michael A. Davis
Department of Poultry Science
Auburn University
Auburn, AL

Paul L. Dawson
Department of Food Science
Clemson University
Clemson, SC

Glenn W. Froning
Department of Food Science and Technology
University of Nebraska
Lincoln, NE

Billy M. Hargis
Departments of Veterinary Pathobiology and Poultry Science
Texas A&M University
College Station, TX

Jimmy T. Keeton
Department of Animal Science
Texas A&M University
College Station, TX

Brenda G. Lyon
U.S. Department of Agriculture
Agricultural Research Service
Russell Research Center
Athens, GA

Clyde E. Lyon
U.S. Department of Agriculture
Agricultural Research Service
Russell Research Center
Athens, GA

Shelly R. McKee
Department of Food Science
and Technology
University of Nebraska
Lincoln, NE

William C. Merka
Department of Poultry Science
University of Georgia
Athens, GA

Julie K. Northcutt
Department of Poultry Science
University of Georgia
Athens, GA

Casey M. Owens
Department of Poultry Science
University of Arkansas
Fayetteville, AR

Joe M. Regenstein
Cornell University
Ithaca, NY

Scott M. Russell
Department of Poultry Science
University of Georgia
Athens, GA

Alan R. Sams
Department of Poultry Science
Texas A&M University
College Station, TX

Denise M. Smith
Department of Food Science
and Toxicology
University of Idaho
Moscow, ID

Doug P. Smith
Department of Poultry Science
University of Georgia
Athens, GA

Lei Zhang
Department of Poultry Science
Auburn University
Auburn, AL

Contents

chapter one

Introduction to poultry meat processing

Alan R. Sams

Poultry processing is a complex combination of biology, chemistry, engineering, marketing, and economics. While producing human food is the main goal of poultry processing, related fields include waste management, non-food uses of poultry, and pet/livestock feeds. When considering the global marketplace, poultry refers to any domesticated avian species, and poultry products can range from a slaughtered carcass to a highly refined product such as a frankfurter or nugget. However, because they dominate the market, chicken and turkeys will be the focus of this book. The common classes of commercial poultry are summarized in Table 1.1. The reader should remember that specific numeric processing conditions in this book are for illustrative purposes and that these conditions may vary between processors. The aims of this book are both to instruct the user in what steps/conditions are used for processing poultry and to explain why things are done that way. This approach will enable the reader to evaluate problem situations and develop possible solutions.

Commercial poultry is extremely uniform in appearance and composition. Tightly managed breeding, incubation, rearing, and nutritional regimes have created a bird that is a virtual copy of its siblings. This uniformity has allowed poultry processing plants to develop into highly automated facilities with an efficiency that is unmatched by other livestock processors. With line speeds of 70 to 140 chickens/min, uniformity, automation, and efficiency are recurring themes and have been keys to the success of poultry processing.

Table 1.1 **Common Classes of Commercial Poultry**

Class of poultry	Age (weeks)	Specifications
Cornish hen chicken	<4	≥25% Cornish and <2.0 lb processed
Broiler or fryer chicken	6–8	Most common commercial chicken
Roaster chicken	8–10	Large bird for whole holiday meals or boneless meat
Stewing hen chicken	52+	Breeder hen that no longer produces eggs at an economical rate
Fryer turkey	9–16	Young turkey usually sold whole
Roaster or young hen/tom turkey	16–24	Most common form of turkey; sold whole, in parts, or as boneless meat
Hen/tom turkey	52+	Breeder bird that no longer reproduces at an economical rate

0-8493-0120-3/01/$0.00+$.50

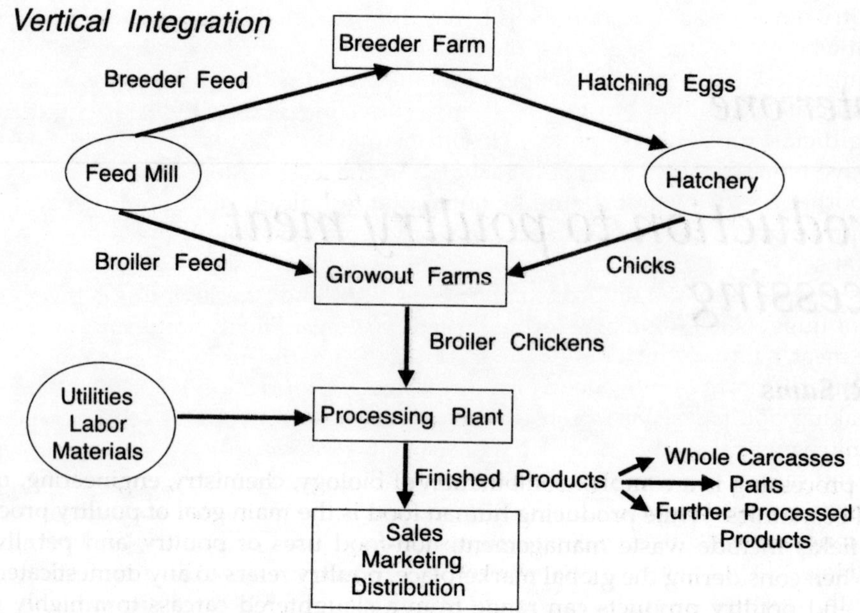

Figure 1.1 Diagram of the material flow between the components of a vertically integrated poultry company.

Poultry companies in the U.S. are vertically integrated. This is a system in which the same entity (e.g., company, cooperative, etc.) owns several (or all) steps of the production process from breeding through processing (Figure 1.1). Vertical intergration ensures maximum efficiency and uniformity. By reducing the number of times a component of the production system (feed, chick, labor, etc.) changes ownership, the profit charged at each level of change can be eliminated. Some poultry companies have taken the concept of vertical integration to a higher level by growing their own grain and purchasing interests in the breeding companies. Improved uniformity is another benefit that results from all parts of the production system having a common goal, a common set of specifications, and a common system of oversight.

The poultry industry is rapidly becoming global. A growing percentage of the U.S. poultry industry revenues come from exports of poultry products, particularly the ones such as dark meat and feet that do not have strong markets in the U.S. As a result, the industry in the U.S. has become keenly aware of the politics and economics of its major customer countries; Russia, Hong Kong/China, Japan, Canada, and Mexico. Although the U.S. is the world leader in poultry production, its industry is still concerned about conditions and any developments in poultry-producing nations with which it competes. Examples of important, competitive advantages in other producer countries include the large grain production in Brazil and the massive potential consumer market developing in China. In an effort to capitalize on some of the production and marketing advantages in various parts of the world, poultry companies based in the U.S. and other countries are establishing production operations in other regions of the world. Another emerging factor in the global marketplace is the development of trading blocks such as the North American Free Trade Agreement (NAFTA), the European Union, and South America's Mercosul. These alliances reduce or eliminate trade tariffs between member nations, standardize many requirements, and regulate trade within and outside of the alliances.

Poultry meat consumption in the U.S. has dramatically increased in recent decades to the point where it has the largest per capita consumption of any meat type. Several factors have contributed to this increased appeal of poultry. First, the fat in poultry is almost exclusively associated with the skin and is easy to remove in response to dietary guidelines for reducing dietary fat. This is contrasted with mammalian meats such as beef and pork, which have more of their fat actually included in the lean sections of the commonly consumed portions. However, it should be noted that technically, lean poultry and lean beef have approximately the same fat and cholesterol contents. The distinction is mainly the ease of fat separation. Second, the industry has been very responsive in developing new products to meet the changing consumer needs. A good example of this is the enormous success of nuggets and similarly formed, fried products. Finally, poultry is an extremely versatile meat, a factor which has possibly contributed to the product development efforts. Poultry meat is more homogeneous in composition, texture, and color than mammalian meat, making poultry easier to consistently formulate into products. When compared to beef, poultry meat also has a milder flavor which is more readily complemented with flavorings and sauces.

Economic production through vertical integration, favorable meat characteristics, and product innovations to meet consumer needs have all contributed to the poultry industry's success. However, the safety of poultry products and the use of water in processing are two issues with which the industry is concerned. Developments in live bird production, processing plant operations, product characteristics, and inspection systems are all being made to reduce bacterial contamination on the product and improve the product's safety. Likewise, the expense and environmental impact of using large quantities of water in processing and then cleaning that water before discharging it have all prompted intense study in these areas. The following chapters will provide the reader with an understanding of these and the many other areas involved in poultry meat processing.

chapter two

Preslaughter factors affecting poultry meat quality

Julie K. Northcutt

Contents

Introduction

Poultry production and processing involve a series of interrelated steps designed to convert domestic birds into ready-to-cook whole carcasses, cut-up carcass parts, or various forms of deboned meat products. The acceptability of poultry muscle as food depends largely upon chemical, physical, and structural changes that occur in muscle as it is converted to meat. During production and management of poultry, antemortem (preslaughter) factors not only exert important effects on muscle growth, composition, and development, but also determine the state of the animal at slaughter. Thus, events that occur both before and after death of poultry influence meat quality.

0-8493-0120-3/01/$0.00+$.50

Antemortem factors affecting quality

According to Fletcher,[1] antemortem factors affecting poultry meat quality may be divided into two categories: those having a long term effect and those having a short term effect. Long term factors are inherent, or they occur over the entire length of the bird's life, such as genetics, physiology, nutrition, management, and disease.[1] These factors will not be discussed in detail in this chapter; however, additional information may be obtained from the cited references.[2–5] Short term factors affecting poultry meat quality are those that occur during the last 24 hours that the bird is alive, such as harvesting (feed and water withdrawal, catching), transportation, plant holding, unloading, shackling, immobilization, stunning, and killing.[1] The remainder of this chapter will focus on addressing these short term antemortem factors, with the exception of immobilization, stunning, and killing, which are discussed in Chapter 3.

Harvesting

Birds are generally reared on litter (wood shavings, rice hulls, peanut hulls, shredded paper, etc.) in enclosed houses, with approximately 20,000 broilers per house or 6000 to 14,000 turkeys per house, depending on house size (Figure 2.1). In the U.S., most birds are grown on a contract basis. Under the terms of the contract, the producer (grower) provides land, labor, housing, equipment, utilities, and litter, while the company provides the birds, feed, and fuel to heat the house. The company then pays the producer according to bird performance.[6,7] Bird age at slaughter depends upon the end product (e.g., whole carcass, cut-up parts, etc.), but the majority of broilers are processed between the ages of 6 and 7 weeks, while turkeys are processed between 14 and 20 weeks of age.

Birds must be "harvested" before they can be processed, and this involves preparing birds for catching or collection, catching birds, and placing birds into containers (coops, crates, etc.). Figure 2.2 shows a schematic of the preslaughter steps including harvesting and up to the point where birds enter the processing plant. Some of the major preslaughter problems that may occur include bird injuries (bruising, broken or dislocated bones, and scratches), bird mortality, and bird weight loss due to feed and water deprivation.[8] These problems are important because they may result in reduced sales from lost or downgraded (not Grade A) products. Bird injuries and carcass defects will be discussed later in the chapter.

Figure 2.1 Typical commercial broiler house.

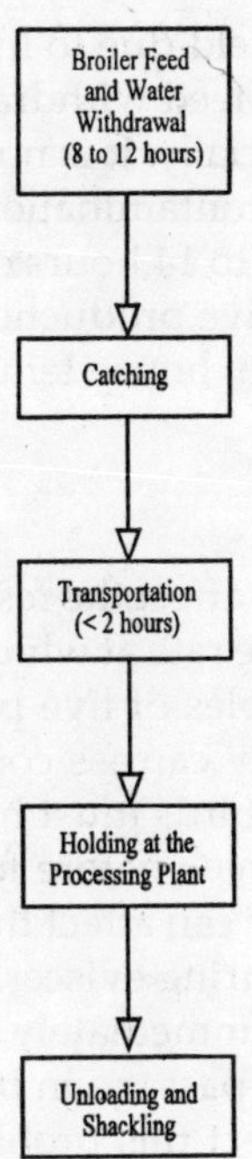

Figure 2.2 Short term preslaughter steps affecting poultry meat quality.

Feed withdrawal

Before birds are caught, loaded, and transported to the processing plant, feed and water are removed to allow time for evacuation of intestinal contents. Removal of feed and water, or feed withdrawal, reduces incidence of carcass fecal contamination which may occur during processing.[9–15] With the USDA's requirement of zero tolerance of carcass fecal contamination in the *Pathogen Reduction/Hazard Analysis and Critical Control Point System (HACCP)* ruling, length of feed withdrawal has become more important to the poultry industry. Zero tolerance of feces means that carcasses contaminated with visible feces are not allowed to enter the immersion chiller. This regulation is discussed in depth in Chapter 5.

Numerous factors influence the effectiveness of a commercial feed withdrawal program, making it extremely difficult to optimize such a program. Before discussing these factors, it is important to have a clear understanding of the definition of feed withdrawal, and the precise goals of a feed withdrawal program. Feed withdrawal refers to the *total length of time* the bird is without feed before processing. This includes the time the birds are in the grow-out house without feed, as well as the time the birds are in transit and in the live hold area at the processing plant.[16]

Length of feed withdrawal is important because it affects carcass contamination and yield, grower payments, processing plant line efficiency, and product safety and quality. Ideally, the length of feed withdrawal before processing should be the shortest amount of time required for the birds' digestive tracts to become empty.[9–14,16] However, this time varies because of differences in house environmental conditions and management practices which affect bird eating patterns. Recommended length of time off feed for broilers before processing is between 8 to 12 hours, while 6 to 12 hours is recommended for turkeys. These time periods are optimal because research has shown this is when the majority of the birds in the flock will have properly evacuated.[9,17] However, the withdrawal time is not so great that there is loss of excessive carcass yield due to live weight loss. Although 8 to 12 hours (broilers) and 6 to 12 hours (turkeys) of feed withdrawal is recommended, a variety of feed withdrawal schedules is used commercially. It is not uncommon to have some plants processing broilers with minimal carcass contamination using a 7 to 8 hour feed withdrawal schedule, while other plants require 12 to 14 hours of feed withdrawal to achieve the same results. For optimal feed withdrawal, live production management practices surrounding bird grow-out must be considered (e.g., house temperature, litter moisture, type of feed, house lighting, etc.).

Live production management

Live production management practices affect the results of feed withdrawal by altering the birds' eating patterns or by changing the rate at which feed passes through the bird's digestive tracts. Table 2.1 gives some examples of live production-related factors which affect broiler feed withdrawal, and ultimately carcass contamination. In order for a feed withdrawal program to work as designed, birds must have normal feed consumption pattern and normal feed passage during the week before feed withdrawal. Variation in bird size (uniformity) within a flock or over time can affect the efficiency of processing plant equipment, specifically at the vent opener during evisceration. Changes in lighting or temperature regimes (hot or cold), a disruption immediately after feed is removed, and the stressors of catching and holding can slow feed passage in birds. When the rate of feed passage is slowed, it may not be possible to correct this problem simply by holding the birds for a longer period of time before processing.[15,16] However, it is best for plants to process flocks with the potential for considerable contamination at the end of a shift when more time could be spent correcting the contamination problems.

Lighting and cooping

Lighting (intensity and duration) and cooping have been found to affect bird activity, and activity of birds affects the rate of feed passage.[11] Under continuous light and access to water, 60 to 70% of the intestinal contents will be evacuated during the first 4 to 6 hours of feed withdrawal (Figure 2.3).[17] However, when birds are exposed to darkness, or after birds are cooped, the evacuation rate is much slower. Research has shown that after a 2-hour feed withdrawal period, broilers in a dark environment had more feed in their crops than broilers in lighted environments (Table 2.2). After 4 hours of feed withdrawal, lighting made no difference in crop contents, except when it was combined with cooping. Cooped broilers held in darkness for 2 hours had more than twice as much feed in their crops than cooped broilers held in the light (Table 2.2). In addition, after 4 hours of feed withdrawal, there was

Table 2.1 Live Production-Related Factors Contributing to Carcass Contamination

- Lack of uniformity in flocks processed
- Differences in bird sizes over time or between shifts
- Excessively long plant holding time and conditions
- Communication problems with growers and catch crews
- Frequent feed outages, especially during the week prior to market
- Time of last feed and target amount of feed left in pans at feed withdrawal
- Policy on fate of left over feed in pans
- Excessive grower activity in house during feed withdrawal
- Extremes in house temperature during feed withdrawal

Source: Modified from Bilgili, S. F., *Broiler Ind.*, 61(11), 30, 1998, and Northcutt, J. K. and Savage, S. I., *Broiler Ind.*, 59 (9), 24, 1996.

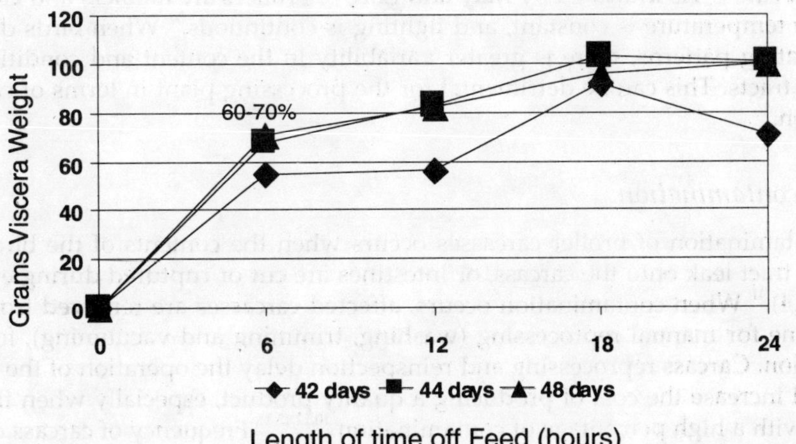

***Figure* 2.3** Effects of length of feed withdrawal on broiler viscera weight. (From Buhr, R. J., Northcutt, J. K., Lyon, C. E., and Rowland, G. N., *Poult. Sci.*, 77, 758, 1998. With permission.)

twice as much feed within the crops of broilers held in darkness compared to crops of broilers held in light.[11] For this reason, poultry companies usually leave birds in the grow-out house on litter with water, but not feed, for 2 to 5 hours before catching. It has been suggested that 4 hours of water consumption for broilers and 2 hours of water consumption for turkeys is optimal after feed withdrawal to allow feed passage from the crop. Longer time on water may cause excessive moisture in the intestinal tract, which increases the likelihood of carcass contamination during evisceration.

Environmental temperature

In addition to lighting and cooping, environmental temperatures have been shown to affect digestive tract clearance of broilers during feed withdrawal.[11,18] This may be related to the consumption of less feed during hot weather in conjunction with reduced bird activity. During the fall and spring when daily temperatures vary widely, birds may gorge

***Table* 2.2 Effects of Lighting and Cooping on the Crop Contents of 45-Day-Old Broilers; Weight of Crop Contents Following Feed Withdrawal**

Holding conditions	Lighting	2 hours (grams)	4 hours (grams)	6 hours (grams)	8 hours (grams)
Litter	Light	13.8[b,c]	2.3[b]	0.6[a]	0.2[a]
Litter	Dark	29.2[a]	4.0[b]	3.1[a]	0.5[a]
Cooped	Light	11.8[c]	6.0[b]	0.4[a]	2.1[a]
Cooped	Dark	21.0[b]	17.0[a]	3.5[a]	1.4[a]

[a–c]Means within a feed withdrawal time with no common superscript are significantly different.

Source: From May, J. D., Lott, B. D., and Deaton, J. W., *Poult. Sci.*, 69, 1681, 1990. With permission.

themselves in the evening after the sun goes down and temperatures begin to decline. If birds have gorged immediately before feed withdrawal, a normal withdrawal period may not be long enough.[16] Birds grown during cold weather with house temperatures below 15.5°C also retain feed in their digestive tracts longer, and the birds are often too cold to stand and eat.[11,18] As indicated by May and Lott,[19] "broilers are nibblers and eat regularly when the temperature is constant, and lighting is continuous." When birds do not have normal eating patterns, there is greater variability in the content and condition of their digestive tracts. This can be detrimental for the processing plant in terms of carcass contamination.

Carcass contamination

Fecal contamination of broiler carcasses occurs when the contents of the bird's crop or digestive tract leak onto the carcass, or intestines are cut or ruptured during evisceration (Figure 2.4).[11] When contamination occurs, affected carcasses are removed from the processing line for manual reprocessing (washing, trimming and vacuuming), followed by reinspection. Carcass reprocessing and reinspection delay the operation of the processing plant and increase the cost of producing a quality product, especially when flocks come through with a high percentage of contamination.[10,12,13,16] Frequency of carcass contamination depends upon the amount of material present in the digestive tract, the condition of the digesta (partially digested food and feces) remaining in the intestines (watery or firm), the integrity of the intestines, and the efficiency of the eviscerating equipment and plant personnel.[15,16]

To study the relationship between feed withdrawal and digestive tract contents, a study was conducted in which the intestinal tracts of 50 to 125 broilers from each of 3 different commercial plants in the U.S. were evaluated. The contents of the crop and gizzard were noted upon dissection, and gizzard bile was reported on a percentage basis. Intestinal shape was observed and recorded as: (1) round and containing feed; (2) flat and void of feed; or (3) round and containing intestinal gas. Table 2.3 shows the results of this study, and a discussion of the findings appears in the next sections.[14]

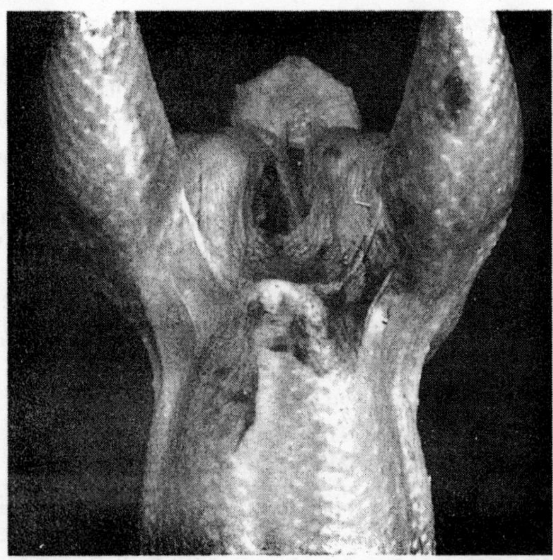

Figure 2.4 Fecal contamination of a broiler carcass.

Table 2.3 **Viscera Contents After Feed Withdrawal**

Time off feed (hours)	Crop contents	Gizzard contents	Intestinal shape	Sloughing of intestinal mucus	Gizzard bile (%)
0–3	Feed	Watery feed	Round	No sloughing	0
9	Water	Litter	Flat	Mild sloughing	30
12	Empty	Litter	Flat	Sloughing	30
14	Empty	Litter	Flat and round	Sloughing to heavy sloughing	35
16–19	Empty	Litter and feces	Flat and round	Sloughing to heavy sloughing	40–70

Source: From Northcutt, J. K., Savage, S. I., and Vest, L. R., *Poult. Sci.*, 76, 410, 1997. With permission.

Short feed withdrawal

When the length of feed withdrawal is too short (less than 6 to 7 hours for broilers, 4 to 5 hours for turkeys), the birds' digestive tracts will be full of feed at slaughter, and the intestines will be large and rounded (Table 2.3). For full-fed birds, the intestines take up a great deal of space in the abdominal cavity, such that the duodenal loop is positioned close to where the vent is opened for evisceration (Figure 2.5). For this reason, the feed-filled intestines are easily cut during vent opening. In addition, processing birds that are full of feed increases the likelihood that the force of evisceration will cause intestinal material to leak out onto the carcass.[14,16,19]

Long feed withdrawal

When the length of feed withdrawal is too long (greater than 13 to 14 hours), a number of problems may occur that increase the likelihood of carcass contamination. Mucus from the intestinal lining will be passed with feces (intestinal sloughing), possibly causing a loss of intestinal integrity. Weaker intestines have a higher incidence of intestinal tearing during evisceration. Figure 2.6 shows intestinal strength data of broilers after various feed withdrawal periods.[21] Intestinal strength of broilers has been found to be approximately 10%

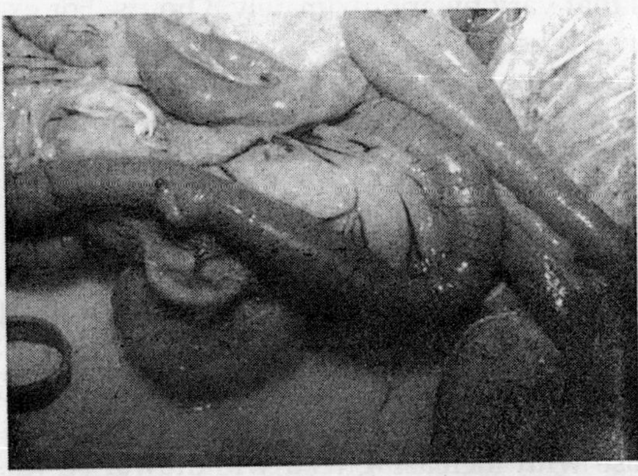

Figure 2.5 Large and rounded intestine from a full fed bird.

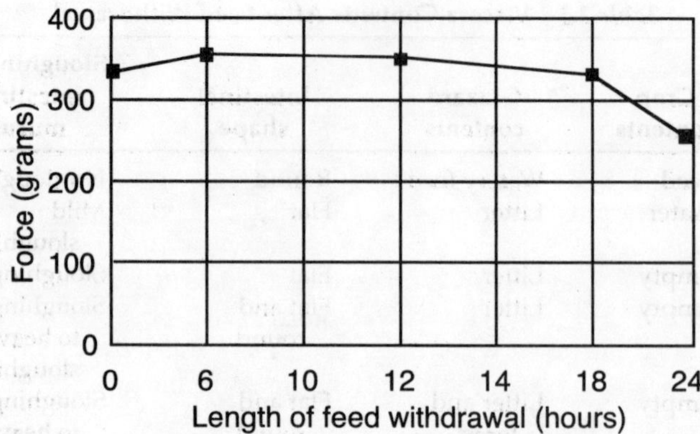

Figure 2.6 Intestinal strength of broilers held without feed for various times before processing. (From Bilgili, S. F. and Hess, J. B., *J. Appl. Poult. Res.*, 6, 279, 1997. With permission.)

lower when broilers were without feed for 14 or more hours before processing as compared to full-fed broilers. Moreover, male birds were reported to have stronger intestines than female birds.[21] In addition to weaker intestines, longer feed withdrawal times often result in bile contamination of carcasses because continuous bile is produced, and the gallbladder becomes enlarged. Enlarged gallbladders may be broken more frequently during evisceration than smaller gallbladders.[14,18,22,23] When the gallbladder reaches maximum capacity, excess bile backs up into the liver and also releases into the intestines and gizzard with antiperistalsis (Table 2.3). This can alter the appearance of the liver and may alter liver flavor. As a result of the bile, the gizzard lining will have a green appearance, indicating the feed withdrawal may be excessive (Table 2.3).[14]

During feed withdrawal, birds consume anything that is available, including litter and fecal material. Thus, there is a mixture of feed, litter, water, and feces in the digestive tract of broilers during the early withdrawal periods. Because of the presence of the other material (residual feed, water, and litter), feces is not easy to identify in the bird's digestive tract until the bird has been without feed for more than 14 hours (Table 2.3). Consumption of fecal material should be avoided because it increases the potential for carcass contamination in the plant, and it may affect the plant's ability to meet the USDA established microbiological standards for poultry.[14,16,18]

Because not every bird eats at the same time, the plant will be processing birds on feed withdrawal schedules that vary by approximately 3 hours. For example, if the target is a 12-hour feed withdrawal schedule for broilers, some birds have just eaten before feed is removed, while others birds ate 2 to 3 hours earlier. In a house of 20,000 birds, a catch crew of 10 will take 2 to 3 hours to empty the house. In a plant running 140 birds per minute, it will take approximately 46 minutes to process the birds on one truck (~6000 birds). The three trucks needed to catch all of the birds in one house will require approximately 2½ hours to slaughter. Because schedules will vary by 3 hours of the target, it is possible to be in the feed withdrawal range where the intestines begin to weaken.[20]

According to Hess and Bilgili,[23] the effect of feed withdrawal on intestinal strength varies with season. Experimental trials were conducted using 51- to 52-day old broilers grown in open-sided (curtain) houses. Force to tear broiler intestines was 15% higher in the winter than in the summer. Moreover, intestinal strength measured during the winter did not decrease with increasing feed withdrawal as was observed during the summer.

Feed withdrawal and microbiological implications

Of particular interest to processing plants as well as the USDA is microbiological contamination of products, especially if the contaminating bacteria are pathogenic. Recent studies have demonstrated that length of feed withdrawal has an effect on pathogenic bacteria in a bird's digestive tract. Byrd et al.[24] reported that feed withdrawal caused a significant increase in *Campylobacter* positive crop samples, with 25% positive crops before feed withdrawal and 62.4% positive crops after feed withdrawal. Corrier et al.[25] reported similar findings for *Salmonella* contaminated crops which increased from 1.9% before feed withdrawal to 10% at the end of feed withdrawal. Stern et al.[26] observed a fivefold increase in *Campylobacter* positive carcasses when they compared full-fed broilers held on litter to broilers held without feed in coops for 16 to 18 hours. Humphrey et al.[27] found broilers held for 24 hours without feed had higher levels of *Salmonella* in their crops, but the speed with which the remaining sections of the intestine were colonized with *Salmonella* was reduced when compared to full-fed broilers. It was suggested that the normal microflora of the crop, specifically lactobacilli that produce lactic acid, changed during feed withdrawal, reducing competitive bacteria and allowing proliferation of *Salmonella*. Hinton et al.[28] reported similar findings when broilers were held without feed for 6, 12, 18, or 24 hours. Broilers held without feed had higher crop pH than full-fed broilers (full-fed crop pH of 5.5 versus 12 hour withdrawal crop pH 6.5). This increase in crop pH may create a more favorable environment for pathogenic bacteria to grow, whereas the lower pH of a full-fed broiler would be a more undesirable environment.

Live shrink and carcass yield

Weight lost by birds during the time period between feed withdrawal and slaughter is referred to as "live shrink." Live shrink is important because it has a significant economic impact on carcass yield. The rate of live shrink has been reported to vary between 0.18% body weight per hour of withdrawal to 0.42% per hour.[9,11,22] For both broilers and turkeys, live shrink during the first 5 to 6 hours of feed withdrawal ranges from 0.3 to 0.6% of the live weight per hour of feed withdrawal. Buhr and Northcutt[22] reported that after the first 5 to 6 hours of feed withdrawal, live shrink was between 0.25 and 0.35% of the bird's body weight per hour of feed withdrawal, with the higher loss for male broilers and the lower loss for female broilers. Comparable results have been found for turkeys (0.2 to 0.4% per hour).[29] In addition to gender, they indicated that live shrink depended upon bird age, grow-out house temperature, eating patterns before feed withdrawal, and preslaughter holding conditions (cooping time and holding temperature). With live shrink, a broiler of market age held off feed for an extra 3 hours before processing (e.g., 15 instead of 12 hours), will weigh approximately 14 grams less than the same broiler processed 3 hours earlier. For turkeys, the loss is even greater. A 16 week old turkey hen held without feed for 3 extra hours (10 hours versus 7 hours of feed withdrawal) would weigh approximately 55 grams less than the same hen 3 hours earlier. This is a combination of 3 hours of less feed for growth and live shrink. In an operation that processes 250,000 broilers a day (average size of a U.S. broiler processing plant), for 5 days a week, an extra 3 hours of feed withdrawal could equate to reducing the live weight processed each week by 3500 kg. For a turkey processing plant (average size approximately 60,000 birds per day), 3 extra hours of feed withdrawal would reduce the live weight processed by 16,500 kg/week. This does not mean that birds given no feed withdrawal will have the highest carcass yields. In fact, birds full of feed that weigh the same as birds held off feed have lower carcass yields because their initial weight includes the digestive tract contents. Research has shown that carcass yield is greatest for broilers held off feed for 6 hours prior to processing; however, in reality, a 6-hour feed withdrawal schedule would be too difficult to manage, and contamination levels would be too high.[18,22]

Feed withdrawal and biological implications

Early research on livestock demonstrated that feed withdrawal resulted in decreased levels of muscle glycogen. In poultry, Murray and Rosenberg[30] reported that breast and thigh muscle glycogen decreased by 0.27 and 0.22%, respectively, after a 16-hour feed withdrawal period. Shrimpton[31] reported reduced muscle glycogen levels in broilers following a 24-hour feed withdrawal period. Warriss et al.[32] found that liver glycogen levels were negligible in broilers after 6 hours of feed withdrawal, and leg muscle glycogen continued to decrease with longer feed withdrawal times. Warriss[8] also reported that transportation of broilers affected liver and leg muscle glycogen. He suggested that holding broilers at the processing plant for more than 1 hour resulted in higher ultimate breast muscle pH (5.84 versus 5.78). These results imply that breast muscle glycogen was depleted during holding at the plant, and glycogen depletion typically occurs when birds are active or stressed.

Kotula and Wang[33] reported that increasing length of feed withdrawal resulted in decreased pH and glycogen levels in breast, thigh, and liver at the time of death for male broilers. For breast muscle, initial pH (<3 min postmortem) ranged from 6.97 for full-fed broilers to 6.36 for broilers off feed for 36 hours. Breast muscle glycogen declined from 7.0 to 3.5 mg/g after 36 hours of feed withdrawal. Thigh muscle followed a similar trend. These same researchers found no difference in final muscle pH (24 hours postmortem) due to feed withdrawal; however, muscle glycogen levels were significantly lower in both breast and thigh from broilers held off feed for longer periods of time before processing.

Injuries associated with catching and cooping

Nearly all broilers are caught and loaded into coops or transport containers by hand. A typical auto-dump coop (module) is shown in Figure 2.7. Catch crews usually consist of 7 to 10 people operating at a rate of approximately 1000 birds per hour. Crew members carry birds upside down by one leg with 5 to 7 birds in each hand, and place approximately 20 birds (depending upon the age and season) in each level of the auto-dump coop. Because this method of catching and loading has been associated with animal welfare problems, poor worker conditions, high labor costs, and carcass damage, several attempts have been made to develop alternative methods for catching broilers.[34–36] Scott[34,35] and Lacy and Czarick[36] have published excellent review articles on handling and mechanical harvesting of broilers.

Irrespective of the method of catching (manual or mechanical), broilers are subjected to handling which not only can result in fear and stress, but may also result in injuries. These injuries are typically bruising and dislocated or broken bones. A bruise generally results from a surface injury where the impact force does not pierce the skin, but instead ruptures cells and capillaries beneath the skin (Figure 2.8).[37,38] This results in the characteristic tissue discoloration which can appear on the broiler within seconds after the injury. The areas of the broiler most frequently bruised are the breast, wings, and legs. It has been estimated that 90 to 95% of the bruises found on broilers occur during the last 12 hours prior to processing,[39] with the grower responsible for approximately 35% of the bruises, and the catch crew approximately 40%, the remainder occurring during transport, unloading, and shackling. Some bruising may even occur during the first few seconds after neck cutting (within 10 seconds) before the bird's blood pressure reaches zero.

In the late 1950s and early 1960s, a group of researchers at the University of Georgia

Figure 2.7 Typical auto-dump coops (modules) being unloaded from the truck.

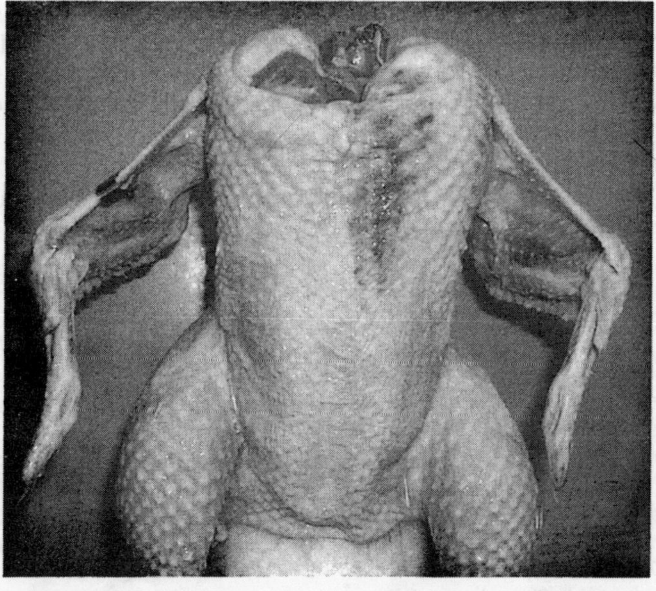

Figure 2.8 Bruising caused by ruptured cells and capillaries beneath the skin.

began to investigate the effects of bruising on poultry and livestock. M. Hamdy, K. May, and co-workers[38–40] suggested that the age of a bruise could be estimated using the color of the bruise. They found that initially after an injury, bruises were red with moderate tissue swelling. Over time, color of bruises changed from red to various shades of purple, yellow, green, and orange before returning to normal. Bruises were reported to heal in broilers within 3 to 5 days depending upon the environmental temperatures, where longer time to heal was required for those birds housed in cooler environments (30°C vs. 21.1°C).[38–40] Similar studies have been conducted by Northcutt and Buhr[41,42] and Northcutt et al.[43] with emphasis on bruise color development, histological tissue damage, and functional properties of poultry meat during further processing.

Bilgili and Horton[44] conducted a year-long field study to evaluate the influence of live production factors on broiler carcass quality and grade. These researchers found that older, heavier broilers had more bruises, leg problems, breast blisters, and broken or dislocated bones. In addition, a positive correlation was found between flock age and birds dead-on-arrival (DOA) at the processing plant. Bird placement density, or the amount of space allowed per bird in the house, influenced broiler bruising, with a higher incidence of bruises occurring when space was limited.

Another contributing factor to broiler bruising is the presence of mycotoxins (toxic metabolite produced by fungi) in grains and feeds. Aflatoxin has been found to increase the birds susceptibility to bruising by increasing capillary fragility and reducing shear strength of skeletal muscle. As little as 0.625 μg of dietary aflatoxin produced extensive hemorrhaging in muscles and internal organs.[45] Additional information on mycotoxicosis and bruising may be found in articles published by Tung et al.[45] and Hoerr.[46]

Summary

Poultry meat quality is affected by numerous antemortem factors, in particular those occurring during the last 24 hours that the bird is alive. These short term factors influence carcass yield (live shrink), carcass defects (bruising, broken/dislocated bones), carcass microbiological contamination, and muscle metabolic capabilities. There is even evidence to suggest that stressful conditions during harvesting, such as catching and cooping, affect the postmortem muscle functional properties. Current issues associated with food-borne illnesses have forced poultry companies to pay even more attention to live production than before to satisfy the "farm-to-table" food safety initiative. These issues will continue to be priorities for the USDA and poultry companies.

References

1. Fletcher, D. L., Antemortem factors related to meat quality, *Proceedings of the 10th European Symposium on the Quality of Poultry Meat*, Beekbergen, The Netherlands, 1991, 11.
2. Moran, E. T., Jr., Live production factors influencing yield and quality of poultry meat, in *Poultry Science Symposium Series*, Volume 25, Richardson, R. I. and Mead, G. C., Eds., CABI Publishing, Oxon, U.K., 1999.
3. Calnek, B. W., Barnes, H. J., Beard, C. W., Reid, W. M., and Yonder, H. W., Jr., Eds., *Diseases in Poultry*, 9th edition, Iowa State University Press, Ames, IA, 1991.
4. National Research Council, *Nutrient Requirements of Poultry*, 9th ed., National Academy Press, Washington, DC, 1994.
5. Sturkie, P. D., Ed., *Avian Physiology*, 4th ed., Springer-Verlag, New York, 1986.
6. Cunningham, D. L., Contract broiler grower returns: A long-term assessment, *J. Appl. Poultry Res.*, 6, 267, 1997.
7. Cunningham, D. L., Poultry production systems in Georgia, costs and returns analysis, unpublished annual reports, Extension Poultry Science, The University of Georgia, Athens, GA,

1990–1996.
8. Warriss, P. D., Wilkins, L. J., and Knowles, T. G., The influence of ante-mortem handling on poultry meat quality, in *Poultry Science Symposium Series,* Volume 25, Richardson, R. I. and Mead, G. C., Eds., CABI Publishing, Oxon, U.K., 1999.
9. Wabeck, C. J., Feed and water withdrawal time relationship to processing yield and potential fecal contamination of broilers, *Poult. Sci.,* 51, 1119, 1972.
10. Bilgili, S. F., Research note: effect of feed and water withdrawal on shear strength of broiler gastrointestinal tract, *Poult. Sci.,* 67, 845, 1988.
11. May, J. D., Lott, B. D., and Deaton, J. W., The effect of light and environmental temperature on broiler digestive tract contents after feed withdrawal, *Poult. Sci.,* 69, 1681, 1990.
12. Papa, C. M. and Dickens, J. A., Lower gut contents and defecatory responses of broiler chickens as affected by feed withdrawal and electrical treatment at slaughter, *Poult. Sci.,* 68, 1478, 1989.
13. Benoff, F. H., The "live-shrink" trap: catch weights a must, *Broiler Ind.,* 41(1), 56, 1982.
14. Northcutt, J. K., Savage, S. I., and Vest, L. R., Relationship between feed withdrawal and viscera condition of broilers, *Poult. Sci.,* 76, 410, 1997.
15. Bilgili, S. F., "Zero Tolerance" begins at the farm, *Broiler Ind.,* 61(11), 30, 1998.
16. Northcutt, J. K. and Savage, S. I., Preparing to process, *Broiler Ind.,* 59 (9), 24, 1996.
17. Veerkamp, C. H., Fasting and yields of broilers, *Poult. Sci.,* 65, 1299, 1986.
18. May, J. D., Branton, S. L., Deaton, J. W., and Simmons, J. D., Effect of environment temperature and feeding regimen on quantity of digestive tract contents of broilers, *Poult. Sci.,* 67, 64, 1988.
19. May, J. D. and Lott, B. D., Effect of periodic feeding and photoperiod on anticipation of feed withdrawal, *Poult. Sci.,* 71, 951, 1992.
20. Northcutt, J. K. and Buhr, R. J., Maintaining broiler meat yields: longer feed withdrawal can be costly, *Broiler Ind.,* 60(12), 28, 1997.
21. Bilgili, S. F. and Hess, J. B., Tensile strength of broiler intestines as influenced by age and feed withdrawal, *J. Appl. Poult. Res.,* 6, 279, 1997.
22. Buhr, R. J., Northcutt, J. K., Lyon, C. E., and Rowland, G. N., Influence of time off feed on broiler viscera weight, diameter, and shear, *Poult. Sci.,* 77, 758, 1998.
23. Hess, J. B. and Bilgili, S. F., How summer feed withdrawal impacts processing, *Broiler Ind.,* 61(8), 24, 1998.
24. Byrd, J. A., Corrier, D. E., Hume, M. E., Bailey, R. H., Stanker, L. H., and Hargis, B. M., Incidence of *Campylobacter* in crops of preharvest market-aged broiler chickens, *Poult. Sci.,* 77, 1303, 1998.
25. Corrier, D. E., Byrd, J. A., Hargis, B. M., Hume, M. E., Bailey, R. H., and Stanker, L. H., Presence of *Salmonella* in the crop and ceca of broiler chickens before and after preslaughter feed withdrawal, *Poult. Sci.,* 78, 45, 1999.
26. Stern, N. J., Clavero, M. R. S., Bailey, J. S., Cox, N. A., and Robach, M. C., *Campylobacter* spp. in broilers on the farm and after transport, *Poult. Sci.,* 74, 937, 1995.
27. Humphrey, T. J., Baskerville, A., Whitehead, A., Rowe, B., and Henley, A., Influence of feeding patterns on the artificial infection of laying hens with *Salmonella enteritidis* phage type 4, *Vet. Rec.,* 132, 407, 1993.
28. Hinton, A., Jr., Buhr, R. J., and Ingram, K., Feed withdrawal and carcass microbiological counts, *Proc. Georgia Poult. Conf.,* Athens, GA, September 30, 1998.
29. Duke, G. E., Basha, M., and Noll, S., Optimum duration of feed and water removal prior to processing in order to reduce the potential for fecal contamination in turkeys, *Poult. Sci.,* 76, 516, 1997.
30. Murray, H. C. and Rosenberg, M. M., Studies on blood sugar and glycogen levels in chickens, *Poult. Sci.,* 32, 805, 1953.
31. Shrimpton, D. H., Some causes of toughness in broilers (young roasting chickens). I. Packing stations procedure, its influence on the chemical changes associated with rigor mortis and on the tenderness of the flesh, *Br. Poult. Sci.,* 1, 101, 1960.
32. Warriss, P. D., Kestin, S. C., Brown, S. N., and Bevis, E. A., Depletion of glycogen reserves in fasting broiler chickens, *Br. Poult. Sci.,* 29, 149, 1988.
33. Kotula, K. L. and Wang, Y., Characterization of broiler meat quality factors as influenced by feed withdrawal time, *J. Appl. Poult. Res.,* 3, 103, 1994.

34. Scott, G. B., Poultry handling: a review of mechanical devices and their effect on bird welfare, *World's Poult. Sci. J.*, 49, 44, 1993.
35. Scott, G. B., Catching and handling of broiler chickens, *Proc. 9th Eur. Poult. Conf.*, Glasgow, U.K., II, 1994, 411.
36. Lacy, M. P. and Czarick, M.. Mechanical harvesting of broilers, *Poult. Sci.*, 77, 1794, 1998.
37. McCarthy, P.A., Brown, W.E., and Hamdy, M.K., Microbiological studies of bruised tissues, *J. Food Sci.*, 28, 245, 1963.
38. May, K. N. and Hamdy, M. K., Bruising of poultry: a review, *World's Poult. Sci. J.*, 22(4), 316, 1966.
39. Hamdy, M. K., Kunkle, L. E., and Deatherage, F. E., Bruised tissue II. Determination of the age of a bruise, *J. Anim. Sci.*, 16, 490, 1957.
40. Hamdy, M. K., May, K. N., Flanagan, W. P., and Powers, J. J., Determination of the age of bruises in chicken broilers, *Poult. Sci.*, 40, 787, 1961.
41. Northcutt, J. K. and Buhr, R. J., Management guide to broiler bruising, *Broiler Ind.*, 61(10), 18, 1998.
42. Northcutt, J. K., Buhr, R. J., and Rowland, G. N., Relationship of the age of a broiler bruise, skin appearance, and tissue histological characteristics, *J. Appl. Poult. Res.*, 9, 13, 2000.
43. Northcutt, J. K, Smith, D. P., and Buhr, R. J., Effects of bruising and marination on broiler breast fillet surface appearance and cook yield, *J. Appl. Poult. Res.*, 9, 21, 2000.
44. Bilgili, S. F. and Horton, A. B., Influence of production factors on broiler carcass quality and grade, in *Proceedings of the XII European Symposium on the Quality of Poultry Meat*, Zaragoza, Spain, 1995, 13.
45. Tung, H-T, Smith, J. W., Hamilton, P. B., Aflatoxicosis and bruising in the chicken, *Poult. Sci.*, 50, 795, 1971.
46. Hoerr, F. J., Mycotoxicoses, in *Diseases of Poultry*, Calnek, B. W., Barnes, H. J., Beard, C. W., Reid, W. M., and Yonder, H. W., Jr., Eds., Iowa State University Press, Ames, IA, 1991, 884.

chapter three

First processing: slaughter through chilling

Alan R. Sams

Contents

Introduction

The processing plant is a highly coordinated system of mechanized operations that kill the birds, remove the inedible portions of the carcasses, and package/preserve the edible portions of the carcass for distribution to the consumer. The efficiency of processing is largely dependent on the uniformity of the birds, so that each machine can do a repeated movement with little or no adjustment between birds. Another important factor is the logistical coordination of carcass flow and production lines so that adequate birds are present to make maximum use of the personnel and equipment. These fixed costs are incurred by the plant regardless of the presence of birds, and therefore need to be paid by the production of poultry meat. This necessitates that every shackle be occupied to produce the maximum amount of product.

0-8493-0120-3/01/$0.00+$.50

Slaughter

Unloading

After their arrival at the processing plant, the birds are unloaded for processing. The coops of birds are removed from the truck and "dumped" onto a conveyor or placed in a position for them to be manually unloaded. The "dumper" can be a source of carcass damage such as bruising and broken bones, because the birds are allowed to freely fall one or more meters to the conveyor belt below. Minimizing this distance can reduce the damage. Manual unloading can also cause carcass damage if the birds are handled roughly. Proper training and supervision are critical to minimizing damage. When the birds are manually unloaded from the coops, they are usually directly hung on a shackle and not placed on a separate conveyor belt. Because of bird size and numbers, dumpers have become the industry norm in the U.S., with manual unloading still occurring in some other parts of the world. Because of their large size and poor body control, turkeys are usually still unloaded manually worldwide.

The ergonomics and safety of the unloading process has become an issue in recent years. Coops and/or workers are on platforms of adjustable heights, maintaining the birds at an optimal position to minimize the bending and lifting required by the worker. The industry has determined that such ergonomic innovations can yield benefits from reduced medical claims and better worker performance/retention. Proper ventilation is also important in the unloading and hanging areas to further improve worker welfare. These are particularly dusty areas, and respiratory health of the workers can be a concern. The hanging areas have traditionally been dark, lit only with "black lights" or dim red lights. This darkening was thought to calm the birds, reducing their struggle against hanging and thereby reducing the damage to their bodies during handling. However, processing companies are increasingly realizing that this benefit may not be as great as once thought, and that improving the working environment with brighter lighting and ventilation more than offsets any increase in carcass damage.

Stunning

The first step in humane slaughter is "stunning" to render the bird unconscious prior to killing. Several methods have been developed to accomplish this goal. The most common and one of the simplest is electric shock. While hanging by their feet, the heads of the birds contact a saline solution (approximately 1% NaCl) that is charged so that an electrical current flows through the bird to the shackle line which serves as the ground (Figure 3.1). A proper electrical stun will produce about 60 to 90 sec of unconsciousness during which the bird is unable to stand or right itself when removed from the shackle and placed on the floor. This is a suggested method of evaluating the effectiveness of the stun. Immediately after contact, the legs are extended, the wings are tight against the body, and the neck is arched. Several seconds after leaving the stunner contact, the bird's posture relaxes and the body becomes almost limp. In addition to humane slaughter, there are other benefits to be gained from proper stunning, such as immobilization for improved killing machine efficiency, more complete blood loss, and better feather removal during picking. Inadequate stunning can result in carcass defects such as incomplete bleeding, while excessive stunning can cause quality defects such as broken clavicles (wishbones) and hemorrhages from ruptured arteries and capillaries. Some commercial poultry is not stunned because some cultures specifically prohibit preslaughter stunning, requiring the birds to be conscious when slaughtered (see Chapter 17).

Figure 3.1 Electrical stunner cabinet containing an electrode covered by a saline solution. Bird movement is from left to right.

There are different conditions used for electrical stunning, depending on the region of the world. Although poultry is not required by law to be stunned before slaughter in the U.S., virtually all commercial poultry is stunned for humane, efficiency, and quality reasons. The birds receive 10–20 mA per broiler and 20–40 mA per turkey for 10 to 12 seconds. These conditions yield adequate time of unconsciousness for the neck to be cut and sufficient blood to be lost so as to kill the bird before it regains consciousness. In most European countries, laws require poultry to be stunned, and with much higher amperages (90 + mA per broiler and 100.+ mA per turkeys for 4–6 seconds). These laws and high amperages are intended for humane treatment to ensure that the birds are irreversibly stunned so that there is no chance they will be able to recover and sense any discomfort. Essentially, these European electrical stunning conditions kill the bird by electrocution and cardiac arrest, stopping blood flow to the brain. Thus, death is by loss of blood supply to the brain for both stunning conditions, but one is by removal of blood and the other is by stopping blood flow to the brain. The harsher European electrical conditions also result in higher incidences of hemorrhaging and broken bones.[1,2]

Other methods of stunning have been developed to replace electrical stunning in areas such as Europe, which require the higher electrical conditions. Exposing the birds to gases to induce either anesthesia or anoxia is one method in commercial use. Carbon dioxide is an anesthetic gas used to induce rapid unconsciousness by altering the pH of the cerebrospinal fluid.[3] Argon and nitrogen are inert gases that displace the air and cause unconsciousness through lack of oxygen.[4,5] There are two main types of gas stunning systems for poultry. First, systems using mixtures of carbon dioxide (10 to 40%) and air (60 to 90%) are shorter duration (30 to 45 sec) and intended to render the bird unconscious but alive for the killing machine. Systems using mixtures of argon (55 to 70%), nitrogen (0 to 15%), and carbon dioxide (30%) are longer duration (2 to 3 min) and intended to render the bird dead at the time of neck cutting. Thus, the carbon dioxide systems would be most analogous with the low amperage, reversible electrical stun, while the argon stun would be most analogous with the higher amperage, irreversible electrical stunning. However, both gas stunning procedures reduce carcass damage relative to the high amperage but not the low amperage electrical stunning.[2] This is because the low amperage electrical stunning has an equally low incidence of carcass damage. An additional note on gas stunning is that these birds are flaccid on the shackles when entering the killing machine. This differs from the stiffer, electrically stunned bird and must be accommodated with minor machine adjustments for bird orientation.

Another stunning system that has received attention for reasons of humane animal treatment is captive bolt stunning.[6] In this stunning method, the head is immobilized and a metal pin or probe is shot into the skull and brain causing immediate and irreversible unconsciousness. The humane and carcass quality effects of this method are still under investigation.

Killing

Within seconds after stunning, the shackle conveyor moves the bird to the killing machine (Figure 3.2). A series of rotating bars grab the wattles and lower neck skin to hold and guide the head into the machine for proper presentation to the cutting blade. The killing machine uses a rotating circular blade to cut the jugular veins and carotid arteries on one or both sides of the neck of the bird. Most killing machines cut both sets of blood vessels by rotating the head from the bird's left to right as it passes over the cutting blade. If the cut is too deep and the spinal nerve cord is cut, the resulting nervous stimulation "sets" the feathers and makes picking more difficult. Conversely, if the cut is too shallow, there will be insufficient bleeding and the residual blood will cause engorged vessels and can discolor the skin. Once the neck has been cut, the bird is allowed to bleed for 2–3 minutes. During this period the bird loses about 30 to 50% of its blood, which eventually causes brain failure and death. If the blood loss is insufficient to cause death or if the neck cut is missed altogether, the bird may be still alive at the end of the bleeding period when it enters the scalder. In this case, the blood rushes to the skin surface in response to the scald water heat, imparting a bright red color to the carcass.

Feather removal

Scalding

Feathers are difficult to remove in their native condition due to their attachment in the follicles. To loosen them, the carcasses are submersed in a bath of hot water which serves to denature the protein structures holding the feathers in place. Two particular combinations of time and temperature have become industry norms and produce quite different effects on the carcass. Scalding at 53.35°C (128°F.) for 120 sec is called "soft scalding" and loosens the feathers without causing appreciable damage to the outer skin layers, the stratum

***Figure* 3.2** Picture of a killing machine showing guide bars and bicycle wheel to keep head in proper alignment for the circular cutting blade below.

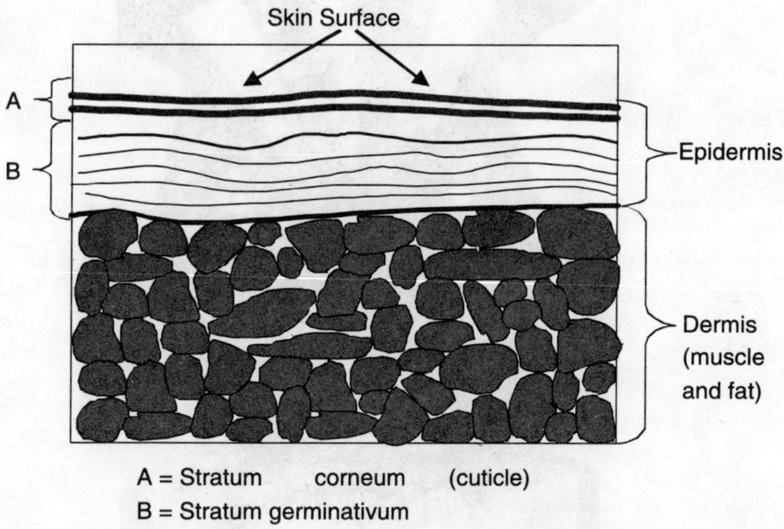

Figure 3.3 Diagram of skin layers. Adapted from Suderman, D. R. and Cunningham, F. E., *J. Food Sci.* 45 (3), 444, 1981.

corneum or "cuticle" (Figure 3.3). Because it leaves this waxy, yellow-pigmented layer of the skin intact, soft scalding is the preferred scalding method for producing fresh poultry with a yellow skin exposed. Such skin color is highly desired in some parts of the world as indicating a healthy bird. If the skin's cuticle will not be exposed or is not pigmented with carotenoids from the feed, the carcasses are usually scalded at 62 to 64°C (145 to 148°F) for 45 sec, a process called "hard scalding." Because it loosens the cuticle, this is a harsher procedure than soft scalding. However, it allows easier feather removal than milder scalding conditions. Once loosened, the outer skin layer and its associated pigmentation is removed by the abrasion of the mechanical pickers. The loss of the waxy cuticle may be beneficial for the processor whose product is destined to be coated and fried. Because of their aqueous basis, fried chicken coatings generally adhere to the skin better in the absence of this waxy, water repellent layer of the skin.

Picking

Picking machines consist of rows of rotating clusters of flexible, ribbed, rubber "fingers" (Figure 3.4). While rotating rapidly, the fingers rub against the carcass and the abrasion pulls out the loosened feathers. By combining a series of these rotating clusters of fingers, each directed at a different region of the carcass, the whole carcass is picked. Picking machines adjusted too close to the bird may cause skin tears in the thigh and breast regions and broken wing, leg, and rib bones. Machines that are too distant may not adequately remove the feathers. Pin feathers are immature feathers that protrude from the skin, still encased in the feather shaft. These pin feathers are difficult to remove with machines and therefore require manual attention. Illustrating the importance of live production issues in processing, a more rapidly feathering bird will have fewer pin feathers when processed. The last step in feather removal is singeing. Carcasses are briefly passed through a flame to burn off the hair-like filoplume structures on the skin because they are aesthetically offensive to consumers and considered a carcass defect.

Before the carcasses leave the picking area, the heads are pulled off of the necks if they have not already come off in the picking machines. The heads, along with the feathers,

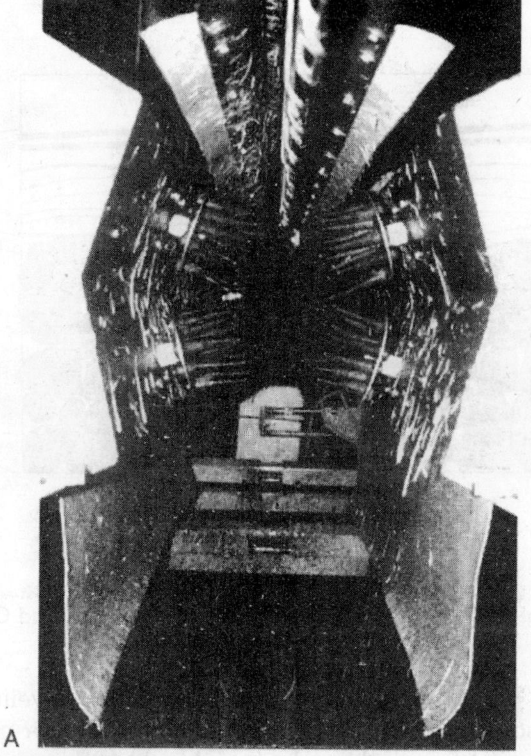

Figure 3.4 Picture of a picker head with the flexible rubber fingers (A). Row of defeathering or "picking" machines with each one targeting its picker heads at a different part of the carcass (B).

blood, and inedible viscera, are called "offal" and are sent to rendering (either in-plant or at a different location), where these materials are ground and cooked into poultry fat and byproduct meal for inclusion in animal feed. The feet are also cut off at the ankle or "hock" joint and sent to be chilled and sorted for sale or inclusion in the giblets. There are usually two categories of feet quality, those that are free of defects and those containing defects such as dark pigmentation or foot pad lesions. The last step before evisceration is to transfer the birds from the kill shackle line to the evisceration shackle line. This is done manually or with a transfer machine. If done manually, this can be a site of bacterial cross-contamination, as one employee handles many birds. This transfer is necessary because one kill line can feed multiple evisceration lines. Separation should be maintained between the live and dead areas of the plant to reduce contamination of the relatively cleaner evisceration room.

Evisceration

Evisceration is the removal of edible and inedible viscera from the carcass. It is a coördinated series of highly automated operations that vary substantially in sequence and design from plant to plant and from one equipment manufacturer to another. Although it is becoming more automated, turkey evisceration is still largely manual worldwide. In broilers, evisceration has three basic objectives: (1) the body cavity is opened by making a cut from the posterior tip of the breastbone to the cloaca (anus); (2) the viscera (primarily the gastrointestinal tract and associated organs, reproductive tract, heart, and lungs) is scooped out; and (3) the edible viscera or "giblets" (heart, liver, and gizzard) are harvested from the extracted viscera, trimmed of adhering tissues, and washed with water. The neck is usually part of the giblets but is collected later, after inspection of the carcass for wholesomeness. Although not technically part of the viscera, the feet (or "paws") have become a valuable product, primarily for export to cultures that use them for human food. In some countries, the paws are included with the packet of giblets sold with the whole carcass.

There are also some countries where a considerable proportion of the poultry is sold without evisceration (Figure 3.5). Broilers processed to only remove the blood and feathers are called "New York dressed" and are even sometimes sold without refrigeration. Cultures preferring this type of product feel that they are fresher because they are whole and are killed and sold within hours (due to short shelf-life). Sometimes these uneviscerated carcasses are held for several days because of the desired "game-like" flavor that develops. Due to the lack of labeling on these carcasses, processors have used adhesive labels on the skin to promote customer loyalty through brand identification (Figure 3.6).

The basic design of most evisceration machines is rotating, vertical cylinders that have ten or more "stations" located around the edge. The shackle line containing the birds wraps around the cylinder and provides the force for rotation so the shackle line and machines move at a coordinated speed. As each shackle and bird contact the cylinder, the bird is grasped and a series of mechanical procedures are performed. When the cylinder has made a complete rotation and completed the series of events, the bird is released from the

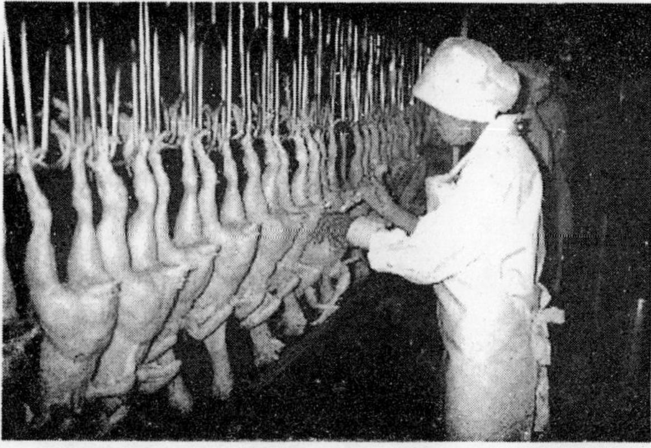

Figure 3.5 New York dressed carcasses ready for market. This employee is puncturing the abdominal skin to release abdominal gas and prevent bloating.

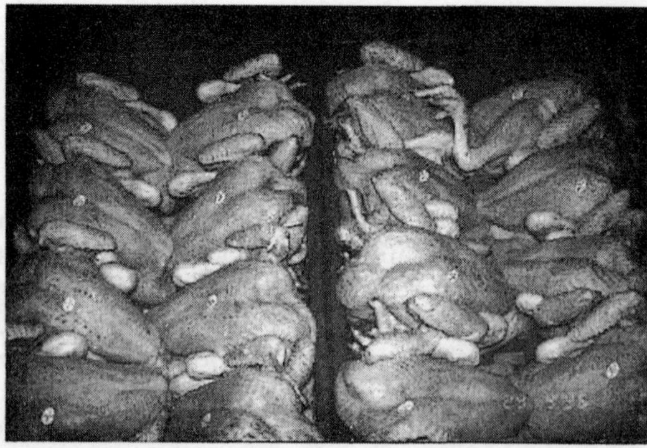

Figure 3.6 Air-chilled New York dressed birds with adhesive labels for brand identification in the marketplace.

machine to travel to the next machine in the sequence. After releasing each bird, machines usually have a washing step for the station before it grasps its next bird. Despite this washing, the fact that approximately every tenth bird contacts the same surface has raised concerns about bacterial cross-contamination between carcasses, and increases the importance of these washing procedures. A typical sequence of evisceration machines, including their common names, is provided in Figure 3.7. In the following discussions, it is important to remember that the birds are hanging by their legs from shackles, and are therefore inverted from the normal, upright, living chicken.

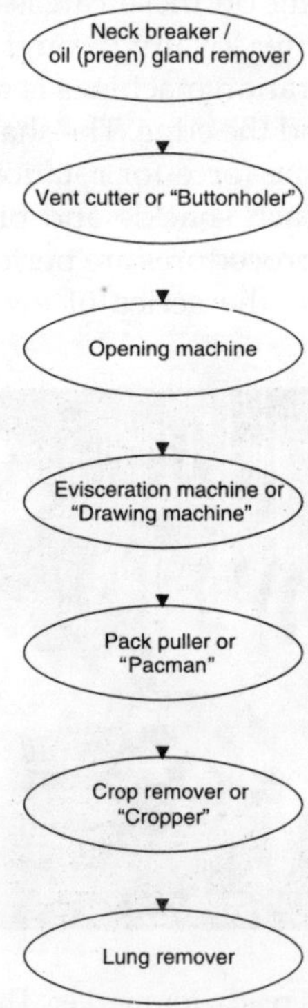

Figure 3.7 Typical sequence and common names for evisceration equipment.

Figure 3.8 Neck breaker showing blade rising from left to right to contact the neck and break the spine.

The neck breaker (Figure 3.8) and oil gland removal can be combined in one machine. A blade pushes against the neck, just anterior to the shoulders. Enough force is applied to break the spine and cut the dorsal skin but not enough to cut the ventral skin or trachea and esophagus. Once the spine is cut, the blade drags the partially severed neck downward and releases it to dangle from the carcass. A blade shaves the preen glade off from the dorsal surface of the tail. This preen gland contains an oily substance that the bird uses to groom its feathers but tastes offensive to humans.

The vent opening machine (vent cutter or "buttonholer") (Figure 3.9) places a probe against the vent opening or anus and then draws a vacuum to grip the surrounding skin. A circular blade then descends and cuts through the immobilized skin around the vent opening, and the probe that is holding the vent opening retracts, pulling the terminal part of the lower intestine out of the bird. The small segment of intestine with the attached vent opening is then released by stopping the vacuum. Poor adjustment of this machine can cause it to cut the intestine, resulting in fecal and bacterial contamination of the carcass. Because all the birds contact the same machine, and that machine has probes and blades

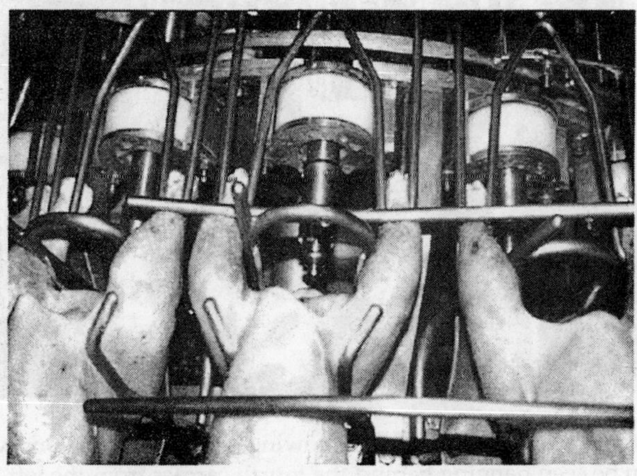

Figure 3.9 Vent opening machine showing probe preparing to contact the vent of the bird.

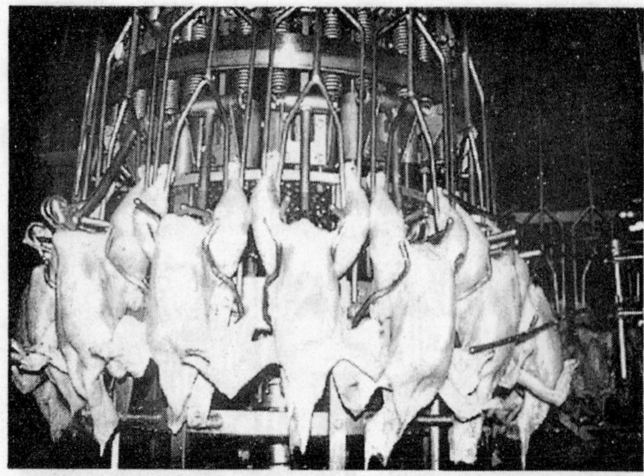

Figure 3.10 Abdominal cavity opening machine showing blade on the left of the machine cutting through abdominal skin.

which cut and are inserted into all of the carcasses, the processing machinery can be a source of cross-contamination, transferring bacteria between carcasses. To reduce this, sanitation of the equipment between birds is important. The stations on the probes on the evisceration equipment are sprayed with chlorinated water, and may also be scrubbed between carcasses.

The next step in evisceration is to increase the opening of the abdominal cavity with the opening machine (Figure 3.10). A large blade is inserted in the abdominal cavity and pressed outward from the spine toward the tip of the keel to cut the skin and enlarge the abdominal opening. This larger opening will allow the evisceration machine, or drawing machine (Figures 3.11 and 3.12), to scoop and draw out the intestinal package of the carcass. The carcass is grasped to immobilize it, and a spoon-shaped scoop is inserted into the body cavity. The scoop travels down along the inside of the breast to a point where it can grab the gizzard and heart. It is then withdrawn, pulling the viscera package out of the bird.

Figure 3.11 Evisceration or drawing machine showing scoop withdrawing viscera from carcass. Visceral "package" can be seen hanging behind the fourth carcass from the left.

Figure 3.12 Evisceration machine showing chain drive of shackle line and machine. The carcasses exiting the machine on the left have the viscera "package" hanging from them in preparation for inspection.

As the birds approach the inspection station (see Chapter 5) their viscera and abdominal fat pads need to be positioned so that the inspector can rapidly view the parts of the bird that need to be evaluated for detection of internal disease, mainly the liver and air sacs. For this positioning purpose, plant employees arrange the viscera in a uniform way to maximize the efficiency of the inspection process. If there is a question about acceptability or if some other treatment is required, birds are hung on a special rack for extra attention. Birds that are deemed unacceptable are placed in designated cans for disposal. Following internal inspection, the viscera package is removed from the carcasses and sent to the giblet harvesting area. The viscera package is removed from the bird by a machine called the pack puller (Figure 3.13). This device inserts a clamping probe into the abdominal cavity, reaching all the way to the neck area of the bird, where it clamps onto the esophagus. As the bird proceeds away from the circular machine, the esophagus and attached viscera package are pulled out of the carcass through the abdominal cavity opening. The viscera is then pumped to a giblet harvesting area where the liver, gizzard, neck, and heart are collected and separated from inedible viscera. In harvesting the giblets, the adhering connective tissue and blood vessels are removed from the heart and liver. The gizzard is split open and

Figure 3.13 Pack pulling machine showing probes that are inserted into the carcasses as the wheel rotates.

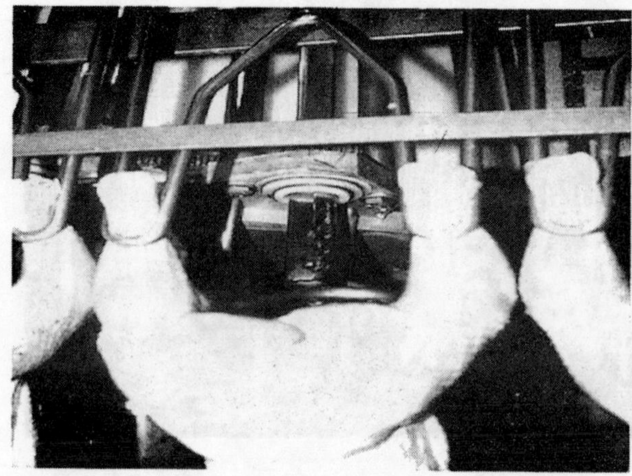

Figure 3.14 Cropper showing serrated probe preparing to enter abdominal opening of the carcass.

the lining is removed by rolling bars containing sharpened edges that grip the tough lining and peel it off the musculature. The inedible viscera are sent to rendering, where they are mixed with other inedible parts of the bird, cooked, and ground into byproduct meal.

The chicken crop is removed with a machine called a cropper (Figure 3.14). This machine inserts a spinning probe with pointed barbs that snags or grabs the crop and pushes it through the space between the neck skin and the spine until it protrudes out of the neck in the location where the head originally was. Once the probe and attached esophagus are protruding out of the carcass, the probe passes a brush and washing station where the adhering crop is removed from the probe before it retracts back through the carcass. As with all equipment possessing probes that enter and then retract through the carcass, sanitation and prevention of bacterial cross-contamination is an important concern.

Finally, a lung removal machine inserts a vacuum probe (Figure 3.15) down into the abdominal opening of the bird and suctions the lungs from the dorsal surface of the rib cage to remove them from the carcass. This can also be done manually with vacuum guns.

Figure 3.15 Vacuum probe of the lung remover machine showing double tubes to extract both lungs from their position embedded against the back of the ribcage.

At some point during the evisceration process, the birds pass through an external inspection station where the surface of the bird is evaluated for surface imperfections such as extensive bruising or other skin lesions. Birds determined by either the interior or exterior inspection stations to need extra trimming or attention are sent to the rework or salvage station where they are trimmed and/or washed as appropriate. Sometimes only part of the bird is of use, so that part is saved and the remainder is condemned for rendering into poultry byproduct meal. Because this rework is a labor-intensive process that removes the birds from the normal product flow, it is expensive and processors strive to keep it to a minimum. Fecal contamination on the surface of carcasses can be cleaned at the rework station or sent through special washing cabinets containing chlorinated water or other antimicrobial compound for "in-line reprocessing."

After the birds have passed inspection but before they are placed in the chilling tank, they pass through an inside/outside (I/O) bird washer which has spray nozzles directed toward the interior and exterior of the bird to remove any adhering material before the birds enter the common chilling tank. The water in the I/O bird washers and the chiller contains chlorine and possibly other antimicrobial materials. Additional sprays or dips may be used before the chiller to rinse the birds with antimicrobial compounds such as trisodium phosphate or acidified sodium chlorite.

The overall dependence of the evisceration process on machinery underscores the importance of machinery maintenance and adjustment for bird size. Poorly adjusted machines are frequent causes of torn skin, broken bones, and ruptured intestines, with resulting fecal contamination of the carcass. Despite the use of machines, there is usually a human for every one or two machines to correct any mistakes the machine makes. People are also important for tasks such as arranging viscera for presentation to inspectors, and harvesting edible viscera. Inspection is an important part of evisceration, and is covered in detail in Chapter 5.

Chilling

The primary objective of chilling poultry is reduction of microbial growth to a level that will maximize both food safety and time available for marketing. Generally, a temperature of 4°C or less is achieved as soon as possible after evisceration (1 to 2 hours postmortem). The U. S. regulations require this temperature to be achieved within 4 hours of death for broiler chickens and 8 hours for turkeys. The two most common methods of chilling poultry are water and air. In addition to differences in the actual procedures, the two chilling methods have different effects on the product. Processors in the U.S. almost exclusively use water chilling, while European processors commonly use air chilling.

Water chilling usually involves multiple stages of tanks. Carcasses are removed from the shackle and slowly pushed through the water by some type of paddle (Figure 3.16) or auger system. The first stage, called a "prechiller," is about 7 to 12°C (45 to 55° F) and lasts for 10 to 15 min. It contains some water discharged from the main chiller, to recycle the refrigeration energy. The main function of the prechiller is to allow water absorption, but it also has some washing and chilling effects on the carcass. At the entrance of the prechiller, the carcass temperature is about 38°C, and the skin lipids are still quite fluid. Water readily penetrates the skin, and to a much lesser extent, the fascia and other subcutaneous tissues. Water absorption is temperature- and time-dependent and is regulated by the U.S. government based on the eventual use of the product. If the product is to be sold in a drainable container, water absorption is limited to 12% of the carcass weight before the prechiller. If the product is to be packaged in a container that does not allow drainage, the limit is 8%. These limits were originally designed to compensate processors for the moisture lost during marketing. A frequent problem is that the actual amount of drip loss varies

Figure 3.16 Paddles in a chiller to move carcasses. Open side of paddle alternates between paddles to agitate the water.

with marketing practices (i.e., temperature, time, and location), and can result in inaccurate label weights with drainable containers. The added water from immersion chilling also adds to shipping costs and can interfere with use of the product in further processing. Allowances for water absorption are occasionally re-evaluated by the government as marketing practices and products change.

After prechilling, the carcasses are about 30 to 35°C when they enter the main chilling tank, the "chiller," which looks generally similar to the prechiller, just larger. The water temperature in the chiller is about 4°C at the entrance and about 1°C at the exit. These low temperatures rapidly reduce the carcass temperature during the 45 to 60 min the carcasses are in the chiller. As the carcass temperature declines, the tissue lipids solidify and seal in the water absorbed in the prechiller. To maximize heat exchange from carcasses to water and maximize cleanliness of the carcasses, water chillers use counter-current flow (Figure 3.17). The carcasses and water flow in opposite directions so that the carcasses are bathed in increasingly cold and clean water throughout the length of the chill tank. An additional practice to enhance chilling rates is injection of air into the bottom of the chill tank (Figure 3.18). The bubbles created as the air moves toward the water surface agitate the water and prevent the formation of thermal layering at the product surface. If the water were not agitated, the water contacting the surface would warm until it reached thermal equilibrium with the carcass temperature. This layer of warmer water would insulate the carcass and reduce further heat exchange from the carcass to the water because the rate of heat exchange between two materials is largely related to the difference in temperatures of the two materials.[7]

Air chilling involves passing the shackle lines of carcasses through large rooms of circulating cold (−7 to 2°C) air for 1 to 3 hours.[8] This can be done with the birds on racks, but it is more efficient and more common to air-chill carcasses on shackles. To enhance cooling,

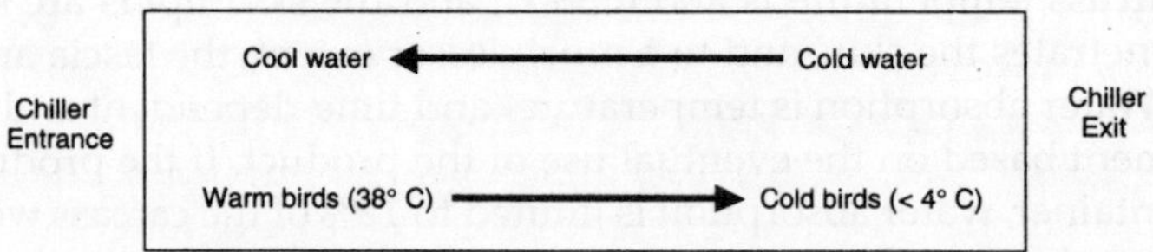

Figure 3.17 Diagram of a counter-current flow chilling system.

Figure 3.18 **Water-immersion chilling tank showing air hoses used to agitate the water with bubbles.**

the product can be sprayed with water, which absorbs heat as it evaporates. Humidity can also be controlled to maximize the air's ability to absorb heat from the carcasses and also the air's ability to evaporate the surface water for evaporative cooling. Air-chilled carcasses can have a dried appearance to their skin, reflecting the drying effect of this chilling method. The dried skin rehydrates, and the appearance usually returns to normal after packaging. Air-chilled poultry carcasses usually exhibit a slight weight loss during chilling. This contrasts with the weight gains resulting from water chilling. Such yield differences contribute to restrictions in international trade between countries using the two different chilling methods.

The two chilling systems have very different effects on the microbial quality of a poultry carcass. Water chilling washes bacteria from the skin and results in carcasses with a generally lower total bacterial load. However, the extensive bird-to-bird contact via water results in a greater potential for spreading bacteria (including pathogens) between carcasses in the water chiller than in an air chiller, where the carcasses are more isolated from each other. The greater potential for pathogen contamination during water chilling is another major factor in trade restrictions between countries using the two different chilling methods. Another microbially-related trade factor is the use of chlorine in product-contact water, such as in chillers. This practice is required in the U.S. and is common in some European countries as an antimicrobial agent, but is banned in other European countries because of the theoretical link to carcinogenesis.

Summary

First processing is critical in the processing plant because it involves the conversion of a living animal to inanimate tissue. The physiological responses on the animal and its living tissues are important to maintaining product quality. In addition to product quality, an overriding theme in processing is the maintenance of product yield, usually interpreted as minimizing losses. Such losses can be in the form of worker productivity with an empty shackle or missed assignment, excessive or unnecessary trimming, or maladjusted equipment making improper cuts. Because first processing is the common funnel in the conversion of live poultry into the variety of products we see in the marketplace, it is critical that all the steps from slaughter through chilling be done correctly and efficiently.

References

1. Fletcher, D. L., Stunning of broilers, *Broiler Ind.*, 56, 40, 1993.
2. Craig, E. W., Fletcher, D. L., and Papinaho, P. A., The effects of antemortem electrical stunning and postmortem electrical stimulation on biochemical and textural properties of broiler breast meat, *Poult. Sci.*, 78, 490, 1999.
3. Eisele, J. H., Eger, E. I., and Muallem, M., Narcotic properties of carbon dioxide in the dog,. *Anesthesiology*, 28, 856, 1967.
4. Mohan Raj, A. B. and Gregory, N. G., Investigation into the batch stunning/killing of chickens using carbon dioxide or argon-induced anoxia, *Res. Vet. Sci.*, 49, 364, 1990.
5. Mohan Raj, A. B., Grey, T. C., Audsely, A. R., and Gregory, N. G, Effect of electrical and gaseous stunning on the carcass and meat quality of broilers, *Br. Poult. Sci.*, 31, 725.
6. Lambooij, E., Pieterse, C., Hillebrand, S. J. W., and Dijksterhuis, G. B., The effects of captive bolt and electrical stunning, and restraining methods on broiler meat quality, *Poult. Sci.*, 78, 600, 1999
7. Singh, R. P. and Heldman, D. R., *Introduction to Food Engineering*, Academic Press, Orlando, FL, 1984.
8. Veerkamp, C. H., Chilling, freezing and thawing, in *Processing of Poultry*, Mead, G. C., Ed., Elsevier Science, England, 1989, 103.

Selected bibliography

Egg and Poultry-Meat Processing, Stadelman, W. J, Olson, V. M., Shemwell, G. A., and Pasch, S., Ellis Horwood Ltd., Chichester, England, 1988

Poultry Products Technology, 3rd edition, Mountney, G. J., and Parkhurst, C. R., The Haworth Press, Inc., Binghamton, New York, 1995.

Meat Science, 5th edition, Lawrie, R. A., Pergamon Press, Inc., Elmsford, New York, 1991.

Muscle Foods: Meat, Poultry, and Seafood Technology, Kinsman, D. M., Kotula, A. W., and Breidenstein, B. C., Eds., Chapman & Hall, New York, 1994.

Poultry Meat Science, Poultry Science Symposium Series, Vol. 25, Richardson, R. I. and Mead, G. C., Eds., CABI Publishing, Oxon, U.K., 1999.

Processing of Poultry, Mead, G. C., Ed., Elsevier Science, New York, 1989.

chapter four

Second processing: parts, deboning, and portion control

Alan R. Sams

Contents

Introduction

Once birds have been converted into processed carcasses and chilled to the required temperature, they can be packaged and marketed whole or they can be converted into some other form such as parts or boneless meat. The many possibilities that exist for a carcass beyond the chiller are grouped together under the area of the plant known as "second processing." Cutting the carcass into parts, deboning meat, and portion control sizing are functions that the poultry industry has begun to do in an effort to save the customer (home consumer, restaurant chain, or supermarket) time. Processors realized that customers were willing to pay for these services, so it has become a major effort in most companies.

Adding value

With the evolution of the modern lifestyle came a shift toward less disposable time and more disposable income. Two-income families and a hectic lifestyle have resulted in a willingness by today's consumer to pay extra for the convenience of partial preparation of the

0-8493-0120-3/01/$0.00+$.50

meal by the processing plant. People did not want to spend their time cutting the carcass into parts before cooking. In addition, consumers became willing to pay for the convenience of buying just the parts of the chicken carcass they wanted. If breast meat was all that was desired, then they could buy a package of just breasts, without the wings and legs they would not eat. These two concepts revolutionized the poultry industry in the 1960s. Even in 2000, the predominant form of chicken marketed in the U.S. is cut up parts. Thus, cutting the carcass into parts was a convenience customers valued and for which they were willing to pay. Restructured and other further processed products are additional ways to change the form of the product and add value, these products will be covered in subsequent chapters. Value can also be added by changing other dimensions of the product such as location or time (where and when it is available). An example of these value additions is the higher price of various food items sold in airports or "convenience" stores.

Parts

Many configurations of parts can be obtained from a carcass. It can be simply cut in two halves, as for grilling, or it can be cut into many pieces. A general summary of the possibilities is provided in Table 4.1. Some foreign markets demand an even more elaborate line of pre-cut parts. For example, some Asian cultures prefer their parts to be cut so that there is minimal hand contact during eating. Figure 4.1 shows a selection of wing products, some having a handle to make eating easier.

Table 4.1. **Commonly Used Configurations for Parts of a Chicken Carcass in the U.S.**

Part	Description
Half carcass	Carcass split evenly into right and left halves
Breast quarter	Anterior right or left quarter containing half of the spine, the ribs, the pectoralis muscles (major and minor), and the attached wing
Leg quarter	Posterior right or left quarter containing half of the spine, the thigh, and the drumstick
Wing	The three segments of the wing with a variable amount of the breast meat (depending on the customer)
Breast	The major and minor pectoralis muscles with or without rib and sternum bones or skin
Thigh	The upper part of the leg containing the femur
Drumstick	The lower part of the leg containing the tibia and fibula
Drumette	The inner portion of the wing
Wing portion	The middle section of the wing, with or without the outer "flipper or wing tip" portion still attached
Whole breast	The anterior half of the carcass without wings, with both breasts still connected in front and with or without the spine connecting them in the back
Keel piece	The pointed posterior tip of the whole breast before splitting (approximately one third of the whole breast)
Breast piece	After the removal of the keel piece from the whole breast, the remaining part is split into right and left halves
Whole leg	Drum and thigh with no spine
Back or strip back	Spine and pelvis, production of quarters puts the back as part of the respective quarters
Breast half or front half	The entire, intact, anterior half of the carcass
Leg half, back half, or saddle	The entire, intact, posterior half of the carcass

Figure 4.1 Selection of wing products, some with "handles."

Cutting the carcass into parts is profitable because it is a way to add value to the product. "Value-added" processing is making some change in the product to increase its appeal to the consumer. This increased appeal is reflected in an increase in price. The price a consumer is willing to pay is a reflection of the value of that product to the consumer. Profitability occurs when one considers that the consumer price increase reflects the added expense of production and an additional margin that reflects the increased, intangible value to the consumer. Cutting a carcass into parts is probably the simplest example of value-added processing for poultry.

Example:

4 lb carcass × \$0.75 = \$3.00

25% breast = 1.00 lb × \$2.50/lb = \$2.50

33% legs = 1.30 lb × \$0.90/lb = \$1.17

14% wings = 0.56 lb × \$1.50/lb = \$0.84

17% back / neck = 0.68 lb × \$0.10/lb = \$0.07

11% giblets = 0.44 lb × \$0.40/lb = \$0.18

\$4.76

Cutting cost = \$0.05/carcass
Value added during cutting = \$4.76 − (3 + 0.05) = \$1.71

In designing parts configurations, it is important to remember that some parts are of greater value and therefore have a greater price and profitability. Therefore, the objective of cutting the carcass into parts is to do it in such a way as to maximize the percentage of the carcass that is put on the most valuable parts. For example, it is more profitable to put as much back meat and bone on the breasts and legs because they are of greater value than the back. This will increase the profit achieved from the overall carcass. This is why it is common to see breasts sold with attached rib meat, and thighs sold with attached pelvis meat.

Grading of poultry will be covered in more detail in Chapter 5, but is also a good example of adding value. A carcass can be downgraded because it has a defect in one of its drumsticks. The single defect decreases the value of the entire carcass. However, if the defective part can be removed by cutting the carcass into parts, then only the bad drumstick will be downgraded, effectively adding value to the remainder of the carcass which will be restored to premium value.

Example:

Whole carcass Grade B (with broken drum)	4.00 lb × \$0.25/lb = \$1.00
. . . compared to . . .	
Broken drum	0.25 lb × \$0.25/lb = \$0.0625
Parts Grade A	3.75 lb × \$1.20/lb = \$4.50

\$1.20 = weighted average Grade A parts value
Virtually the same grading cost for the Grade B carcass and the Grade A parts
Salvaged ("added") value = \$3.5625 = (\$4.50 + \$0.0625) − \$1.00

Grading can add value in another way. The customer gets a certain reassurance of product quality and consistency from purchasing poultry that has been graded. This reassurance has a specific value for which the consumer is willing to pay, over the amount it actually costs the processor to have the product graded.

Cutting poultry carcasses into parts can be done manually with a knife or table saw, or automatically with a wide variety of available machines. With machines, the cuts are made by positioning blades in very specific orientations to the oncoming carcasses. Inherent in such an intricate operation is the need for good maintenance and sharp blades so the cuts are performed accurately. Whatever the method, carcasses are usually cut into two (halves), four (quarters), eight (breasts, wings, thighs, drumsticks), or nine (two breast pieces, a keel piece, wings, thighs, drumsticks) pieces. The latter two configurations are called eight- and nine-piece cuts, and are common in the fried chicken restaurant industry.

Yield

Yield is a measure of efficiency. It can be generally defined as the amount of output obtained for every unit of input, expressed as a percent.

$$\text{yield} = \text{efficiency} = (\text{output}/\text{input}) \times 100$$

There are many different types of yield, and each one has its particular measure of efficiency for use in managing the processing plant. The proportion of the live bird body retained through first processing is called the ready-to-cook yield (RTC). This is a direct measure of the efficiency of catching, transportation, unloading, evisceration, and trimming. Any one of these factors can cause a loss of product and therefore RTC yield. An average value for this yield is 70 to 75%, and is slightly higher when the giblets are with the carcass than when the carcass is sold without giblets (WOG). This 75% can be used in combination with the live bird feed conversion to evaluate the overall efficiency of the vertically integrated company. Because only 75% of the live bird is going to be salable, a more useful measure of live production efficiency would be to calculate feed efficiency as feed consumption per unit of salable product. Such a measure would combine the efficiencies of the live production and processing plant arenas. Of the WOG carcass, about 60% is meat and 40% is bone. Of the 60% of the carcass that is meat, about 60% is light meat and 40% is dark meat. These are general numbers and will vary due to processing plant, bird age, and genetic strain. These numbers are useful for projecting available meat volume for some future operation, such as deboning the next day or future product development.

Another form of yield is examination of the proportion of the carcass that goes to each part (see earlier example on cut up parts). As previously mentioned, this yield classification can be critical to profitability when producing cut up parts.

Aging and deboning

The immense demand for boneless breast meat has not only created a high-priced market for this product, but also a high degree of customer quality sensitivity. Customers are willing to pay for the deboning convenience, but they expect to get consistently high quality for their money. To increase the pressure further, a boneless breast fillet has no skin to hide its blemishes or seal in its juiciness, and no way to keep it tender. Its quality is vulnerable in the processing plant and in the kitchen. Because the blemishes are largely trimmed or sorted out, tenderness and yield are the greatest challenges facing the producers of boneless breast meat. Although both thigh and breast meat are commercially deboned, breast meat will be the focus in this presentation because of its much greater demand in the marketplace and in product formulations.

Aging

Early meat scientists observed that meat deboned soon after death became objectionably tough, and this toughening response lessened with the progression of postmortem time. The practice of holding meat for a period of time between death and deboning became known as "aging" or "maturing." As science learned more of the biochemistry of muscle and meat, it became apparent that the muscle needed to develop rigor mortis before deboning in order to prevent toughening.[1] Although the common early practice was to store the intact carcass overnight or longer, the efficiency and productivity pressures of today's processing plants have forced many processors to reduce this aging time to a minimum. In general, storing the carcass or front half under refrigeration until 4 hr after death (about 2.5 to 3 hr after exiting the chiller) is the minimum required aging time for broiler chicken meat in modern commercial plants. This is covered further in Chapter 7. The aging period is expensive because of the energy, labor, and space that it requires and the reduced meat yield due to water dripping from the muscle during refrigerated storage. Hirschler and Sams[2] estimated that an average-sized broiler processing plant lost 2 to 3% of its breast meat due to aging, which translated into approximately five million U.S. dollars per year. Therefore, reducing the need for aging has been the subject of research for many years.

It should be noted that these tenderness concerns only pertain to whole muscle products such as breast fillets. Because some broiler chicken breast meat and the majority of boneless turkey breast meat is formulated into restructured products, aging beyond the chilling period may not be necessary. Whole fillets are usually the premium products in a company's line, and getting acceptable tenderness with minimum aging time requires an understanding of the process of rigor mortis and its effect on meat.

Rigor mortis

Rigor mortis is the process of cell death.[3,4] When an animal dies, its individual cells remain alive, continuing their metabolism by using energy stored in them. With the loss of blood as an oxygen supply, the cells gradually shift from aerobic (oxygen-dependent) metabolism to anaerobic (oxygen-independent) metabolism. They continue to use energy but make it more slowly because anaerobic metabolism is less efficient than aerobic. This imbalance causes the cell's supply of the primary energy compound, adenosine triphosphate (ATP), to decrease. The production of lactic acid, the end product of this form of metabolism also occurs with the increased anaerobic activity. While lactic acid would be removed by the blood in the living animal, this compound accumulates in the muscle cells of dead animals and causes the cell pH to decrease from near neutrality (7) to a more acidic pH of about 5.7. This decline in pH reduces the activity of some of the ATP-producing enzymes, further reducing the production of ATP (Figure 4.2). The pH reduction during rigor mortis development affects protein functionality and further processed products, as described in subsequent chapters.

Adenosine triphosphate is an important compound in the function of a muscle cell because it not only provides energy for many reactions, but also helps regulate the interactions of the protein fibers involved in contraction. A muscle consists of overlapping protein filaments, thick filaments made of a protein called myosin, and thinner filaments made of a protein called actin (Figure 4.3). These filaments are part of a repeating structure called a sarcomere, which serves as the basic contractile unit of the muscle. One end of each thin filament is anchored in a structure called a "Z-disc" or "Z-line" at one end of the sarcomere, and the other partially overlaps one end of some thick filaments in the middle of the sarcomere. The other end of each thick filament overlaps the thin filaments at the other end of the sarcomere.

When a nerve signal reaches the muscle, it signals the release of calcium from storage vesicles into the fluid surrounding the filaments (Figure 4.3). In the presence of ATP, these calcium ions trigger the ATP to form a bridge between the thin and thick filaments. The ATP molecule then releases its energy, providing the fuel to pull the thin filaments and the ends of the sarcomere (to which they are attached) together. A new ATP molecule is then

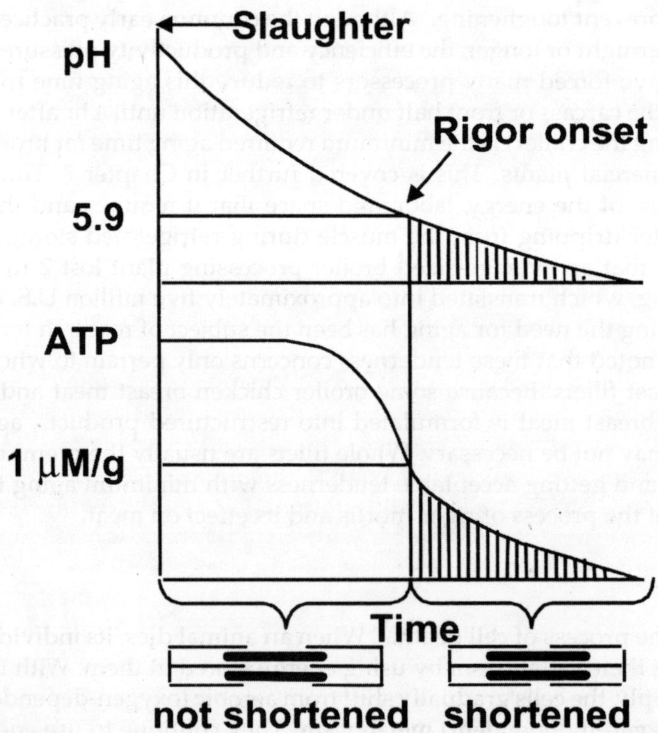

Figure 4.2 Adenosine triphosphate (ATP) and pH decline during rigor mortis development.

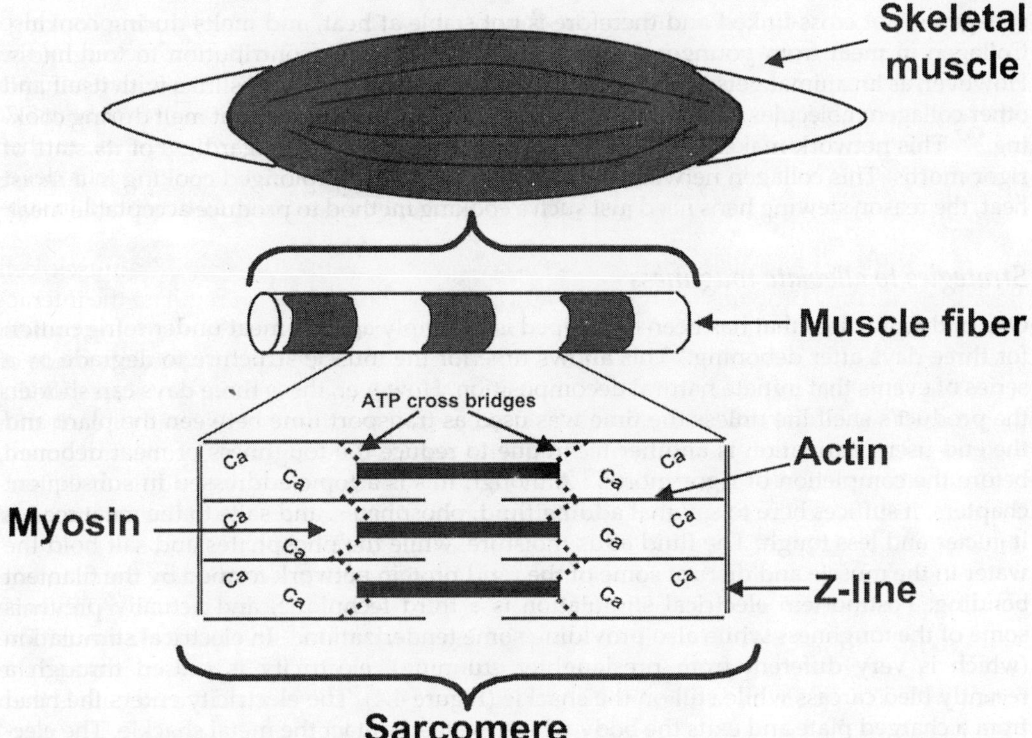

Figure 4.3 Muscle diagram showing sarcomere, filaments, and roles of calcium and ATP in cross bridge formation between the filaments.

needed to break the bond between the filaments and allow the muscle to relax to its original length. So, ATP causes contraction by providing energy and relaxation by breaking the bond between contracted thick and thin filaments. The minimum concentration for ATP to function in these roles is about 1 μM ATP/g muscle (Figure 4.2).[3] Therefore, when a muscle cell's ATP concentration falls below this level, it is no longer responsive to nervous or other stimuli and is in rigor mortis.

Cutting and deboning the muscle before rigor mortis is developed will cause a nervous signal response in the muscle and cause it to contract. Furthermore, the extent of the muscle's contraction is no longer limited by skeletal restraints, so the degree of shortening is greater for the free muscle. Additionally, when the muscle is removed from the carcass, it cools more rapidly because it no longer has the insulating skin cover and surrounding muscles. When muscles chill rapidly, the calcium storage vesicles leak. If this happens early enough after death, there can be sufficient ATP still present to initiate contraction and sarcomere shortening, a process called "cold shortening."[4] Overlap of the contractile filaments is important to toughness because meat with more overlap (shorter sarcomeres) is more dense and has more filaments per cross-sectional area for teeth to cut through during biting. Also, shorter sarcomeres have less fluid space in them, and therefore less fluid.[5] Less fluid means less juiciness, a characteristic that contributes to the toughness sensation. Deboning is not the only stimulus that can induce shortening and toughness in pre-rigor muscle. Cooking meat before rigor mortis is developed will also induce toughness.

The toughness due to the overlapping contractile filaments ("contractile toughness") is not to be confused with toughness of meat from older animals. This other toughness is primarily due to cross-linking of the connective tissue protein, collagen. In young animals, the collagen is not cross-linked and therefore is not stable at heat, and melts during cooking. Collagen in meat from younger animals provides very little contribution to toughness. However, as an animal gets older, the collagen forms heat-stable crosslinks with itself and other collagen molecules, forming a heat resistant network that does not melt during cooking.[4,6,7] This network makes the meat from older animals tough, regardless of its state of rigor mortis. This collagen network only breaks down with prolonged cooking in a moist heat, the reason stewing hens need just such a cooking method to produce acceptable meat.

Strategies to alleviate toughness

One of the practices that has been developed is to simply age the meat under refrigeration for three days after deboning.[8] This allows time for the muscle structure to degrade by a series of events that initiate natural decomposition. However, these three days can shorten the product's shelf life unless the time was used as transport time between the plant and the end user. Marination is another technique to reduce the toughness of meat deboned before the completion of rigor mortis.[9] Although this is a topic addressed in subsequent chapters, it suffices here to say that adding fluid, phosphates, and salts to the meat makes it juicier and less tough. The fluid adds moisture, while the phosphates and salt hold the water in the muscle and disrupt some of the rigid protein network formed by the filament bonding. Postmortem electrical stimulation is a third technique, and actually prevents some of the toughness while also providing some tenderization.[10] In electrical stimulation (which is very different from preslaughter stunning), electricity is pulsed through a recently bled carcass while still on the shackle (Figure 4.4). The electricity enters the head from a charged plate and exits the body where the feet contact the metal shackle. The electrical characteristics and timing cause two effects. One is that the pulses exercise the muscle, accelerating the depletion of ATP and developing rigor mortis earlier. The other effect is that the pulses cause such forceful contractions that the filaments are torn, reducing the integrity of the protein network responsible for toughness of the muscle. An important consideration in aging and using any of these strategies to reduce aging is that a target tenderness level be established. A company needs to evaluate its customers' desires and determine what level of tenderness is needed. The processors can then select or combine these techniques for reducing aging.

Portion control and uniformity

Portion control is an important concept in food service, the segment of the food industry known as the hotel, restaurant, and institutional industry (HRI). This segment has been growing rapidly in response to the increasing trend of the U.S. consumer eating more meals away from home. Portion control is an attempt to have uniform portion size, appearance, and quality so that each customer gets the same amount of food, there is less picking from a buffet line (leaves extremely large and small portions behind), and it allows more accurate food supply and cost projections. Portion control is also important in retail food marketing because the home consumer prefers uniformly sized portions because of their cooking (time, endpoint temperature, method) and serving (appearance, doneness) consistency.

The production of cut up parts was helped by the natural suitability of each part as an individual meal portion. This is particularly important in the fried chicken restaurant industry in which each piece needs to have approximately the same amount of meat so that each customer gets the same amount of product, regardless of the part they eat. This was partially the origin of the nine-piece cut. The breast of the commercial broiler was a disproportionately large proportion of the carcass and was therefore not an appropriate

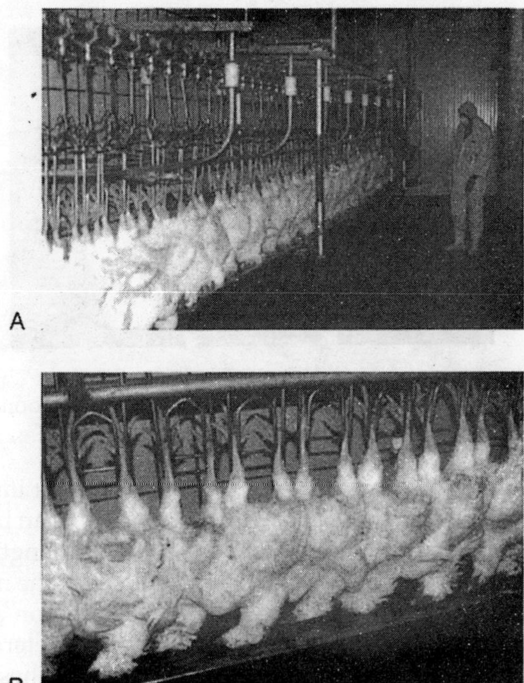

Figure 4.4 (A) Postmortem electrical stimulator in a broiler processing plant. (B) Closer view of electrical stimulator showing contact of birds' heads with the charged plate.

portion size relative to the other pieces. The solution was to cut the breast into three pieces, and to cut the wing off so as to include some of the breast meat on the wings. This distributed the carcass more evenly to all the component pieces. Further control of portions is usually obtained by sorting the carcass by weight into ranges specified by the customer. This results in all the parts of a given type (e.g., thighs) being about the same size. The scales for this weighing are usually in line with the overhead shackle line. When the shackle containing a bird of the desired weight passes over the scale, the carcass is released into a bin for cutting into parts. It should be noted that portioning could also begin with live bird. Most processors will use specific genetic strains and/or slaughter birds at a specific age to produce the largest percentage of their birds in the range specified by the customer.

Because of its high value and specialized uses, boneless breast meat is portioned in many different ways. The first thing that can be done is to trim off the connective tissue, membranes, and fat from the edges. This intricate operation is still done manually in most cases (Figure 4.5). Depending on the amount of muscle tissue this trimmed material (or "trim") contains, it can be used in some restructured product like a nugget or patty. Certainly, this trimming is a way of adding value to the fillet because it is a convenience service that reduces customer waste and increases uniformity of the product. The price increase for a trimmed fillet will be enough to cover the added production expense plus the profit associated with the added value of convenience. After trimming, the fillets can be sorted according to weight. The fillets pass down a high-speed conveyor and over a series of scales, each scale set for a specific weight range. When the fillet crosses a scale set for that fillet's weight, the scale signals a computer to use a lever to push the fillet into a bin. The computer can also keep track of productivity and inventory as it gets a signal for every fillet.

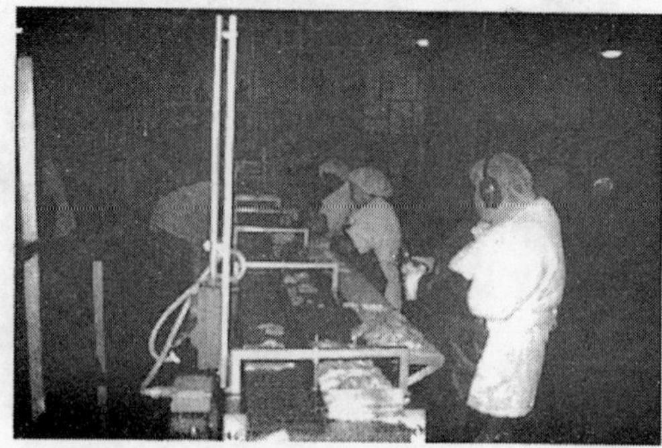

Figure 4.5 Trimming and portion sizing line in plant producing boneless breast fillets.

More refined portioning can be achieved by cutting the fillet into smaller pieces with vertical or horizontal cuts. A specific length, width, and height can be created with combinations of these cutting procedures. The vertical cuts shape the length and the width of the piece. This can be done by hand or more quickly by a machine. The machine uses a camera and computer to create a digital image of the fillet passing by on a conveyor. Using the specification for the desired size of the portion, the computer determines the best way to cut the fillet. It then directs a cutting device to make the cut. These devices usually use water knives, which are highly focused, high-pressure sprays of water. These jet streams of water easily cut through the soft muscle tissue. Frequently, the center is the premium part of the fillet, with the outer edges used with other trim in restructured products.

The horizontal cut controls the thickness of the fillet and is sometimes called a "slitter" cut. For this cut, the fillet rides between two plates or conveyor belts toward a horizontal blade or water knife (Figure 4.6). The plate or the second belt immobilizes the fillet as the knife cuts it. The distance between the blade and the belts or plates determines the thickness of the resulting fillet, and is set according to customer specifications. The portion of the fillet possessing the original outer surface of the muscle (away from the sternum) is

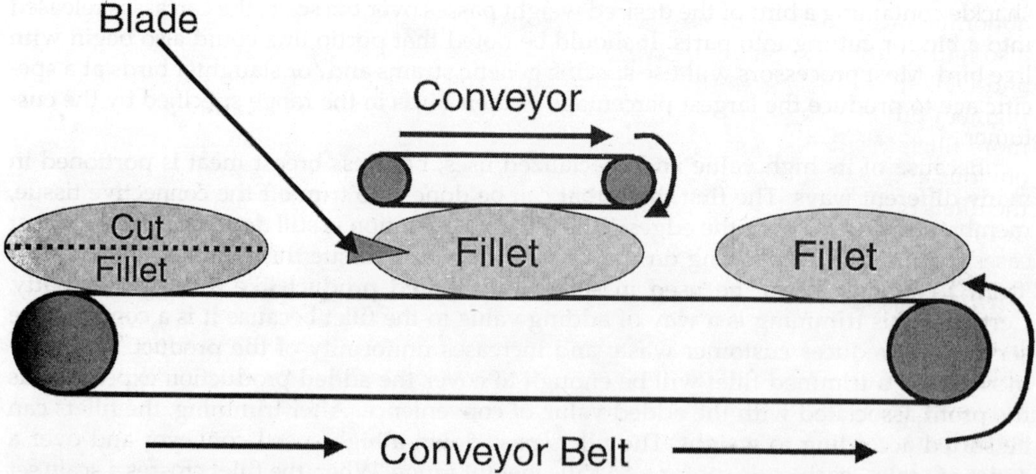

Figure 4.6 Schematic diagram of a "slitter" machine showing opposing conveyor belts and product being cut.

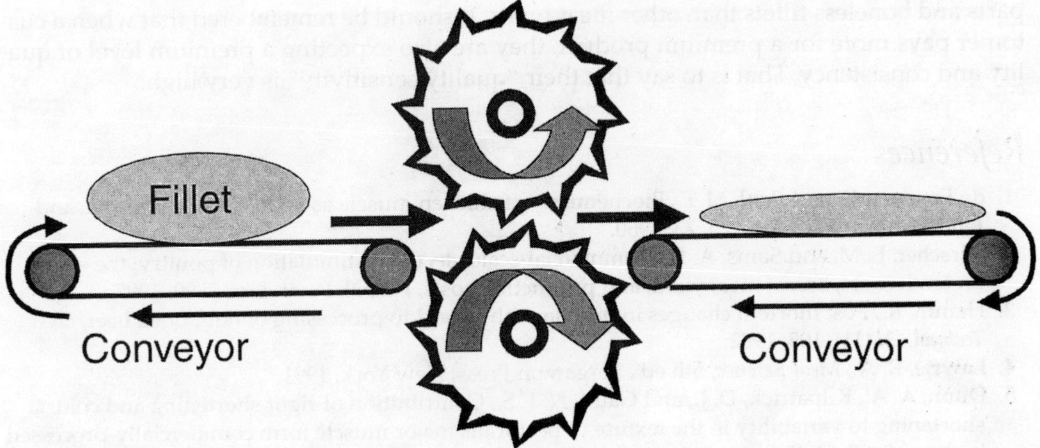

Figure 4.7 Schematic diagram of a "bridging" (or "cubing" or "flattening") machine showing opposing rollers with flattened meat.

the premium piece because it most resembles the original muscle. Although a chicken fillet is only slittered once, a turkey fillet can be slitter cut multiple times because of its greater thickness.

A final way to control the dimensions of the fillet is to flatten it through a process called "bridging" or "cubing." The fillet passes between two closely spaced rollers that have knobs protruding from their surfaces (Figure 4.7). The fillet is squeezed between the rollers, reducing its thickness and increasing its length and width. The knobs puncture the epimysial connective tissue layer, increase surface area for marinade absorption, and improve tenderness by physically disrupting the muscle structure.

In addition to size, color is another uniformity concern for poultry processors. Genetics, preslaughter heat stress, and chilling rate have all been associated with paleness and reduced water-holding capacity.[11–14] Some processors are sorting out the pale fillets from further processing because their poor functional performance can cause variation in product quality. Pale, soft, and exudative (PSE) and dark, firm, and dry (DFD) are both abnormal meat color conditions related to muscle metabolism.[4] Variation in meat color is also important to skinless, boneless breast meat in retail packages.[15,16] In addition to PSE and DFD meat, the concentration of muscle pigment proteins like myoglobin and blood pigment proteins like hemoglobin can alter meat color. Because there are usually four or more fillets displayed side by side in a package, variation in one or more fillets is very obvious and leads to rejection of the entire package by consumers. Many processors sort the fillets by color and only package uniform appearing units together.

Summary

In contrast to the first processing area of the plant, second processing involves a considerable amount of manual labor to perform the intricate cutting, trimming, and portioning. However, second processing is where much of the value is added to the profit and therefore where most of the plant's profit is derived. Production of parts and boneless breast meat are the major functions of modern processing plants and are excellent examples of adding value to the processed carcass. Portion control is a growing segment of poultry processing because poultry carcasses are more easily made into consumer-ready portions of parts and boneless fillets than other meat types. It should be remembered that when a customer pays more for a premium product, they are also expecting a premium level of quality and consistency. That is to say that their "quality sensitivity" is very high.

References

1. de Fremery, R. and Pool, M. F., Biochemistry of chicken muscle as related to rigor mortis and tenderization, *Food Res.*, 25, 73, 1960.
2. Hirscher, E. M. and Sams, A. R., Commercial-scale electrical stimulation of poultry: the effects on tenderness, breast meat yield, and production costs, *J. Appl. Poult. Res.*, 7, 99, 1997.
3. Hamm, R., Post mortem changes in muscle with regard to processing of hot-boned beef, *Food Technol.*, 36(11), 105, 1982.
4. Lawrie, R. A., *Meat Science*, 5th ed., Pergamon Press, New York, 1991.
5. Dunn, A. A., Kilpatrick, D. J., and Gault, N. F. S., Contribution of rigor shortening and cold shortening to variability in the texture of pectoralis major muscle form commercially-processed broilers, *Br. Poult. Sci.*, 36, 401, 1995.
6. Nakamura, R., Sekoguchi, S., and Sato, Y., The contribution of intramuscular collagen to the tenderness of meat from chickens with different ages, *Poult. Sci.*, 54, 1604, 1975.
7. Light, N. D. and Bailey, A. J., Molecular structure and stabilization of the collagen fibre, in *Biology of Collagen*, Vudik, A. and Vuust, J., Eds, Academic Press, New York, 1980.
8. McKee, S. R., Hirschler, E. M., and Sams, A. R., Physical and biochemical effects of broiler breast tenderization by aging after pre-rigor deboning, *J. Food Sci.*, 65(5), 959, 1997.
9. Lyon, C. E., Lyon, B. G., and Dickens, J. A., Effects of carcass stimulation, deboning time, and marination on color and texture of broiler breast meat, *J. Appl. Poult. Res.*, 7, 53, 1998.
10. Sams, A. R., Commercial implementation of postmortem electrical stimulation, *Poult. Sci.*, 78, 290, 1999.
11. McKee, S. R. and Sams, A. R., The effect of seasonal heat stress on rigor development and the incidence of pale, exudative turkey meat, *Poult. Sci.*, 76, 1616, 1997.
12. McKee, S. R. and Sams, A. R., Rigor mortis development at elevated temperatures induces pale exudative turkey meat characteristics, *Poult. Sci.*, 77, 169, 1998.
13. Wang, L.-J., Byrem, T. M., Zarosley, J., Booren, A. M., and Strasburg, G. M., Skeletal muscle calcium channel ryanodine binding activity in genetically unimproved and commercial turkey populations, *Poult. Sci.*, 78, 792, 1999.
14. Owens, C. M., McKee, S. R., Matthews, N. S., and Sams, A. R., The development of pale, exudative meat in two genetic lines of turkeys subjected to heat stress and its prediction by halothane screening, *Poult. Sci.*, 79, 430, 2000.
15. Boulianne, M. and King, A. J., Biochemical and color characteristics of skinless boneless pale chicken breast, *Poult. Sci.*, 74, 1693, 1995.
16. Allen, C. D., Fletcher, D. L., Northcutt, J. K., and Russell, S. M., The relationship of broiler breast color to meat quality and shelf-life, *Poult. Sci.*, 77, 361, 1998.

Selected bibliography

Egg and Poultry-Meat Processing, Stadelman, W. J, Olson, V. M., Shemwell, G. A., and Pasch, S., Ellis Horwood Ltd., Chichester, England, 1988.

Poultry Products Technology, 3rd ed., Mountney, G. J. and Parkhurst, C. R., The Haworth Press, Binghamton, New York, 1995.

Meat Science, 5th ed., Lawrie, R. A., Pergamon Press, Elmsford, New York, 1991.

Muscle Foods: Meat, Poultry, and Seafood Technology, Kinsman, D. M., Kotula, A. W., and Breidenstein, B. C., Eds., Chapman & Hall, New York, 1994.

Poultry Meat Science, Poultry Science Symposium Series, Vol. 25, Richardson, R. I. and Mead, G. C., Eds., CABI Publishing, Oxon, U.K., 1999.

Processing of Poultry, Mead, G. C., Ed., Elsevier Science, New York, 1989.

chapter five

Poultry meat inspection and grading

Sacit F. Bilgili

Contents

Summary

The Food Safety and Inspection Service (FSIS) and the Agricultural Marketing Service (AMS) are two branches of the United States Department of Agriculture (USDA) responsible by law for administering poultry inspection and grading activities, respectively. In the U.S., poultry inspection involves examination of each bird to determine its wholesomeness and fitness for human food, maintenance of sanitary standards, and supervision of the

0-8493-0120-3/01/$0.00+$.50

preparation, slaughter, processing, packaging, and labeling of poultry products at approved facilities. Inspection of poultry destined for interstate or foreign commerce is mandatory, and the associated costs are paid by the USDA.

Grading, on the other hand, is the classifying and sorting of poultry and poultry products according to groups of conditions and quality characteristics. Since grading activities are voluntary and must be compensated by the user, poultry plants can establish and use their own quality standards (i.e, plant grades). Grading services provided by the AMS utilize nationally standardized quality criteria, trained USDA personnel, and USDA grade marks.

Federal food safety and quality standards and regulations, as they apply to meat and poultry products, have changed dramatically since their inception. This transformation process is expected to continue in the future to effectively address the changing nature of meat and poultry products and consumer needs.

History of meat and poultry inspection

Humans have consumed animal meat from the earliest times. The consumption of certain types of meat was prohibited by various ancient cults and religions. Many early civilizations regulated the slaughter and handling of animals and performed meat inspections.[1] The consumption of meat from diseased or ill animals was considered harmful and was prohibited in Europe beginning in the 12th century. Although the sale of unwholesome and contaminated meat was a serious offense in early European cultures, the first formal and modern legislation on meat inspection was not passed until 1835 in England.

In the U.S., the production and marketing of farm animals for food was a local enterprise during the colonial period. Meat inspection was rudimentary and was performed within the auspices of the farmers, butchers, and consumers. As the population and settlements grew, marketing distances between producers and consumers increased. With the development of transportation systems, interstate and foreign commerce in meat and meat products emerged. In the early 1880s, Europeans regarded American meat as unwholesome, and restricted imports.[2]

The first meat inspection law in the U.S. was the Meat Inspection Act, enacted in 1890 primarily to regain the confidence of Europeans in American beef, and to ensure that exports met European requirements. This act provided for limited inspection of exported meat and was not effective in restoring the confidence of export markets, despite further amendments in 1891 and 1895. Many foreign governments continued to reject U.S. inspection certificates on exported meat products.[3]

The unsanitary conditions in Chicago's meat packing houses exposed by Upton Sinclair's book, *The Jungle,*[4] raised public concern about unsanitary processing conditions. Hence, following federal investigations ordered by Theodore Roosevelt, Congress passed the Meat Inspection Act (MIA) of 1906. This act, which represented one of the nation's first consumer protection measures, required mandatory inspection of meat and meat products sold in interstate and foreign commerce. In addition to establishing sanitation requirements for plants, the MIA required inspection of cattle, hogs, sheep, and goats for disease at the time of slaughter, inspection of processed products for harmful additives, and examination of all labels for truthfulness and accuracy.

In the early 1900s, the poultry industry was small and represented a secondary occupation for farmers who raised birds for personal consumption. Chickens and turkeys were also produced on small farms and sold, either live or slaughtered, at local markets. Since poultry was considered a minor meat product, it was not included in 1906 legislation.

There were no standardized methods for the production, slaughter, and marketing of poultry and poultry products. The public need for poultry meat was easily met by small-scale farm production. Live or New York-dressed (only blood and feathers removed) poultry was the common market form, usually processed and inspected by the housewife for any signs of abnormalities, spoilage, or evidence of unwholesomeness.[5]

As poultry production and consumption increased, purchasers began to demand inspection of live and slaughtered poultry. At that time, New York City served as a major distribution point for poultry. An outbreak of avian influenza caused the New York Live Poultry Commission Merchants Association to begin inspection of live poultry in 1924.[2] The need to ensure product wholesomeness led many cities, counties, and states to initiate their own inspection programs. In 1926, the Federal Poultry Inspection Service (FPIS) was established under a joint agreement between the USDA, the New York Live Poultry Commission Merchants Association, and the Greater New York Live Poultry Chamber of Commerce to assist localities in their inspection programs. Since most poultry was processed and shipped as New York-dressed, inspections were performed at the point of delivery. FPIS was authorized to conduct its own voluntary postmortem inspection, and provided inspection of eviscerated poultry at the request of purchasers. FPIS also issued wholesomeness certificates for canned poultry products, required by local and foreign governments. A voluntary inspection program was created for poultry by the USDA in 1927, with only one commercial slaughter plant operating at the time. In 1938, commercial slaughter operations were restricted to packing plants, eliminating the common practice of "on-the-farm slaughter" of animals from other premises.[6]

World War II increased the military demand for poultry products. Military purchasers asked the USDA to supply the inspection and certification services necessary for processors to meet their specifications. As market preference shifted from live poultry to New York-dressed poultry and then to ready-to-cook (RTC) poultry, the USDA modified its inspection and certification program. Point-of-delivery inspection was not satisfactory for RTC poultry, since it did not include evaluation of slaughter and evisceration conditions. The wartime poultry needs of the military were met by plants surveyed and found to meet military sanitation requirements. Thereafter, the USDA required evisceration and canning plants to process New York-dressed poultry purchased only from plants that met USDA sanitation requirements.[5]

The development and formalization of procedures for conducting antemortem and postmortem inspections at the dressing plants accelerated the trend toward consolidation of dressing and evisceration activities within a single plant. At this time, the poultry inspection activities of FPIS were limited to assuring wholesomeness and promoting sales of poultry products in jurisdictions that required certification.[5]

In 1957, Congress established a mandatory program under the Poultry Products Inspection Act (PPIA). The term "poultry" was defined as any live or slaughtered domesticated birds such as chickens, turkeys, ducks, geese, or guineas. Game birds, including pigeons and squabs, and ratites (i.e., ostriches, emus, and rheas) were not covered by this act. The PPIA required several kinds of inspection for poultry and poultry products in interstate and foreign commerce:[5]

- Inspection of birds prior to slaughter (antemortem inspection)
- Inspection of each carcass after slaughter and before processing
- Inspection of plant facilities to ensure sanitary conditions
- Inspection of all slaughtering and processing operations
- Verification of truthfulness and accuracy of product labeling
- Inspection of imported poultry products at the point of entry

The responsibility of implementing PPIA remained with the USDA's Agricultural Marketing Service (AMS), which had administered the voluntary inspection program. The Humane Slaughter Act of 1958 brought about requirements for humane slaughter of animals used for products that were sold to federal agencies. In 1962, Congress passed the Talmadge-Aiken Act, which established cooperative agreements permitting state employees to carry out inspection activities in about 300 plants. These plants were considered "federally inspected," and thus allowed to sell their products in interstate commerce.

The Wholesome Meat Act of 1967 represents the first major amendment to the MIA. This act extended inspection and enforcement requirements to meat products in intrastate commerce. It also strengthened the regulation of imported meat and formalized the federal-state cooperative inspection program, in which USDA provided funds for state inspection programs but required them to be "at least equal to" the federal inspection program. During this time, about 16% of the chickens processed in the U.S. were not inspected by USDA because they were not marketed across states lines, and 31 states did not establish their own programs to cover the inspection of such poultry. The Wholesome Poultry Products Act of 1968 similarly required inspection of virtually all poultry sold to consumers by covering the inspection activities either through state or federal programs. The act provided for federal technical assistance and up to 50% of the funding for state-approved inspection programs. If states choose to end their inspection programs or cannot maintain a standard equal to USDA, FSIS must assume responsibility for inspection. The Wholesome Poultry Products Act amended the PPIA but did not change the federal antemortem and postmortem inspection processes. Poultry inspection laws have not changed significantly since 1968, despite a tremendous increase in the quantity of poultry inspected (Table 5.1).

In 1978, the Humane Methods of Slaughter Act amended the previous inspection acts to require that meat inspected and approved be produced only from livestock slaughtered in accordance with humane methods. The provisions of this act were also extended to state-inspected plants and foreign plants exporting to the U.S. However, it is important to note that poultry and poultry products were not included in the scope of this act.

Table 5.1 **History of USDA-Inspected Poultry Plant**

Year	Number of plants	Live weight inspected (lb)
1927	1	—
1928	7	3.2 million
1940	35	76 million
1954	260	1 billion
1958	268	2 billion
1964	201	6.6 billion
1975	154	13.7 billion
1981	371	20 billion
1988	528	26 billion
1991	508	34 billion
1996	459	44 billion

Source: Modified from NRC, *Poultry Inspection: The Basis for a Risk-Assessment Approach,* Committee on Public Health Risk Assessment of Poultry Inspection Programs, National Academy Press, Washington, D. C., 1987.

The FSIS was established within the USDA by the Secretary of Agriculture in 1981 to regulate meat, poultry, and egg product industries under one agency. Since then, public health-related meat and poultry inspection activities have been administered by the FSIS.

The effectiveness of mandatory bird-by-bird organoleptic inspection had been frequently questioned by consumer groups, industry, and the scientific community. In the early 1980s, FSIS asked the National Research Council (NRC) to evaluate the scientific basis of meat and poultry inspection programs. The NRC reports, released in 1985[7] and specifically for poultry in 1987,[8] basically recommended the establishment of risk assessment-based meat and poultry inspection programs to address public health concerns.

Congress enacted discretionary inspection authority in 1986, which allowed the FSIS to vary the type and nature of inspection in processing plants based on the compliance history of the plant, commitment of plant management, and product type. The Performance Based Inspection System (PBIS) allowed computer generation of inspection schedules, and tasks for each plant and plant process based on risks. Each plant's compliance history is documented through the deficiency classification guide.

In 1988, the National Advisory Committee on Microbiological Criteria for Foods (NACMCF), and in 1990 the National Advisory Committee on Meat and Poultry Inspection (NACMPI), were established to provide advice and recommendations to the Secretary of Agriculture on issues pertaining to microbiological criteria and inspection programs, respectively, assess safety and wholesomeness of meat and poultry.[9,10]

A major milestone in the NRC recommendations was reached when the Pathogen Reduction and Hazard Analysis Critical Control Point (HACCP) system's final rule was published on July 26, 1999.[11] Under the new regulations, which were phased in between 1997 and 2000, each meat and poultry plant was required to develop a written HACCP plan to systematically address all significant hazards associated with its products. Regulatory performance standards were also introduced to reduce *Salmonella* in raw meat and poultry. In addition to the establishment of written plant Sanitation Standard Operating Procedures (SSOPs), microbial testing (generic *Escherichia coli*) was also required to monitor process control and verify the effectiveness of reducing fecal contamination during slaughter operations.[12] The scope of the PBIS was extended nationally in 1996, with a computer-based system for organizing inspection requirements, scheduling inspection activities, and recording noncompliance history was introduced for each processing establishment. Progressive Enforcement Action (PEA), a plant performance tracking system with increasingly severe enforcement actions, was also implemented to enforce compliance with the regulatory requirements.

Poultry inspection

The FSIS implements the provisions of the inspection act through Poultry Inspection Regulations Part 381.1 through 381.311. These detailed and prescribed regulations are designed to be enforceable and valid in a court of law, and written to meet the following four basic objectives:

1. That poultry is processed under sanitary conditions in an approved facility
2. Inspected for wholesomeness (i.e., suitable for human food)
3. Free from adulteration
4. Properly (truthfully and informatively) labeled

The overall food safety responsibilities of FSIS, which are administered and enforced through a national network of veterinarians and inspectors, include the following list:[12]

- Antemortem and postmortem inspection of poultry and other animals intended for human food and further processing of meat and poultry products
- Provide pathological, microbiological, chemical, and other necessary examinations of meat and poultry products for disease, infections, extraneous materials, drug and other chemical residues, or other kinds of adulteration
- Conduct emergency responses, including retention, detention, or voluntary recall of meat, poultry, and egg products containing adulterants
- Conduct epidemiological investigations of food-borne health hazards and disease outbreaks
- Monitor the effectiveness of state inspection programs to assure equivalence to those under federal acts
- Implement cooperative strategies to control food safety hazards associated with animal production practices
- Monitor foreign inspection systems and facilities that export meat, poultry, and egg products to the U.S. to assure equivalence to U. S. standards
- Reinspect imported meat and poultry products at ports of entry, and egg products at destination
- Provide public information to ensure the safe handling of meat, poultry, and egg products by food handlers and consumers
- Coordinate U.S. representation and participation in the Codex Alimentarius Commission activities

About 8000 Inspection Operations employees carry out the inspection laws in over 6000 meat, poultry, and other slaughter and processing plants in the U.S. and U.S. Territories. Some 250,000 different meat and poultry products, including other prepared foods that contain at least 2% or more cooked poultry or at least 3% raw meat, fall under FSIS inspection.[13]

There are eight primary public health-related inspection activities conducted by the FSIS in processing plants:

1. Antemortem inspection
2. Postmortem inspection
3. Condemnation and final disposition
4. Sanitary slaughter and dressing
5. Poultry chilling
6. Plant sanitation
7. Carcass reinspection
8. Residue monitoring

Antemortem inspection

The FSIS inspectors examine and observe the animals, on a discretionary basis, prior to slaughter for signs of disease and other abnormal conditions. Most companies augment this process by providing the FSIS early data on probable disease conditions that may be present in market-age flocks. Suspect flocks are segregated from healthy poultry and slaughtered under separate arrangements. Dead and dying animals are prevented from being slaughtered during antemortem inspection (Table 5.2). Antemortem inspection

Table 5.2 **Poultry Slaughter: Federally Inspected Poultry for 1999**

	Chickens		**Turkeys**		
	Young	**Mature**[1]	**Young**	**Old**[1]	**Ducks**
Head inspected (×1000)	8,098,247	175,542	263,315	1,951	23,318
Average live weight (lb)	4.99	5.20	25.4	26.6	6.60
Condemnation[2] (%):					
Antemortem	0.42	1.7	0.26	0.99	0.16
Postmortem	1.5	6.1	1.97	6.7	2.2

[1] Fully mature breeder birds.

[2] Calculated based on pounds condemned as percent of pounds of live weight inspected.

Source: Modified from USDA, Poultry Slaughter, 02.01.00, National Agricultural Statistics Service (NASS), Agriculture Statistics Board, U. S. Department of Agriculture, Washington, D. C., 2000.

accounts for a small portion of FSIS inspection activities, since through vertical integration nearly all broilers produced in the U.S. are reared under closely monitored conditions.

Postmortem inspection

Bird-by-bird inspection of carcasses is required by law for all poultry slaughtered in a federally inspected establishment. The FSIS line inspectors examine external and internal surfaces of the carcasses and internal organs after evisceration for disease conditions and contamination that would make all or part of the carcass unfit for human consumption (Table 5.2). Veterinarians supervise the line inspectors to assure uniformity in the inspection process and provide expertise in detecting disease conditions.

There are currently three inspection systems available for broiler processing plants:

1. The Streamlined Inspection System (SIS) allows maximum evisceration line speeds of 70 birds per minute (bpm), with two line inspectors per evisceration line, each examining alternate birds.
2. The New Enhanced Line Speed (NELS) allows maximum evisceration line speed of 91 bpm, with three inspectors per evisceration line, each examining every third bird.
3. Two new evisceration systems approved by the FSIS are Maestro (Meyn Poultry Processing, Gainesville, GA) and Nu-Tech (Stork Gamco, Inc., Gainesville, GA). Another, (Sani-Vis, Johnson Food Equipment Inc., Kansas City, KS) is under field testing by the FSIS for approval. These new systems allow evisceration line speeds of up to 140 bpm, with four line inspectors per line, each examining every fourth bird.

In all of these systems, eviscerated carcass and the corresponding viscera must be presented in such a way that the entire carcass, including internal and external surfaces, and all internal organs, can be rapidly but thoroughly inspected. Hocks must be cut for inspection so that synovial membranes and tendons can be inspected. With the conventional evisceration systems, viscera remains attached to the carcass to facilitate inspection. However, to achieve and maintain the SIS or NELS evisceration line speeds, carcass and visceral organs must be properly presented for inspection. To facilitate inspection, plant workers (presenters) complete the mechanical evisceration process by separating the viscera from the adhering fat tissues and opening the body cavity for ease of inspection. The new evisceration systems involve physical separation of the viscera from the carcass to reduce likelihood of contamination with the digestive tract contents. Separated viscera are then presented, either in a pan or shackle, parallel with the carcass for inspection. The new evisceration systems provide significant labor savings, since separation of viscera eliminates the need for presenters at each inspection station. Depending upon the system used, mechanical devices called selectors thrust every second, third, or fourth carcass and associated viscera at each inspection station.

Line inspectors are also provided with a plant helper for removing condemned carcasses, viscera, and parts from the evisceration line, retaining questionable carcasses for veterinary disposition, segregating contaminated carcasses (ingesta, fecal, retained yolk, bile) and those carcasses destined for off-line salvage, marking carcass parts for later trim by plant workers down the evisceration line, and for recording whole carcass condemnation causes on inspection tally sheets for each lot. Carcasses and viscera that are allowed to remain on the evisceration line are assumed provisionally inspection-passed. After automatic or manual giblet (heart, liver, and gizzard) separation, the carcasses pass through the final plant trim station, where plant employees remove marked portions of the carcass found unwholesome during inspection and with other localized defects such as breast blisters, broken or dislocated bones, hemorrhagic skin discolorations and bruises, skin sores and scabs, etc.

Inspection regulations provide detailed descriptions of conditions (i.e., physical layout of inspection stations (Figure 5.1), line speed control mechanisms, proper carcass presentation techniques, hand-washing and recording facilities, etc.) under which postmortem carcass inspection is conducted. The USDA inspection mark (Figure 5.2) is required on consumer packages and shipping containers of federally inspected poultry and poultry products.

Condemnation and final disposition

Based on postmortem bird-by-bird examination by line inspectors, carcasses are classified either as:

1. Inspection-passed—carcasses and or viscera with no apparent signs of disease or unwholesome condition remain on the evisceration line but may require trimming of minor lesions to meet RTC requirements prior to chilling.
2. Trimmed/salvaged/washed and passed—carcasses exhibiting localized or discrete lesions are either trimmed by the inspector helpers or affected parts marked for later trim on the line. Carcasses with localized disease involvement (mild airsacculitis and cellulitis) and those contaminated with extraneous material (gall, retained yolk, and digestive tract contents) are usually marked, removed from the evisceration line, and hung on a separate line for salvage. Salvage or reprocessing of these carcasses is performed at an off-line station and involves the use of approved procedures and re-inspection prior to chilling (Figure 5.3).[14]
3. Retained for disposition by the veterinarian—carcasses with questionable lesions are retained at the inspection station for evaluation by the Inspector in Charge (IIC).
4. Condemned as whole carcasses[15–17]—condemned carcasses are recorded on inspection tally sheets and classified either as field- or plant-related condemnations.
   - Field causes of whole bird condemnations:

     Tuberculosis: condemnations due to tuberculosis are almost nonexistent in broilers because of their young market age. It has been occasionally observed in older fowl.

Figure 5.1 Inspection station with carcass hung back for further examination, "condemed" cans in foreground, and clipboard for recording defect frequency above shackle line.

Leukosis: this condition usually refers to tumor development due to Marek's disease, caused by a herpes virus. Marek's disease can cause tumors in visceral organs (spleen, liver, reproductive organs, muscles, bones, and nerves), as well as skin (feather follicle enlargement due to lymphocytic infiltration); (Figure 5.4).

Septicemia/toxemia: a general condemnation category that includes clear signs of systemic disease involvement. Affected carcasses are often emaciated due to

Figure 5.2 USDA inspection mark for wholesomeness.

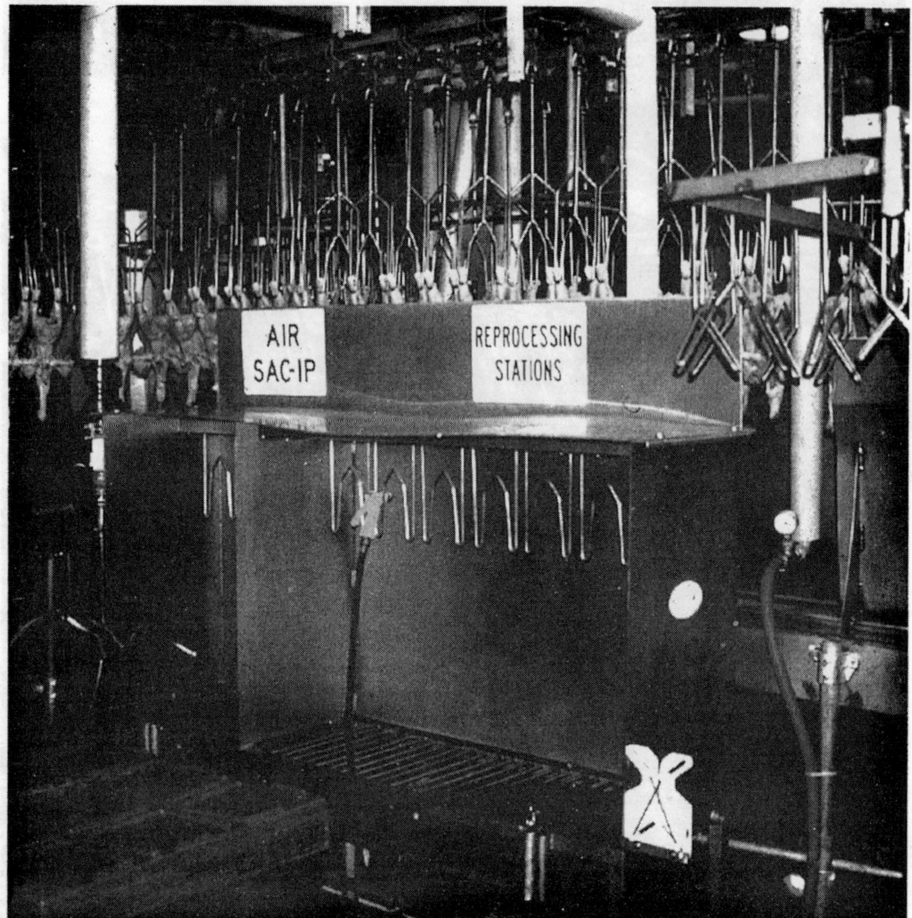

Figure 5.3 Off-line reprocessing station with washing area in cabinet and boxes for holding salvaged parts in the foreground.

severe muscle loss and appear blue or dark discolored because of dehydration. Typically, the fat depots are brown-discolored and may include pinpoint hemorrhages around the coronary region of the heart.

Airsacculitis: these carcasses show exudate or pus involving the respiratory system and the airsacs. Carcasses showing signs of systemic disease are condemned (Figure 5.5).

Synovitis: carcasses with reddened, inflamed, and swollen hock joints, especially with exudate, are considered to have synovitis. Those with enlarged or ruptured tendons, with or without green discoloration of the area, are usually affected by tendonitis or tenosynovitis. Regardless of the condition, the affected legs are usually trimmed. Whole carcasses are condemned only if signs of systemic infection is evident (Figure 5.6).

Cellulitis: cellulitis or infectious process (IP) refers to inflammation of the tissue underlying the skin of birds. The lesions, typically found on the pelvic area of the carcass, are generally localized in nature and can be trimmed. Only those carcasses with diffuse lesions are condemned as whole carcasses (Figure 5.7).

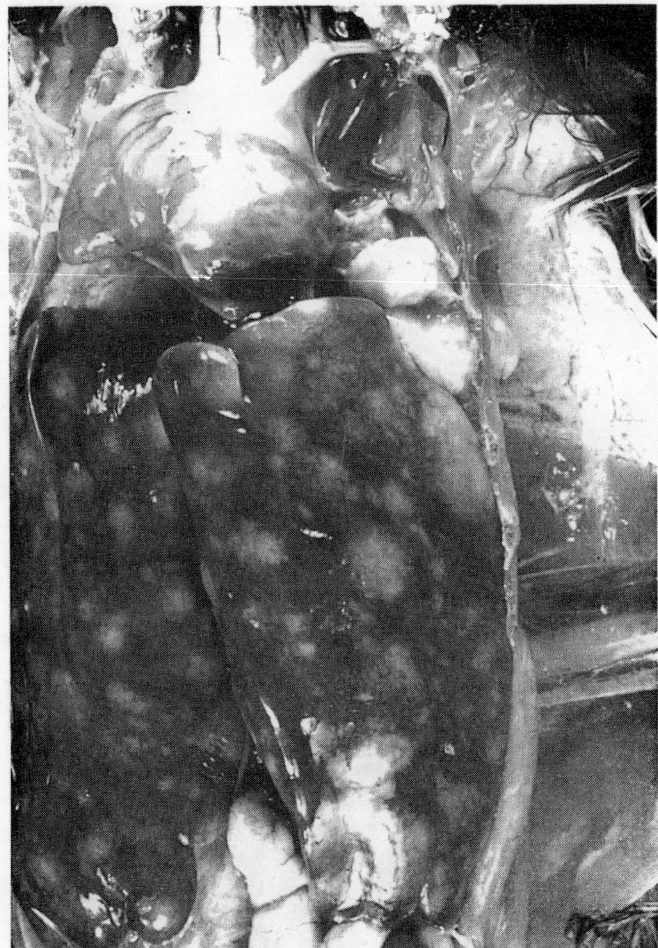

Figure 5.4 Leukosis showing liver tumors.

Tumors: the most common tumor in young chickens is squamous cell carcinoma, which actually regresses with age. The tumors are seen as craters on the skin after defeathering. At present, carcasses with two or more tumors are condemned (Figure 5.8).

Bruises: fresh or old hemorrhages (bruises) are usually trimmed on the line to remove the affected portions. Only those carcasses with severe lesions are condemned as a whole for this condition.

- Plant causes of whole bird condemnations:

Cadavers: birds that have died from causes other than loss of blood during slaughter. The carcasses and viscera appear cherry red in color, and may have a foul odor. Usually blood accumulates in lower regions (i.e., neck) of the carcass.

Contamination/mutilation: carcasses that are contaminated with extraneous material (oil, paint, grease, etc.), those which cannot be inspected because of excessive contamination with digestive tract contents, those that are mutilated by equipment and have no salvageable parts, and those that fall into the drain are condemned as a whole.

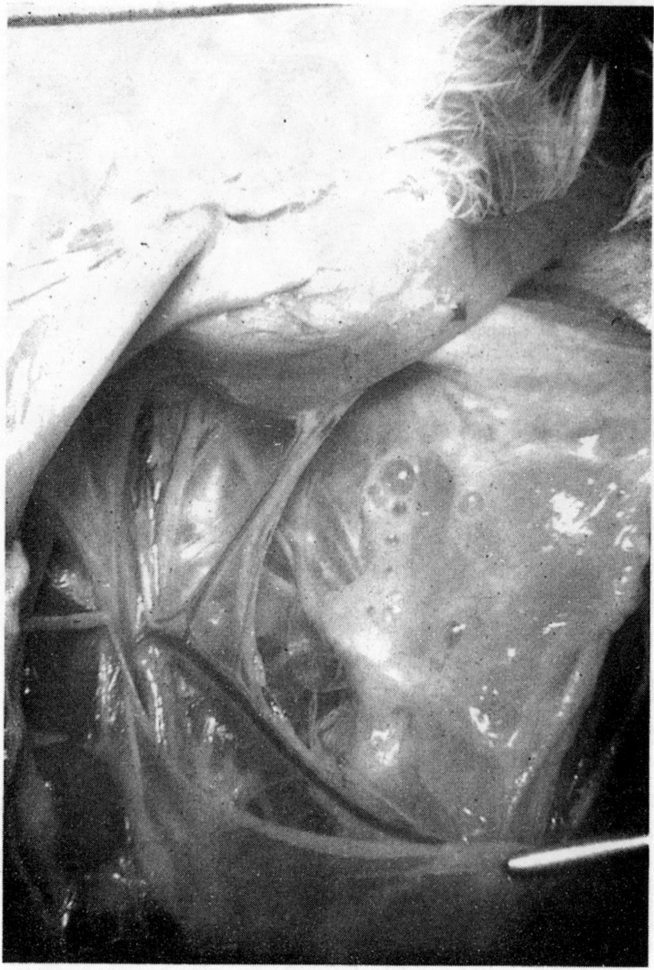

Figure 5.5 Airsacculitis showing cloudy air sac membranes.

Overscalding: a carcass with a cooked appearance, with white color deep down the pectoral muscles is considered an overscald. This condition is typically observed after stoppage of the picking line due to mechanical problems or equipment breakdown.

Sanitary slaughter and dressing

Preventing fecal contamination of the carcasses from spillage of digestive tract contents or smearing of fecal material on edible meat surfaces is the single most important aspect of sanitary slaughter and dressing regulations. The FSIS inspectors monitor picking and evisceration operations to assure minimal contamination of the product. Carcasses contaminated with extraneous material are removed from the evisceration line at the inspection stations and sent to a separate station for reprocessing. Such carcasses must be reprocessed by a combination of vacuuming, trimming, and washing with water containing 20 ppm chlorine and reinspected prior to chilling. The FSIS has recently modified Finished Product Standards to introduce a "Zero Fecal Tolerance" policy for carcasses entering the chiller.[18]

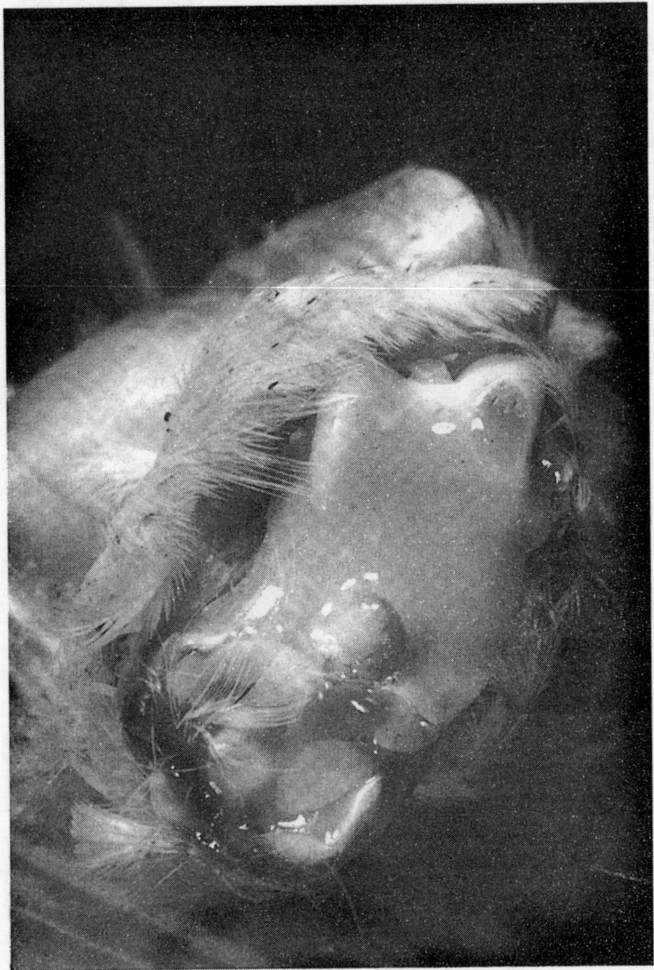

Figure 5.6 Synovitis with cloudy fluid accumulation in the hock joint.

Poultry chilling

After inspection, giblet harvest, and trim, carcasses classified as RTC are promptly cooled to inhibit bacterial growth in immersion chillers. The internal temperature of the meat must be reduced to below 4.4°C within 4 hours (for 4-lb broiler), 6 hours (4- to 8-lb) and 8 hours (>8-lb broiler or turkey), unless they are to be frozen or cooked immediately. Similarly, giblets (heart, liver, gizzard, and necks) are chilled in separate immersion chillers to <4.4°C within 2 hours. The FSIS also monitors scalder and chiller water overflow (3 and 2 gallons per bird entering the tank, respectively), monitors chill water and chilled product temperatures, and regulates the extent of moisture uptake. Moisture uptake limits are currently being re-evaluated by the FSIS.

Plant sanitation

FSIS inspectors constantly monitor the facilities and equipment for proper sanitation. The structural aspects of the premises, water supply, waste handling system, slaughter and processing equipment, personnel facilities and practices, ice and dry storage areas, coolers and

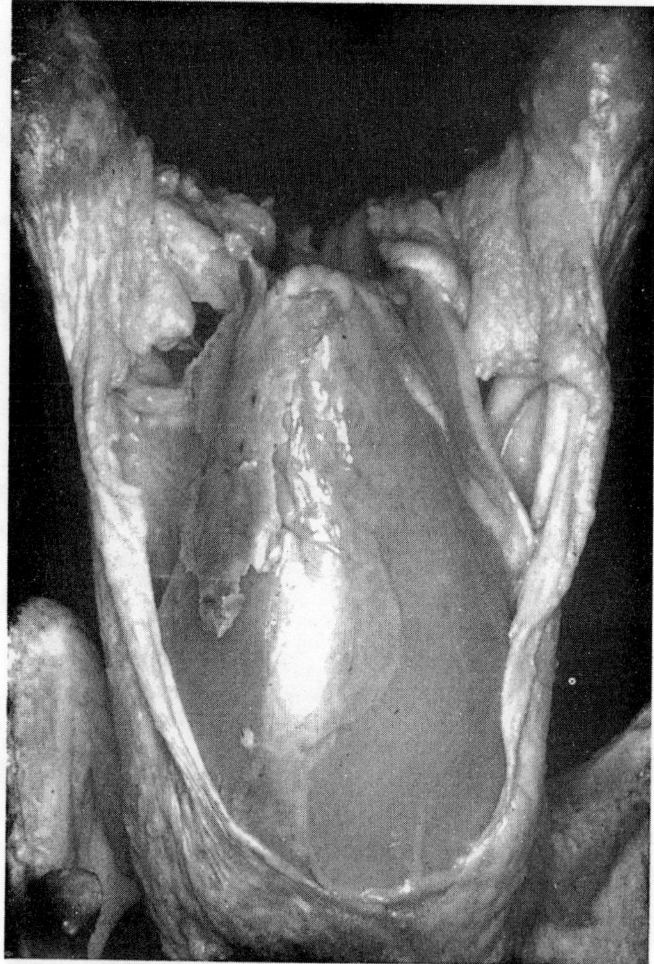

Figure 5.7 Cellulitis or infectious process showing plaque formation between skin and breast muscle.

freezers, pest control programs, and other hygiene-related features of the plant environment are monitored. Sanitation activities also include preoperational inspections, as well as maintenance of sanitary conditions during slaughter and processing. Since 1997, with the introduction of the Pathogen Reduction and HACCP ruling,[18] each plant is required to develop and implement Sanitation Standard Operating Procedures (SSOPs). These detailed written sanitation programs describe the sanitation procedures and their use frequency, assign responsibilities, and indicate corrective actions and record-keeping activities.

Carcass reinspection

After the standard postmortem inspection and trim/salvage operations, samples of carcasses are systematically reinspected (pre- and postchill) by the plant quality control personnel and FSIS inspectors. Processing defects (ingesta, excessive feathers, remnants of viscera, bile stains), carcass trim defects, and chilling conditions (temperature and extraneous material such as metal particles or grease) are re-evaluated with separate on-line

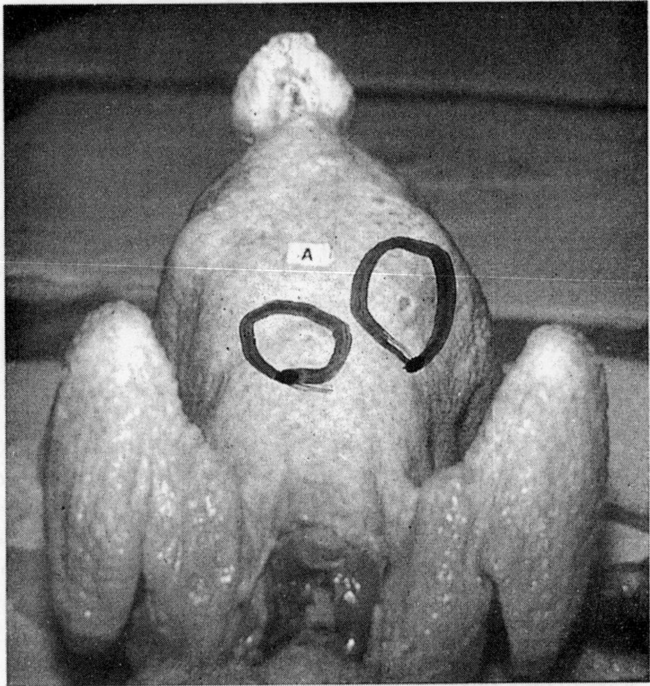

Figure 5.8 Squamous cell carcinomas (circled) on skin surface.

tests under the Finished Product Standards (FPS) system. In this system, defects are recorded on successive samples based on a Cumulative Sum System (CUSUM), which replaced the Acceptable Quality Level (AQL) standards developed in 1973. A sample of carcasses are also examined prior to chilling to assure that the zero-fecal standard is met. In addition, salvaged product and carcasses that are reprocessed for fecal contamination are also reinspected prior to chilling.

Residue monitoring

The FSIS inspection activities also include monitoring for drug and chemical residues in animal tissues resulting from the improper use of or accidental exposure to pesticides, herbicides, animal drugs, and controlled feed additives, as well as from industrial accidents which may contaminate animal feeds or the environment. Under the umbrella of the National Residue Program, the FSIS, Food and Drug Administration (FDA) and Environmental Protection Agency (EPA) cooperate in determining the presence and the level of chemicals found in poultry products. The FDA and EPA prescribe the conditions under which approved chemicals and drugs are approved and used in the poultry production system. Tissue samples are randomly utilized to screen for number of chemicals on the basis of their documented toxicity, exposure, and persistence levels.[19]

Other inspection activities

The FSIS also utilizes plant-operated voluntary Total Quality Control (TQC) programs to monitor further processing facilities since many companies utilize various process control systems to assure consistency in their products, to comply with the FSIS requirements, and to adhere to their own quality standards. In addition to in-plant inspection activities, FSIS monitors the meat and poultry supply for wholesomeness and labeling accuracy throughout the food chain (warehouses, brokers, distributors, retail chains, etc.). Products found unsafe for human consumption are removed from the marketing chain through detention, seizure, or recalls. Although the vast majority of plants regulated by FSIS comply with the inspection laws, FSIS uses several enforcement tools (warning letters, criminal prosecution, injunctions, withdrawal of inspection, and plant closing) when violations occur. The FSIS is also responsible for assuring the safety of imported meat and poultry, which must meet the same standards as domestic products. For a country to be eligible to export to the U.S., it must impose inspection requirements at least equal to those enforced in the U.S. Finally, imported products are reinspected by FSIS when they enter the U.S.

The pathogen reduction and HACCP program

On July 25, 1996, FSIS issued the Pathogen Reduction-Hazard Analysis Critical Control Point (HACCP) System's Final Rule.[18] This program introduced four major changes in the meat and poultry inspection program:

1. Requirement for all meat and poultry plants to develop and implement written SSOPs
2. Set pathogen reduction performance standards for *Salmonella* for slaughter plants and for plants producing raw ground meat products
2. Established microbial testing for generic *E. coli* to verify the adequacy of process controls for preventing fecal contamination
3. Mandated all meat and poultry plants to develop and implement product specific HACCP plans to improve the safety of their products

The requirements contained in the final rule were phased in over a period of four years. The requirement for SSOPs in all plants, and generic *E. coli* testing in slaughter plants, became effective in January 1997. The requirements for HACCP and *Salmonella* performance standards were phased in based on plant size over a three-year period. Immediately after the introduction of Pathogen Reduction-HACCP Final Rule, FSIS modified the FPS to introduce the zero tolerance standard for visible fecal material on pre-chill carcasses and carcass parts.[20] FSIS considers fecal material a vehicle for pathogens. Since microbiological contamination is reasonably likely to occur during the slaughter process, plants must adopt controls to prevent fecal contamination and occurrence of pathogens. To verify this, ten carcasses are examined (twice per shift) after the final carcass wash and before chilling for visible fecal material. As with other plant inspection deficiencies, if visible contamination is detected during a test, the nature of the noncompliance is documented (NR) and the plant is notified to take corrective actions consistent with the HACCP plan.[21]

Sanitation standard operating procedures (SSOPs)

Maintenance of sanitary conditions during meat and poultry processing is essential for achieving food safety. FSIS requires that all meat and poultry establishments develop, maintain, and adhere to SSOPs. Unsanitary facilities and equipment, poor food handling conditions, improper personal hygiene, and other unhygienic practices create an environment conducive to product contamination with microorganisms, including pathogens. Traditionally, FSIS has enforced sanitation requirements through very prescriptive regulations, detailed guides, and direct hands-on involvement by inspectors in day-to-day preoperational and operational procedures. The SSOP requirements were developed to shift the sanitation responsibility and accountability from FSIS inspectors to the slaughter establishments. The SSOP Final Rule[22] was considered a significant move by the FSIS to reform, reorganize, and recodify meat and poultry inspection regulations.

The SSOPs are developed, documented, implemented, and maintained by each plant. The general requirements of SSOPs are:

1. SSOPs must describe all sanitation procedures conducted daily, prior to and during operations, sufficient to prevent direct product contamination and adulteration.
2. The SSOPs must specify frequency and identify employee(s) responsible for the implementation and maintenance of sanitation procedures.
3. Appropriate corrective action(s), including restoration of the sanitary conditions, prevention of recurrence, disposition of the affected product(s), re-evaluation, and modification of the sanitation activities must be documented.
4. Daily sanitation records will be maintained, up to six months, to document the implementation and monitoring of the SSOPs. Such records will be initialed and dated by the responsible employee(s).
5. The SSOPs will be signed and dated by an individual with overall authority for the establishment.

The FSIS verifies the adequacy and effectiveness of each SSOP by reviewing its daily implementation and written documentation, or by direct observation and testing to assess the sanitary conditions in the plant. The SSOPs and Good Manufacturing Practices (GMPs)[23] are considered essential prerequisites to HACCP system.

Hazard analysis critical control points (HACCP) system

HACCP is a straightforward, logical, and systematic process control system that focuses on prevention of food-borne hazards. Both the principles of the system and its applications throughout the food chain have been extensively described[24–26] and endorsed by many national and international organizations including the National Academy of Sciences,[7] International Commission on Microbiological Specifications for Foods,[27] Codex Alimentarius Commission,[28] and the NACMCF.[29] There are seven widely accepted HACCP principles:

- Principle 1: Conduct a Hazard Analysis. Prepare a list of steps in the process where significant hazards are reasonably likely to occur and describe the preventative measures that can be applied to control those hazards. A food safety hazard is defined as "any biological, chemical, or physical property that may cause a food to be unsafe for human consumption." Food safety hazards might be expected to arise from natural toxins, microbiological contamination, chemical contamination, pesticides, drug residues, zoonotic diseases, decomposition, parasites, unapproved use of direct or indirect food or color additives, and physical (i.e., metal, glass, plastic) hazards. The hazard analysis includes food safety hazards that can occur before, during, and after entry into the establishment.
- Principle 2: Identify Critical Control Points (CCPs) in the process. The CCP is defined as "a point, step, or procedure in a food process at which control can be applied and, as a result, a food safety hazard can be prevented, eliminated, or reduced to acceptable levels." Several "CCP Decision Trees" were developed to separate CCPs from other process control points.[29]

- Principle 3: Establish critical limits for preventative measures associated with each identified CCP. The critical limit is defined as "the maximum or minimum value to which a physical, biological, or chemical hazard must be controlled at a critical control point to prevent, eliminate, or reduce to an acceptable level the occurrence of the identified food safety hazard." Critical limits are usually objective and quantitative measurements, such as time/temperature, humidity, water activity, pH, salt concentration, or chlorine level.
- Principle 4: Establish CCP monitoring procedures. Monitoring is a planned sequence of measurements and observations to not only assess whether a CCP is under control, but also to produce an objective record for verification. Monitoring could be done continuously, such as the automatic time/temperature equipment used at a cooking step, or it could be non-continuous, where the measurements are taken at prescribed frequency. The establishment of monitoring frequency depends on the process and may require the use of statistically based sampling schemes.
- Principle 5: Establish corrective actions. Develop procedures to be taken when monitoring indicates that there is a deviation (i.e., a failure to meet a critical limit) from an established critical limit. The corrective actions are preplanned and must include: disposition of the non-complying product; elimination of the cause of the deviation to prevent recurrence; demonstration that the CCP is under control; and maintenance of records of the corrective actions.
- Principle 6: Establish recordkeeping procedures that document the HACCP system. Typical HACCP system records include Product(s) Description Forms, Product and Ingredient Forms, Process Flow Diagram Forms, Hazard Analysis/Preventative Measures Forms, CCP Determination Forms, Critical Limits/Monitoring/Corrective Action Forms, Verification Forms, and the Master HACCP Plan.
- Principle 7: Establish verification and validation procedures to ascertain that the HACCP system is working correctly. Verification activities include the use of analytical tests or audits to evaluate the accuracy of monitoring procedures, equipment calibrations, microbiological sampling, record reviews, on-site inspections and process audits, and product/environmental sampling. Validation is a broader assessment to demonstrate that the HACCP plan that is put into place can actually prevent, eliminate, or reduce the levels of hazards identified in the process. Scientific literature, experimental research findings, scientifically based requirements, regulatory standards, or information developed by process authorities can be used to validate HACCP plans for specific products.

Microbial testing

In accordance with the Pathogen Reduction/HACCP rule, each federally inspected establishment must conduct microbial testing for generic *E. coli* to monitor process control and for *Salmonella* to meet performance standards established for raw meat products. These criteria and standards were based on the FSIS Nationwide Microbiological Baseline Data Collection Program.[30]

Testing for E. coli

Slaughter plants are required to test for generic (Biotype I) *E. coli* on processed carcasses after chilling to verify that their processes are preventing and removing fecal contamination. Generic *E. coli* is selected because it is an excellent indicator of fecal contamination and it is easy, and relatively inexpensive, to culture and enumerate. The *E. coli* performance criteria are not an enforceable regulatory standard, but provide for an objective point of reference or guideline to assess process control. Each poultry slaughter establishment must have written procedures for testing to include sampling (responsibilities, location, and randomization), handling (collection, sample integrity, and shipping conditions), culture (an approved method with sensitivity to detect <5 cfu/ml rinse fluid), and analysis (tabulation or process control chart). For poultry, whole carcasses are randomly selected after chilling and drip, 1/22,000 (broilers) and 1/3000 (turkeys) and rinsed with an approved diluent in a bag. *E. coli* are then enumerated in carcass rinse fluid (cfu/ml) using one of the approved methods found in Official Methods of Analysis of Association of Official Analytical Chemists (AOAC).[31] The most recent 13 test results are typically displayed on a control chart. The establishment is considered to be operating within the *E. coli* performance criteria if out of the last 13 tests conducted, none exceed the upper limit (1000 cfu/ml) of marginal range (M), and fewer than three are between the lower limit (100 cfu/ml) of marginal range (m) and M.

Testing for Salmonella

Poultry plants are required to meet the established national *Salmonella* performance standard for raw products when sampled and tested by FSIS. Unlike the *E. coli* performance criteria, *Salmonella* standard must be achieved by the establishment, not on a lot-by-lot basis, but consistently over time through the use of appropriate process controls. The frequency, timing, and analysis of *Salmonella* tests are done by federal inspectors. The national *Salmonella* performance standard for broilers is 20%. This standard equates to a maximum of 12 *Salmonella* positive samples in a complete set of 51 samples. Plants that fail to meet the performance standard must take corrective actions or FSIS suspends inspection services for that particular product. During the first year of *Salmonella* testing, which covered about 200 large meat and poultry plants from September 1998 to January 1999, national *Salmonella* prevalence level was reduced 10.9%.[32]

HACCP-based inspection models project

Consistent with the Pathogen Reduction/HACCP Final Rule,[11] FSIS is gradually removing many prescriptive "command-and-control" inspection requirements and allowing plants the freedom to develop their HACCP systems to meet food safety standards. Existing food safety standards, *Salmonella* incidence and zero tolerance for fecal contamination, will be supplemented in the near future to include other pathogenic microorganisms (i.e., *Campylobacter jejuni/coli*) and other food safety issues.[33] The HACCP-Based Inspection Models Project (HIMP) is basically a new model for inspection that will replace the existing SIS, NELS, and Maestro/NuTech inspection models by allowing plants to design their process control systems around the "bird-by-bird" inspection or "carcass sorting" process currently carried out by FSIS line inspectors. With this system, many of the organoleptic defects monitored by the inspectors, although considered consumer protection concerns, will be addressed by trained plant personnel. Under the HIMP, 28 volunteer plants (20 chicken, 5 turkey, and 3 hog) will be extending their HACCP plans and other process control systems to assume the responsibility of preventing unsafe and unwholesome meat and poultry from entering the food supply.

Although possibly subject to further modification, diseases and conditions observable during postmortem inspection are categorized according to their food safety (FS) or other consumer protection (OCP) significance. Based on this classification, septicemia/toxemia and fecal contamination are considered two important food safety hazards to public health and carry "zero" performance standard. Other diseases and conditions are aesthetic defects that rarely, if ever, present a direct food-borne risk to the consumers, but are rather considered unacceptable components of meat and poultry products. Currently OCPs are categorized as: animal infectious, neoplastic and degenerative conditions (cellulitis, airsacculitis, tuberculosis, synovitis and tenosynovitis, visceral infections, leukosis, carcinoma and sarcomas, and ascites); miscellaneous non-degenerative conditions (bruises, breast blisters, overscalding, mutilation, skin sores and scabs, and cadavers); and contamination with digestive tract contents and dressing defects with (cloaca, intestine, esophagus) and without (feathers, hair, lung, and trachea) contact with digestive tract. FSIS will establish performance standards for food safety and other consumer protection conditions that each pilot plant must meet. Under the HIMP, prevention of unacceptable levels of OCPs will be monitored and documented with the use of Statistical Process Control methods and control charts.

Plants will carry out these HIMP activities under FSIS inspectors oversight and verification inspection, where samples of products will be examined by FSIS inspectors to assure that meat and poultry products meet FSIS requirements. FSIS inspectors will continue to have the authority to stop or slow evisceration lines as appropriate, retain product that may be adulterated or misbranded, withhold the marks of inspection (Figure 5.2), and reject facilities, equipment or any parts of the plant determined to be out of compliance with the inspection regulations. In addition to the HIMP, FSIS is developing an "in-distribution" pilot program to verify that meat and poultry products produced by federally inspected establishments are not adulterated or misbranded as they move in the distribution system to the consumer. FSIS is testing the feasibility of deploying inspection resources to carry out both in-plant and in-distribution activities. Testing of this in-distribution model is expected to begin in 2000.[34]

Poultry grading

Grading is defined as classifying and sorting of poultry and poultry products, shell eggs, and rabbits according to various groups of conditions and quality characteristics. The development of grading standards and regulations to implement the grading services is authorized by the Agricultural Marketing Act of 1946. Federal grade standards evolved slowly from the establishment of the Office of Markets in 1913 until World War II, when the military started requiring consistency in inspection and quality in purchasing foods delivered to the U.S. troops.[35] Today, *Poultry and Egg Market News,* and standardization and grading activities are provided by the Poultry Division of the USDAs Agricultural Marketing Service (AMS). *Poultry and Egg Market News* is a supply, demand, and price report that AMS distributes nationally, tracking the sales of about 50 commodities. In addition to implementing grading services based on official USDA standards, the Poultry Division of AMS is also involved in certification of poultry and shell eggs purchased through food procurement contracts to verify specifications for quantity, quality, condition, formulation, net weight, packaging, storage, and transportation.[35]

The use of grading and certification services is voluntary. In addition to a grading fee, establishments using AMS must also provide space, equipment, lighting, or other facilities as required by the grader or regulations.[36] Grading is performed by a USDA grader assigned to an establishment on a full- or part-time basis, depending upon the volume and nature of products produced. Usually, resident graders are assisted by plant employees to handle large volumes of poultry. However, the resident grader performs final check grading and certification through an appropriate sampling plan. Companies can develop and

Figure 5.9 Official USDA grademark.

use their own grade specifications (i.e., plant grade). However, the letters "U.S." or "USDA" may only be used with a poultry grademark if the poultry has been graded by an authorized USDA grader (Figure 5.9).

All poultry that is graded by the USDA must first be inspected by FSIS for wholesomeness. After the poultry pass inspection, the product is eligible for grading according to official standards of quality. All products, whether in the form of RTC whole carcasses, parts, or further processed products may be graded. The RTC poultry that contain processing defects (excessive protruding feathers, bruises that require trimming, remnants of lungs, trachea, and other organs, extraneous material of any type) or with off-condition (slimy, slippery, putrid, or sour odors) are not graded and must be reworked to remove the defective portions.

For RTC carcasses and parts, the standards of quality include:

1. Conformation: the skeletal deformities that may affect the normal distribution of flesh. Dented, crooked, knobby or V-shaped breasts, deformed wings and legs, and wedge-shaped frame are example defects that detract from normal appearance.
2. Fleshing: amount of flesh is consistent with the carcass and its parts. Most of the flesh is located on the breast, thighs, and drumsticks. There are certainly sex differences in the amount of flesh over the back, with females carrying more flesh than males.
3. Fat covering: a well-developed layer and distribution of fat in the skin. Fat typically accumulates around the feather tracts, but some fat is also deposited between the feather tracts over the back and the hips.
4. Feathers: carcass or its parts must be free of protruding feathers and hairs (down on ducks and geese) to meet the RTC requirement and to be eligible for grading.
5. Exposed flesh: exposed flesh can result from cuts, tears, and trims on the carcasses. It detracts from the appearance of the product and may lower the eating quality by allowing the flesh to dry out during storage and cooking.
6. Discolorations: lightly shaded areas on the skin due to incomplete bleeding or hemorrhaging that are free of clots. Dark red, blue, or green discolored bruises must be removed before grading.
7. Disjointed or broken bones and missing parts: carcass and parts free of broken and disjointed bones.
8. Freezing defects: darkening and dehydration of poultry skin or the surface of skinless products due to freezing and storage (i.e., freezer burns).
9. Accuracy of cut: when parts are separated at a joint, the joint should be evenly split. Also, a part should only contain the appropriate anatomical tissues. For example, a "whole leg" should contain only a drum and thigh with no spine, while a "leg quarter" should contain a drum, thigh, and half of the spine.

Table 5.3 **Summary of Specifications for "A" Quality Poultry**

Conformation:							
Breastbone				**Normal**			
Back				**Slight curve or dent**			
Legs and Wings				**Slight curve**			
				Normal			
Fleshing:		Well fleshed, considering kind and class					
Fat covering:		Well developed layer, especially between heavy feather tracts					
Defeathering:		Turkeys		Ducks and Geese[1]		All Other Poultry	
Protruding feathers		(less than or equal to 3/4 in.)		(less than or equal to 1/2 in.)		(less than or equal to 1/2 in.)	
(Feather length)		Carcass	Parts	Carcass	Parts	Carcass	Parts
		4	2	8	4	4	2
Exposed Flesh:[2]		Carcass		Large Carcass Parts[3] (halves, front and rear halves)		Other Parts[3]	
Weight Range							
Minimum	Maximum	Breast and Legs	Elsewhere	Breast and Legs	Elsewhere		
None	2 lb	1/4 in.	1 in.	1/4 in.	1/2 in.	1/4 in.	
Over 2 lb	6 lb	1/4 in.	1 1/2 in.	1/4 in.	3/4 in.	1/4 in.	
Over 6 lb	16 lb	1/2 in.	2 in.	1/2 in.	1 in.	1/2 in.	
Over 16 lb	None	1/2 in.	3 in.	1/2 in.	1 1/2 in.	1/2 in.	
Discolorations:		Lightly Shaded			Moderately Shaded[4]		
Carcass							
		Breast and Legs		Elsewhere	Hock of Leg		Elsewhere
None	2 lb	3/4 in.		1 1/4 in.	1/4 in		5/8 in.
Over 2 lb	6 lb	1 in.		2 in.	1/2 in		1 in.
Over 6 lb	16 lb	1 1/2 in.		2 1/2 in.	3/4 in.		1 1/4 in.
Over 16 lb	None	2 in.		3 in.	1 in.		1 1/2 in.

Discolorations: Large Carcass Parts (halves, front and rear halves)		Lightly Shaded		Moderately Shaded[4]	
		Breast and Legs	Elsewhere	Hock of Leg	Elsewhere
None	2 lb	1/2 in.	1 in.	1/4 in.	1/2 in.
Over 2 lb	6 lb	3/4 in.	1 1/2 in.	3/8 in.	3/4 in.
Over 6 lb	16 lb	1 in.	2 in.	1/2 in.	1 in.
Over 16 lb	None	1 1/4 in.	2 1/2 in.	5/8 in.	1 1/4 in.

Discolorations: Other Parts		Lightly Shaded	Moderately Shaded[4]
None	2 lb	1/2 in.	1/4 in
Over 2 lb	6 lb	3/4 in.	3/8 in.
Over 6 lb	16 lb	1 in.	1/2 in.
Over 16 lb	None	1 1/4 in.	5/8 in.

Disjointed and Broken Bones:	Carcass—1 disjointed and no broken bones Parts—Thighs with back portion, legs or leg quarters may have femur disjointed from the hip joint. Other parts—none.
Missing Parts:	Wing tip and tail
Freezing Defects:	Slight darkening on back drumstick. Overall bright appearance. Occasional pock marks due to drying. Occasional small areas of clear pinkish, or reddish colored ice

[1] For ducks and geese, hair or down is permitted on the carcass or part.

[2] Maximum aggregate area of all exposed flesh.

[3] For all parts, trimming of skin along the edge is allowed, provided at least 75% of the normal skin cover associated with the part remains attached.

[4] Moderately shaded discolorations and discolorations due to flesh bruising are free of clots and limited to areas other than the breast and leg except for the area adjacent to the hock.

Source: Modified from USDA Poultry Grading Manual, U. S. Department of Agriculture, *Agriculture Handbook No. 31,* 1998.

In assessing these quality standards, location, severity, and total aggregate area of defects must be taken into account, in addition to the class (species), market age, and sex of poultry. There are no grade standards for giblets (heart, liver and gizzard), detached necks and tails, wing tips, and skin. Grade standards for boneless-skinless breasts include presence of bones, tendons, and cartilage, discolorations and blood clots, and other product-specific factors.

The consumer grades for whole carcass and parts are U.S. Grades A, B, and C. Summary of specifications for A grades are presented in Table 5.3.[37] Lower grade whole carcasses (B and C) are often cut up, since parts from these carcasses may qualify as Grade A and therefore be of higher value. Proportion of downgrading by class of poultry (Figure 5.10) and the extent of non-RTC defects are also summarized for 1999 (Table 5.4).

In modern processing operations, poultry is initially graded by plant employees trained and authorized by the USDA and monitored by an official resident grader. Most carcasses and parts are graded on the production line or after a cooler. When monitoring the graded product, the resident grader utilizes the Acceptable Quality Level (AQL) procedure on a subsample of product to determine the cumulative scores of defective, under-grade birds or parts present. When AQL tests indicate excessive defects, the product is classified as "USDA retained" until the product has been reworked to meet the grade criteria.

Since products vary in complexity and detail, procurement programs for further processed products may contain additional specifications on preparation and processing, metal detection, freezing, packaging and labeling, test weights, portion control, temperatures, storage, and transportation.

Standards and grades are used extensively throughout the marketing system in the U.S. and provide common language for buyers and sellers of poultry and poultry products. Commercial operations also use the standards and grades as a basis for their own product specifications, for advertising, and for establishing brand names.

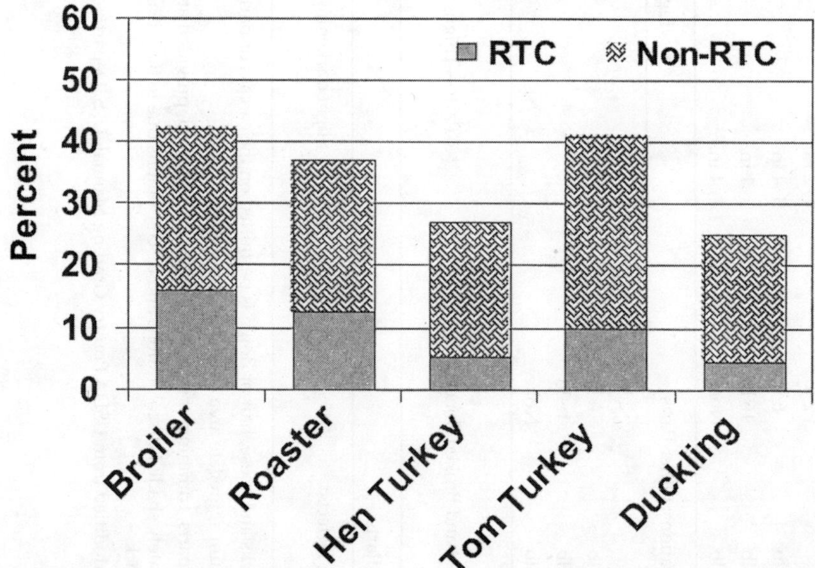

Figure 5.10 Frequency of downgrading by class of poultry for 1999. RTC = ready-to-cook criteria, non-RTC = downgrades due to plant operation malfunction.

Table 5.4 **Proportion of Post-Chill Non-RTC Defects by Class**

	Product				
			Young turkey		
	Frying chicken	**Roaster**	**Hen**	**Tom**	**Duckling**
			%		
Conformation	0.1	0.2	0.1	0.1	0
Bruised breast	1.3	1.4	1.2	1.4	0.3
Drum	1.8	1.1	0.7	0.8	0.9
Thigh	0.2	0.1	0.2	0.3	0.3
Wing	2.6	1.9	1.5	1.7	0.8
Shoulder	0.8	0.6	0.5	0.3	0.1
Total bruises	6.7	5.0	4.0	4.6	2.3
Discolorations	1.1	0.8	0.5	0.8	6.7
Disjointed bones	0.2	0.6	0.2	0	0
Broken leg	2.1	2.3	0.2	0.2	0.4
Wing	1.4	2.8	0.3	0.8	0
Total trim	2.2	1.0	4.2	8.5	1.8
Total exposed flesh	3.5	2.0	3.6	2.7	1.3
Missing wings	6.7	8.9	5.4	8.0	3.9
Drum	0.6	0.3	1.2	1.4	0.6
Leg	0.8	0.4	1.1	1.1	1.5

Source: Modified from USDA, *Grade Yield Survey,* Agricultural Marketing Service, Poultry Grading Branch Marketing and Regulatory Programs, U. S. Department of Agriculture, Washington, D. C., 2000.

References

1. Forrest, J. C., Aberle, E. D., Hedrick, H. B., Judge, M. D., and Merkel, R. A., Meat inspection, in *Principles of Meat Science*, Freeman, W. H., Ed., San Francisco, 316, 1975.
2. Libby, J. A., History, in *Meat Hygiene*, Fourth edition, Lea & Febiger, Philadelphia, 1, 1975.
3. Olsson, P. C. and Johnson, D. R., Meat and poultry inspection: Wholesomeness, integrity, and productivity, in *Seventy-Fifth Anniversary Commemorative Volume of Food and Drug Law*, Food and Drug Law Institute, ed., Food and Drug Law Institute, Washington, D.C., 1984, 220.
4. Sinclair, Jr., U. B., *The Jungle*, Doubleday, New York, 1906.
5. NRC, *Poultry Inspection: The Basis for a Risk-Assessment Approach*, Committee on Public Health Risk Assessment of Poultry Inspection Programs, National Academy Press, Washington, D.C., 1987.
6. CAST, *Foods from Animals: Quantity, Quality and Safety*, Report No. 82, Council for Agricultural Science and Technology, Ames, IA, 1980.
7. NRC, *Meat and Poultry Inspection: The Scientific Basis of the Nation's Program*, Report of the Committee on the Scientific Basis of the Nation's Meat and Poultry Inspection Program, Food and Nutrition Board, National Academy Press, Washington, D.C., 1985.
8. NRC, *Poultry Inspection: The Basis for a Risk-Assessment Approach*, Report of the Committee on Public Health Risk Assessment of Poultry Inspection Programs, Food and Nutrition Board, National Academy Press, Washington, D.C., 1987.
9. USDA, *Meat and Poultry Inspection*, Report of the Secretary of Agriculture to the U.S. Congress, Food Safety and Inspection Service, U.S. Department of Agriculture, Washington, D.C., 1988.
10. USDA, *Meat and Poultry Inspection*, Report of the Secretary of Agriculture to the U.S. Congress, Food Safety and Inspection Service, U.S. Department of Agriculture, Washington, D.C., 1991.
11. USDA, Pathogen Reduction; Hazard Analysis and Critical Control Point (HACCP) Systems; Final Rule, *Fed. Regis.*, 9 CFR Part 304, 1996.

12. USDA, *Meat and Poultry Inspection,* Report of the Secretary of Agriculture to the U.S. Congress, Food Safety and Inspection Service, U.S. Department of Agriculture, Washington, D.C., 1996.
13. USDA, *Food Safety and Inspection Service,* 9. Food Safety, Agriculture Fact Book, U.S. Department of Agriculture, Washington, D.C., 1998, 1.
14. USDA, *Guidelines for Offline Salvage of Poultry Parts,* Meat and Poultry Inspection Technical Services, Food Safety and Inspection Service, U. S. Department of Agriculture, U.S. Government Printing Office:0-617-013, Washington, D.C., 1988.
15. Ewing, M., Exley, S., Page, K., and Brown, T., Understanding the disposition of broiler carcasses. *Broiler Ind.,* March 28, 1977.
16. USDA, *Meat and Poultry Inspection Regulations,* 381.81-381.92, Title 9, Chapter III, Subchapter C, Code of Federal Regulations, Food Safety and Inspection Service, U. S. Department of Agriculture, Washington, D. C., 1970.
17. USDA, *Meat and Poultry Inspection Manual,* Food Safety and Inspection Service, U. S. Department of Agriculture, Washington, D.C., 1987.
18. USDA, *Pathogen Reduction; Hazard Analysis Critical Control Point (HACCP) Systems,* Final Rule, 9 CFR Part 304, et al., Food Safety and Inspection Service, U. S. Department of Agriculture, Washington, D.C., 1996.
19. USDA, *FSIS Facts: The National Residue Program,* FSIS-18, Food Safety and Inspection Service, U.S. Department of Agriculture, Washington, D.C., 1984.
20. USDA, *Enhanced Poultry Inspection; Revision of Finished Product Standards with Respect to Fecal Contamination,* Docket No.94-016F, Food Safety and Inspection Service, U. S. Department of Agriculture, Washington, D.C., 1996
21. USDA, *Poultry Post-Mortem Inspection and Reinspection-Enforcing the Zero Tolerance for Visible Fecal Material,* Food Safety and Inspection Service, Directive 6150.1, U. S. Department of Agriculture, Washington, D.C., 1998.
22. USDA, *Sanitation,* 9 CFR part 416, Food Safety and Inspection Service, U. S. Department of Agriculture, Washington, D. C., 1996.
23. DHHS, *Current Good Manufacturing Practice in Manufacturing, Packing, or Holding Human Food,* 21 CFR Part 110, Food and Drug Administration, Washington, D.C., 1996.
24. Bauman, H., HACCP: Concept, development, and application, *Food Technol.,* 44(5),156, 1990.
25. Adams, C., Use of HACCP in meat and poultry inspection, *Food Technol.,* 44(5), 169, 1990.
26. Stevenson, K. E., Implementing HACCP in the food industry. *Food Technol.,* 44(5), 179, 1990.
27. ICMSF, *Microorganisms in Foods 4. Application of Hazard Analysis Critical Control Point (HACCP) System to Ensure Microbiological Safety and Quality,* Blackwell Scientific, Boston, MA, 1988.
28. CAC, Report of the 24th session of the Codex Committee on Food Hygiene — Alinorm 88/13A. Codex Alimentarius Commission, Rome, 1988.
29. NACMCF, Hazard Analysis Critical Control Point System, *Int. J. Food Microbiol.,* 16, 1, 1992.
30. USDA, *Nationwide Broiler Chicken Microbiological Baseline Data Collection Program,* Food Safety and Inspection Service, U. S. Department of Agriculture, Washington, D.C., 1996.
31. AOAC, *Official Methods of Analysis,* 16th edition, Official Analytical Chemists International, Gaithersburg, MD, 1995.
32. USDA, *One-Year Progress Report on Salmonella Testing for Raw Meat and Poultry Products,* Food Safety and Inspection Service, Backgrounders, U. S. Department of Agriculture, Washington, D.C., 1999, 1.
33. USDA, *HACCP-Based Inspection Models,* Backgrounders, Food Safety and Inspection Service, U. S. Department of Agriculture, Washington, D.C., 1998, 1.
34. USDA, *HACCP-Based Inspection Models Project: Models Phase,* pages 1–5, Backgrounders, Food Safety and Inspection Service, U. S. Department of Agriculture, Washington, D.C., 1999, 1.
35. USDA, *United States Classes, Standards, and Grades for Poultry,* AMS 70.200 *et. seq.* Agricultural Marketing Service, Poultry Programs, U. S. Department of Agriculture, Washington, D.C., 1998.
36. Brant, A. W., Goble, J. W., Hamann, J. A., Wabeck, C. J., and Walters, R. E., Guidelines for establishing and operating broiler processing plants, U.S. Department of Agriculture, Agriculture Handbook No. 581, 1982.
37. USDA, Poultry Grading Manual, U.S. Department of Agriculture, Agriculture Handbook No. 31, 1998.

chapter six

Packaging

Paul L. Dawson

Contents

0-8493-0120-3/01/$0.00+$.50

Introduction

The poultry industry is concentrated in that there are approximately 60 to 70 processors in the U.S., and nearly 50% of production is handled by five major processors. Since there are a limited number of poultry cuts and a relatively small number of processors with sizeable operations, poultry is typically packaged in small consumer portions at a central processing location. Retailers absorb a cost for this service but can eliminate labor, equipment, packaging inventories, contamination problems, and the inefficiencies of a small packaging operation.

Before discussing specific packaging systems for poultry, the function of the package and packaging materials will be reviewed. As a final introductory statement, one must remember that packaging can increase the shelf life of poultry meat, however, it cannot improve the quality of the product. Thus, good manufacturing and handling practices must be used to maintain a high quality packaged product.

Basic concepts — package function

The functions of a food package can be divided into four areas: containment, information, convenience, and protection.[1] *Containment* includes the holding of a product without necessarily protecting it. Holding multiple pieces of chicken parts such as legs, thighs, wings, or breasts allows for them to be sold in various volumes or combinations. *Information* is both a governmental regulation and marketing tool. The package carries the nutritional labeling, proper handling practices, product information, and identifiers required by law. The package also contains the product price, claims, and cooking suggestions as well as package recycling messages. *Convenience* is a function of the package. Single serving sizes of sliced meat and microwaveable packages allow for cooking/reheating and consumption of the product in a part of the package. *Protection* is the most important package function, protecting the product from microorganisms, rodents, dust, external contaminants, humidity, light, and oxygen. The package should also protect the product from tampering and physical damage during handling. Unpackaged meat would quickly dehydrate, therefore the package must prevent moisture loss. Poultry meat having higher pigment concentrations must be protected from loss of the bright red oxymyoglobin color. Products such as ground leg meat and comminuted meat are packaged in films with high oxygen permeability to maintain the oxymyoglobin state. Comminuted meat is packaged in paperboard boxes that allow for oxygen presence and prevent light from contacting the meat surface.

The function of the packaging materials can also be categorized according to its function within the total package system. These functions categorize the material as to whether it contributes *barrier, strength,* or *sealing* properties. For example, aluminum foil is often added as a layer for its barrier properties, excluding light and gases. Aluminum foil also provides a surface for application of graphics. Polyester (polyethylene terephthalate,

PETE) is often added for its strength. Polyethylene is an excellent sealing agent and is used as the sole sealing agent or in combination with other materials in a composite. A general summary of the packaging material functions is:

- **Barriers:** aluminum foil, ethylene vinyl alcohol (EVOH), vinylidene chloride copolymer (Saran), polyvinyl chloride (PVC), and acrilonitrile (Barex)
- **Strength properties:** polyester, PETE, nylon, polypropylene (PP)
- **Sealing agents:** polyethylene (PE), Surlyn, polystyrene (PS).

General packaging practices

In practice, poultry meat packaging can be categorized into primary, secondary, and tertiary levels. The primary package is the one with which most consumers are familiar, and is the food contact surface. The primary package will carry the labeling and any additional consumer information. The most common primary package for poultry meat is polymer (plastic) film wrap or overwrap. However, the primary package can also be a metal in the case of canned cooked and retorted meat or poultry meat-containing soups and sauces. The primary package can contain a mixture of materials such a paper, foil, and cellophane that alters the package's properties or allows for special graphics. It may be flexible, semi-rigid, or rigid. Flexible packaging materials include "plastics" (polymers), paper, or a thin laminate; semi-rigid materials include thermoformed polymers, aluminum foil, or paperboard; rigid materials are thick polymers, metal, or glass.[2] Plastic is a generic term used for a family of polymers. These polymers have a relatively simple chain structure when compared to the structures of most food components.

The secondary package is an outer box, case, or wrapper that contains or unitizes several single primary packages together. The secondary package does not contact the food surface but serves to protect the primary packages from breakage, damage, dirt, and soiling during distribution. The secondary package is often a cardboard box containing many tray packs of chicken parts that are pre-labeled and priced. More sophisticated meat packaging systems may have the secondary package unit gas flushed with an inert gas, with the fresh meat in the primary package surrounded with a high gas-permeable film which will allow the meat to "bloom" when removed from the secondary package, but will allow the inert gas into the package prior to removal. This system will reduce microbial growth during shipping and holding but allow for proper color development at the retail store.

The tertiary package holds several secondary packages in shipping loads, such as pallet-sized units. Stretch wrap is often utilized to stabilize the pallet during loading, unloading, and shipping.

Packaging materials

There are relatively few different types of food packaging materials. However, there are many different variations within some material types, and many combinations of materials are utilized. The materials used to package meat products include fiber-based (paper, paperboard), glass, and metal. In addition, nearly all poultry packages have plastics as either coatings, linings, overwraps, or bags. The most common plastic materials, their use and properties are summarized in Table 6.1.

Paper, paperboard, and fiberboard

Paper, paperboard, and fiberboard differ in their relative thickness, paper being the thinnest, paperboard thicker, paper sheet more rigid, and fiberboard made by combining

Table 6.1 **Plastics Used for Packaging Meat Products**

Polymer type	Use	Features
Ionomer	Heat-seal layer	Resists seal contamination
Nylon (uncoated)	Films, thermoformed trays	Also used as bone guards
Nylon (PVdC[1] coated)	Films, thermoformed trays	
PETE[1] (uncoated)	Films, trays	Good clarity
PETE[1] (PVdC coated)	Films	
LDPE[1]	Bags, wraps	Low cost, low gas barrier
LLDPE[1]	Heat-seal layer	Good clarity
EVA[1]-LDPE copolymer	Seal layer, films, wraps	Heat shrinkable
PP[1] (non-oriented)	Semi-rigid containers	
PVC[1]	Fresh meat wrap	Gas transmission rate depends on plasticization
PVdC	Barrier layer	Barrier less affected by moisture

[1]PVdC, polyvinyldienechloride; PETE, polyethylene terephthalate (polyester); LDPE, low density polyethylene; EVA, ethylene vinyl acetate; LLDPE, linear low density polyethylene; PP, polypropylene; PVC, polyvinyl chloride.

layers of paper. The material used for secondary shipping cartons of poultry meat is most often corrugated paperboard, named because of the wavy inner layer of paperboard that adds strength. Although this material is commonly referred to as "cardboard," "corrugated paperboard" is the term used in the packaging industry. These secondary paperboard boxes are sometimes produced from wood pulp and reprocessed paper, which is bleached and coated or impregnated with waxes, resins, lacquers, or plastics. The added layer improves the package's resistance to high humidity and improves wet strength, grease resistance, appearance, and barrier properties. Acid treatment of paper pulp can result in glassine paper with high oil and water resistance. The acid modifies the cellulose, giving rise to long wood pulp fibers that also add strength to the paper.

Metals

Metals used for canned poultry meat include steel and aluminum. The steel can has greater strength and resistance to denting, while the aluminum can is lightweight and resistant to atmospheric corrosion. The steel can was at one time coated with tin to prevent corrosion at the food contact surface, however, this layer is now a steel alloy such as a chromium alloy, which is much cheaper than tin. The metal can is also coated with an additional organic layer on both the inside and outside can surfaces. This further protects the can from corrosion by the food constituents and also protects the food from contamination by the metal, particularly from metal-catalyzed degradative reactions. Phenolic compounds are used in this organic layer for meat spreads, while modified epons are used for other meat-containing products. Aluminum foil can be used in flexible pouches and is often combined with plastics and paper in layers. Foil offers a complete barrier to light, oxygen, and water vapor.

Plastics (polymers)

Plastics (polymers) comprise by far the most common packaging material for poultry meat products due to their versatility, cost, and convenience.

Polyethylene (PE)

$$—CH_2—CH_2—CH_2—CH_2—CH_2—$$

The molecular structure of PE is a $(CH_2)_n$ with short side ethylene chains located along the main ethylene chain, which prevents close stacking and results in a less dense structure. There are three major types of PE which differ in their structure, properties, and manufacturing processes. These three types are high density polyethylene (HDPE), low density polyethylene (LDPE), and linear low density polyethylene (LLDPE). LDPE and LLDPE have different molecular structures but have similar densities (0.910 to 0.925 g/cm^3). LDPE and HDPE differ in the length of side chains and thus, also differ in the overall density of the film. The HDPE is more dense, less clear, stronger, and stiffer than LDPE. The LLDPE is produced under higher pressure and results in a film with similar density to LDPE but with the strength and toughness of HDPE. HDPE also forms a good seal at relatively lower temperatures and has better grease and heat resistance than LDPE. LLDPE is stiffer and has a higher range of heat sealability than LDPE and is being used as a laminate layer as well as in bags and stretch wraps.

Polypropylene (PP)

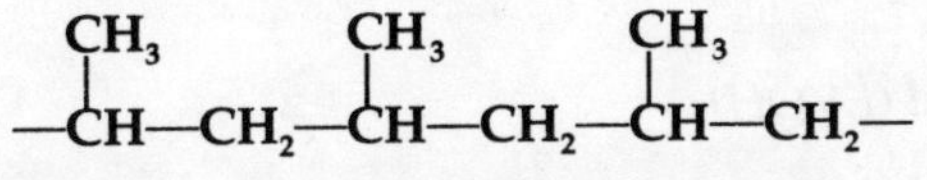

$$\text{—}\underset{|}{\overset{CH_3}{}}\!CH\text{—}CH_2\text{—}\overset{CH_3}{\underset{}{CH}}\text{—}CH_2\text{—}\overset{CH_3}{CH}\text{—}CH_2\text{—}$$

The structure of PP is a carbon chain with every other side group being a methyl (CH_3) instead of a hydrogen as with PE. This structure results in a harder and more resilient polymer than HDPE with a permeability to water vapor and gases between those of LDPE and HDPE. The structure of PP can be varied several ways, including oriented or non-oriented, and can be extruded and coated to become heat sealable and change other film properties. The main application for meat packaging is in cook-in products, due to PP's high heat tolerance and impermeability to moisture during water bath or steam cooking.

Ionomers (Surlyn)

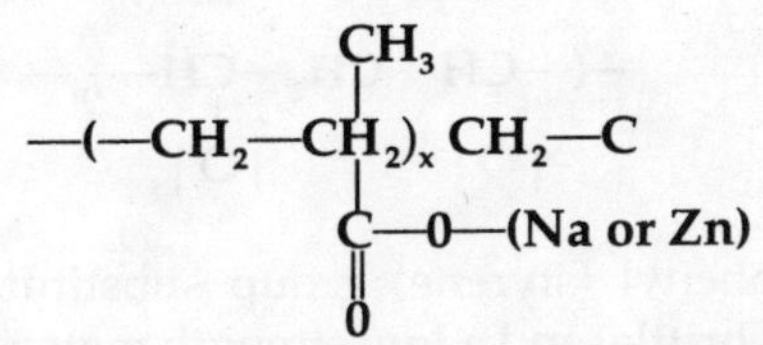

$$\text{—(—}CH_2\text{—}CH_2)_x\ CH_2\text{—}\underset{\underset{\underset{O}{\|}}{C\text{—}O\text{—}(Na\ or\ Zn)}}{\overset{CH_3}{C}}$$

Ionomers are polymers that have been copolymerized with an acid. Some part of the acid remains in the film structure in the form of an ammonium salt or metal, usually zinc or sodium. The incorporation of these ions increases the lipophilic nature of the polymer. The films are flexible, tough, and transparent, and are excellent heat-sealing agents. In meat packaging, ionomers are used as the food contact and heat sealing surface in laminated materials. They have a wide heat sealing range, possess good grease resistance, and will adhere well to most other packaging material, including aluminum foil.

Polyvinyl chloride (PVC)

$$\text{—(—}CH_2\text{—}ClCH\text{—}CH_2\text{—)}_n\text{—}$$

This polymer is similar in structure to PE except for having a chloride substituted for a hydrogen at alternating ethylenes. PVC is difficult to process since it begins to break down at about 80°C. PVC is ideal for stretch and shrink retail packages where high oxygen and water vapor permeability (compared to PVdC) is desired and only limited shelf life is required. It is often used as in-store packaging for deli meat, fresh meats, and cured meat products.

Polyvinylidene chloride (PVdC)

$$—(—CH_2—CCl_2—)_n—$$

Saran is a trade name for PVdC, and has an additional chloride atom included in the ethylene molecule compared to the PVC structure. This film is also clear and strong, with low permeability to gas and moisture. It is used as a layer in multilayer material for pouches, bags, and thermoformed packages for meat, where it functions as an oxygen and water vapor barrier. PVdC can be heat-sealed, is printable, and can withstand cooking or retorting. It is used to package frankfurters, luncheon meats, hams, or wherever modified atmosphere packaging is preferred.

Ethylene vinyl alcohol (EVOH)

$$(\ldots CH_2—CH_2—CH_2—\underset{|}{\underset{OH}{CH}}—CH_2—CH_2—CH_2—\underset{|}{\underset{OH}{CH}}—)_n—$$

This film is an excellent oxygen barrier, however, it is hydrophilic. Thus, its oxygen permeability will fluctuate with high humidity. The hydroxyl groups in the polymer backbone make EVOH water soluble and disrupted by high humidity. To improve moisture resistance, EVOH is placed between layers of PP, PE, and/or PETE.

Polystyrene (PS)

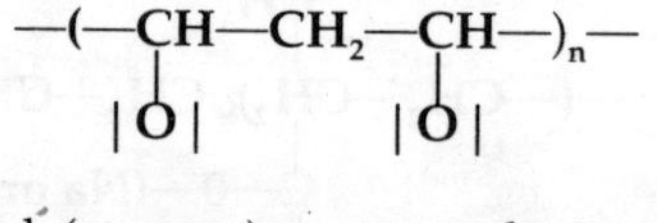

$$—(—\underset{|O|}{CH}—CH_2—\underset{|O|}{CH}—)_n—$$

The PS structure has a phenyl (styrene) group substituted for a hydrogen in the PE structure. PS is clear, hard, brittle, and a low strength material. It is used as disposable containers as well as packaging films. PS can be foamed to form expanded polystyrene (EPS) (Styrofoam), which is used as the tray in tray-packed poultry meat. Both the clear and foamed thermoformed trays have high oxygen permeability. High impact polystyrene (HIPS) has good tensile strength and stiffness. Styrene is one of the few materials with the thermal melt strength required to form trays.

Polyamides (nylons)

$$H(—\overset{H}{N}—(CH_2)_n—\overset{H}{N}—\overset{O}{\overset{\|}{C}}—(CH_2)_n—\overset{O}{\overset{\|}{C}})_n\ OH$$

Polyamides (nylons) include polymers formed by condensation of certain amino acids, and this is, therefore, the only food "plastic" containing nitrogen. The nylons are designated with paired numbers, the first number indicating the number of carbon atoms in the amine portion and the second indicating the number of carbon atoms in the carboxylic acid section. They have relatively high melting points and low gas permeability, but they will absorb moisture and lose strength when exposed to moisture. Nylons are used in cook-in-the-film meat applications, sometimes in combination with Surlyn (an ionomer).

Polyesters

$$—(—CH_2—CH_2—O—\overset{\overset{\displaystyle O}{\|}}{C}—\langle o \rangle—\overset{\overset{\displaystyle O}{\|}}{C}—O—)_n—$$

The most common polyester is polyethylene terephthalate (PETE), used in carbonated beverage containers. It has excellent strength, clarity, and heat stability and is used for vacuum packaging and cook-in applications for meat. PETE is strong, clear, and has very low moisture and gas permeability. PETE is also used in sterilizable pouches and boil-in-bag applications.

Polycarbonates (PC)

$$C1—\overset{\overset{\displaystyle O}{\|}}{C}—C1—O—\underline{O}—\overset{\overset{\displaystyle CH_3}{|}}{\underset{\underset{\displaystyle CH_3}{|}}{C}}—\underline{O}—O—\overset{\overset{\displaystyle O}{\|}}{C}—OH$$

The PCs contain polyesters of carbonic acid. They are stiff, transparent, tough, and hard. PCs have high gas permeability and absorb moisture which causes them to lose mechanical properties. Despite relatively high cost, their inertness to food has promoted the use of PCs in plates for oven-treated dinners.

Cellophane

Cellophane is regenerated cellulose film made from trees and manufactured from sheets of wood pulp. The fibrous wood pulp is regenerated into a nonfibrous form, and with the addition of plasticizers obtains the needed degree of flexibility. Cellophane is a good gas and grease barrier but will break down in the presence of moisture, so it is often coated with a hydrophobic layer.

The physical properties of the polymer films are summarized in Table 6.2.

Water vapor and oxygen permeability

The barrier properties of polymers are important for packaging meat products. Packaging material offers resistance to the transfer of gases and water/odor vapors through the package. This resistance is known as the barrier property of the film. Water vapor transmission rate (WVTR), or more accurately, water vapor permeability (WVP) and oxygen transmission rate (OTR) are two critical package material properties that will affect the quality of poultry meat. The WVTR is determined based on water passage through a specified film surface area, while WVP also takes into account the film thickness and relative humidity gradient on either side of the film. Both WVP and OTR are expressed as gas or vapor

Table 6.2 Physical Properties of Packaging Materials

Packaging material	Density (g/ml)	Tensile strength (Kpsi)	Elongation at break (%)	WVP at 100°F and 90% RH (g/100 in.^{2} · 24 h)	OTR at 77°F and 0% RH (ml/100 in.^{2} · 24 h · 1 atm)	Heat seal (temp °F)
HDPE[1]	0.945–0.967	2.5–6	200–600	0.4	100–200	275–310
LDPE[1]	0.91–0.925	1.5–5	200–600	1–2	500	250–350
LLDPE[1]	0.918–0.923	3–8	400–800	1–2	450–600	220–340
EVA[1]	0.93	2–3	500–800	2–3	700–900	150–350
Ionomer	0.94–0.96	3.5–5	300–600	1.5–2	300–450	225–300
PETE[1]	1.3–1.4	25–33	70–130	1–1.5	3–6	275–350
PVC[1]	1.22–1.36	4–8	100–400	2–30	30–600	280–340
PVdC[1]	1.6–1.7	8–16	50–100	05–.3	0.1–1	250–300
EVOH[1]	1.14–1.19	1.2–1.7	120–280	3–6	.01–.02	350–400
PC[1]	1.2	9–11	100–150	12	180–300	400–420
Nylon 6	1.1–1.2	6–24	30–300	22	2.6	400–550
PS[1]	1.0–1.2	5–8	1–30	7–11	350	—
PP[1]	0.90–1.2	4.5–6	100–600	11–12	—	260–290

[1]PVdC, polyvinyldienechloride; PETE, polyethylene terephthalate (polyester); LDPE, low density polyethylene; EVA, ethylene vinyl acetate; LLDPE, linear low density polyethylene; PP, polypropylene; PVC, polyvinyl chloride; EVOH, ethylene vinyl alcohol; PC, polycarbonate; PS, polystyrene; PP, polypropylene.

exchange rates with both the relative humidity and temperature conditions stipulated for a 1 ml thick film at 1 atm of pressure. Several different units have been used to report the WVP and OTR of films.

WVP	OTR
ml/m^2 24 h at 38°C and 90% RH	ml/m^2 24 h at 20°C and 0% RH
ml/m^2 24 h at 25°C and 75% RH	ml/m^2 24 h at 25°C and 50% RH
g/100 in.^{2} 24 h at 100°F and 90% RH	ml/100 in.^{2} 24 h at 77°F and 0% RH

In general, for meat packaging the polymer films can be categorized into relative levels of OTR utilizing the ml/m^2 24 h units as: 0–1200, low; 1200–5000, medium; and 5000 and greater, high. The OTR will change for some materials with changes in temperature and humidity. Nylon and EVOH will change dramatically with changes in relative humidity due to their hydrophilic nature.

Thermal properties of plastics are also important to poultry meat packaging. Most bags and trays used in meat packaging are sealed by fusing two layers of polymer together by the application of heat. Meat packaging often requires a skin-tight finish that is accomplished by shrinking the film wrap with heat.

Packaging material properties can be modified in a variety of ways both during the formation of the individual polymers (additives, orientations, etc.) and by combining multiple layers of polymers to produce the desired properties. Two methods used to produce multilayered polymers are lamination and coextrusion. The method chosen depends upon the materials to be combined. Lamination can be described as gluing two polymers together while coextrusion combines the layers by melting or molding them together.

The package seal is a major control point for preventing contamination of packaged poultry meat. The term hermetic is used to describe a package and seal that is impervious to dust, dirt, bacteria, mold, yeast, and gases. Metal and glass containers are true hermetic containers. Some flexible packaging materials are designed to allow some gas exchange

Figure 6.1 "Ice-pack" broilers packaged in a wax-coated corrugated box, "wet-shipper." (Courtesy of Mountaire Farms, Selbyville, DE.)

and therefore are not hermetic by definition. These types of flexible packaging do not allow microorganisms to impenetrate their structure. The seal is a frequent failure point when failure occurs, and plays an important role in preventing contamination and purge from leaking into retail cases.

Fresh poultry

Current practice

The oldest method of packaging and distributing fresh poultry meat is in a "wet shipper." The wet shipper is a wax-coated corrugated box in which whole birds are placed with ice (Figure 6.1). The "dry shipper" is similar to the wet shipper with ice excluded. More recently, whole carcasses have been placed in polymer bags and sealed or clipped (Figure 6.2). Almost 90% of all chicken parts are packaged directly into consumer portions using highly oxygen permeable polystyrene foam trays with a high oxygen permeable PVC or polymer-based, stretch film overwrap. These include breast, thigh, drum, and wing portions (Figure 6.3). Most of the remaining portion of poultry meat is packaged in bulk ice packs at the central processor, but it ultimately ends up in a similar tray and stretch wrap package at the retail level (Figure 6.4).

Figure 6.2 Whole carcasses packaged in polymer bags with either sealed or clipped closures.

Figure 6.3 Chicken parts packaged in polystyrene tray with an oxygen permeable overwrap film.

In a bulk pack system described by Timmons,[3] the retail packages are placed in corrugated containers with plastic liners. The liner is then gas-flushed with a modified atmosphere and sealed within the corrugated container (Figure 6.5). This system provides approximately five additional days of shelf life, compared to the non-modified atmosphere packaging method. A low oxygen barrier material such as HDPE/LDPE co-extrudate is used to allow release of off odors produced during storage. Due to the concern for buildup of off odors in the package, high barrier film materials have limited use in the poultry industry. Only about 1–2% of poultry meat requires high oxygen barrier packaging, including precooked products. Other requirements of fresh/frozen poultry packages are non-fogging, non-wrinkling, high clarity, puncture resistance, and sealability.

Poultry parts can be deep-chilled or crust-frozen in overwrapped trays by passing the package through a chill tunnel (−40°C or lower) for approximately one hour. This process hardens the surface of the meat without freezing the interior and greatly extends the shelf-life of the product. USDA regulations require that poultry meat must remain above 26°F at the meat core to be labeled "fresh." The typical fresh meat package for retail display is a foam tray overwrapped with a clear film. An absorbent pad is usually placed under the

Figure 6.4 Turkey drum portions packaged at the retail store in a polystyrene tray with an overwrap film

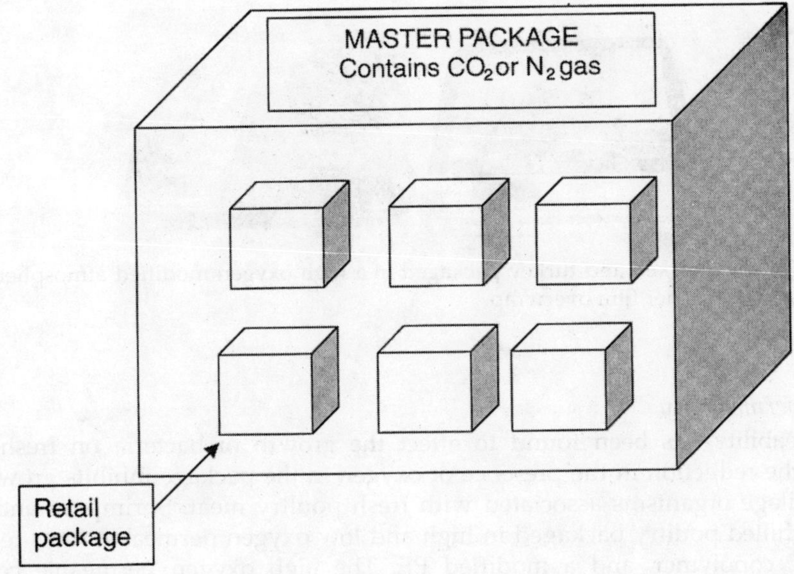

Figure 6.5 Master package containing retail packages in a modified atmosphere.

meat to absorb purge. The pad is comprised of an absorbent material such as cellulose, surrounded with a porous, non-absorbent "plastic." The overwrap film has a relatively high degree of oxygen permeability to allow the raw meat pigment to "bloom." Fresh meat packaging overwrap materials are stretch PVC or stretch-shrink PE, with the trays made from EPS.

Raw poultry meat is highly perishable even when stored under chilled conditions. The growth of psychrotrophic spoilage bacteria is most often the cause of spoilage. While other factors will limit the shelf life of the poultry meat (especially initial bacterial levels), vacuum or modified atmosphere packaging can extend it. Generally, when vacuum or CO_2 atmosphere is combined with chill temperatures, a significant increase in shelf life can be obtained. Furthermore, increasing the CO_2 levels to 80 and 100% can reduce the growth rate of spoilage bacteria on chicken compared to 20% CO_2 and vacuum packaging.

There are generally three methods used to vacuum-package poultry meat, depending upon the type of meat:

1. Whole carcasses are packaged in heat-shrinkable plastic bags with low oxygen permeability, using a rotomatic or chamber with a clip seal or heat-seal system.
2. Cut-up poultry uses a vacuum system prior to heat-sealing the package.
3. Ground poultry uses a thermoforming or horizontal overwrap machine where the meat is placed in a tray, a vacuum is pulled, then the package is gas-flushed before being sealed.

Ground poultry meat requires different packaging due to color stability. Ground turkey breast meat is a popular product and is sometimes mixed with ground turkey thigh meat. Ground chicken thigh meat has a less stable color, so there is currently less on the market. All of these ground products are packaged in high oxygen atmospheres of 70 to 90%, usually in PS foam trays with an overwrap film or lid stock that is a barrier to oxygen (Figure 6.6). The package headspace is usually held at a gas volume to meat/volume ratio of 1:1 or greater.

Figure 6.6 Ground chicken and turkey packaged in a high oxygen modified atmosphere in polystyrene trays with a barrier film overwrap.

Research

Film permeability

Film permeability has been found to affect the growth of bacteria on fresh poultry. Generally, the reduction in the presence of oxygen in the package inhibits growth of the typical spoilage organisms associated with fresh poultry meat. Shrimpton and Barnes[4] evaluated chilled poultry packaged in high and low oxygen permeable films testing PE, PVC/PVdC copolymer, and a modified PE. The high oxygen permeable copolymer delayed the detection of off odors and resulted in a higher concentration of oxygen in the package headspace compared to the other films evaluated. Fluorescent pigment production, lipolytic activity, and proteolytic activity of chicken spoilage bacteria were directly related to the availability of oxygen due to the packaging procedures.[5] The bacterial numbers paralleled the increases measured in biochemical activities. In addition, fresh poultry meat packages must maintain constant moisture content within the package in order to maintain product quality as well as restrict bacterial growth.[6] The use of films with various oxygen permeabilities will affect the growth of bacteria and the color and odor of refrigerated poultry meat.[7] Generally, low OTR films will retard bacterial growth while high OTR films will lower the off odor impact upon opening the package (Figure 6.7).

Vacuum and Modified Atmosphere Packaging (MAP)

Vacuum and modified atmosphere packaging (MAP) has been used to extend the shelf-life of packaged meat for several decades. Several MAP packaging systems for fresh poultry meat exist including: flexible trays with vacuum or gas-flush, rigid trays with lid stock and gas flush, heat-sealable bags with vacuum or gas flush, and master/bulk packaging overwrap for vacuuming or gas flush multiple packages.[8] Carbon dioxide content is critical in MAP to control the growth of aerobic spoilage bacteria. Haines[9] was the first to show an inhibitory effect of CO_2 on aerobic spoilage bacteria. Barnes et al.[10] found that vacuum-packaged, chill-stored poultry lead to the growth of mainly lactic acid bacteria and, in some cases, cold-tolerant coliforms. The use of CO_2-enriched atmospheres for chilled poultry is based on the early work of Ogilvy and Ayres.[11] They found that the ratio of poultry meat shelf life in CO_2 to the shelf life in air could be expressed as a linear function with CO_2 concentration. The CO_2 affected both the lag growth phase and generation time of the bacteria present. A minimum concentration of 20% in the package headspace is required to see a significant improvement in shelf life.[12,13] The growth of pathogens on fresh chicken was inhibited by increasing the concentration of CO_2 with storage at 1.1°C, however, the lactic acid bacteria present were not inhibited, due to their facultative anaerobic abilities.[14] Thomson[15] also found that a high CO_2 atmosphere inhibited the growth of bacteria on poultry compared to chicken packaged with ambient air. Fresh ground or skinless poultry meat is packaged in high oxygen atmosphere (70 to 80%) with the balance of atmosphere being CO_2 to

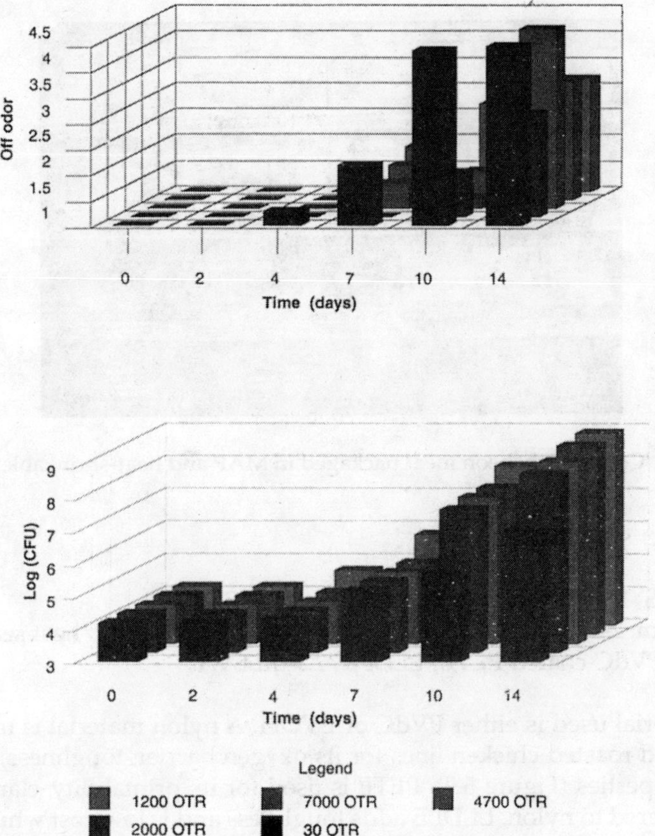

Figure 6.7 The effect of film oxygen transmission rate on the shelf-life of ground chicken leg meat. (From Dawson, P.L., Han, I.Y., Voller, L.M., Clardy, C.B., Martinez, R. M., and Acton, J.C., *Poult. Sci.*, 74, 1381, 1995. With permission.)

maintain color yet limit the growth of spoilage bacteria. A refrigerated shelf life of 14 days is attainable using this system,[8] and slightly longer if accompanied by deep chilling. For retail packages, the CO_2 concentration should be limited to 35% to minimize package collapse and excessive purge. Nitrogen is often used as a filler gas, which minimizes purge without the addition of oxygen.

Processed meat

Current practice

Processed meat products include nitrite-cured meat and non-cured cooked products. Processed meat is typically packaged in heat-shrinkable films such as EVA/PVdC/EVA or nylon/EVOH/ionomer co-extruded materials (Figure 6.8). Also, either nylon or PETE-based film, with a heat-sealable layer (ionomer or EVA) is used for processed poultry meat. Dried meat products stored at room temperature require a high oxygen and moisture barrier film such as PVdC or EVOH. Some dried meats are packaged in aluminum foil (PE laminated) films. The two common barrier tray packages used for poultry are

1. A non-barrier EPS (PS foam) tray overwrapped with a barrier film
2. An EPS (PS foam) tray with a built-in barrier and barrier lidding sealed to the tray

Figure 6.8 Cooked luncheon meat packaged in MAP and heat-shrinkable packages.

The package layer structure is

- Tray: HIPS/PS foam/HIPS/adhesive/barrier film
- Barrier film: 2–3 ml LLDPE/adhesive/PVdC-coated nylon/heat seal coating
- Lidding: PVdC-coated PETE/LLDPE/2.5 ml EVA

The barrier material used is either PVdC or EVOH. A nylon material is used in packages for hot wings and roasted chicken lines for its oxygen barrier, toughness, heat resistance, and forming properties (Figure 6.9). PETE is used for its printability, clarity, and relative lower cost compared to nylon. LLDPE adds toughness and is low cost while EVA provides a heat-sealing layer and proper seal strength upon cooling.

Cook-in-the-bag type products are restructured deli-type meats such as turkey hams, turkey breasts, and turkey rolls that are cooked in the bag after the package is sealed. The advantages of cook-in technology include increased shelf life, higher quality products, and increased product yields. The bag or casing must be capable of withstanding the temperatures required to fully cook the meat. The cook-in bag consists of layers that perform different functions. The water vapor and gas barrier layer is EVOH, however, some degree of adhesion to the meat product is required of the package. The adhesion layers are formed from nylons and/or Surlyn. The adhesion allows for minimum purge after cooking and

Figure 6.9 Roasted chicken packaged in polystyrene foam trays with a heat-shrinkable overwrap film.

storage. Non-adhesion bags may be used for products that are cooked in the package then removed for further processing such as smoking, browning, or addition of other flavors. The cook-in packaging system requires the combination of stuffers, clippers, pumps, and shrink tunnels for a continuous packaging operation.

Research

Frankfurters that were vacuum-packaged did not develop mold after 24 days of refrigerated storage while frankfurters from the same batch that were not vacuum-packaged did develop mold.[16] Natural-casing wieners held under MAP with a blend of 70% nitrogen, and 30% CO_2 were shelf-stable for 30 days.[8] Cured meats packaged in low OTR films with the removal of oxygen from the package will maintain their cured meat color and flavor while inhibiting the growth of spoilage organisms.

Interaction of the meat surface and film sealant layer is somewhat similar to the interaction of the myofibrillar proteins solubilized at the surface of meat tissue particles during preparation of comminuted meat products.[17, 18] Therefore the degree of meat-to-film binding is dependent on the extractable myofibrillar protein in the meat product.[19] Meat-to-film adhesion has been examined with the "peel" test.[19, 20] The inherent problem with this test is whether the force measured is that between meat and film or between meat and meat. Scanning electron micrographs of film surfaces reveal that when ground chicken meat emulsions were exposed to a non-binding film during heating, little or no meat residue appeared (Figure 6.10a). However, when the same emulsion was exposed to a binding film, meat residue adhered to the film surface (Figure 6.10b).[21]

A weak protein solution extracted from chicken breast meat was exposed to three different cook-in films (PE, nylon, and Surlyn-based). The total bound protein and the classes of bound amino acids were determined in samples held in a constant temperature water bath and in a water bath heated in temperature gradients. Protein adhesion occurred in all three film types, however, protein adhesion followed the trend Surlyn > nylon > PE after 60 min of heating at 25.8°C.[22] The amount of bound protein increased with Surlyn with heating from 55 to 80°C while PE and nylon showed little or no increase, respectively. Based on the amino acid class bound to the film, both hydrophobic interactions and hydrogen bonding participate in meat-to-film adhesion (Figure 6.11).

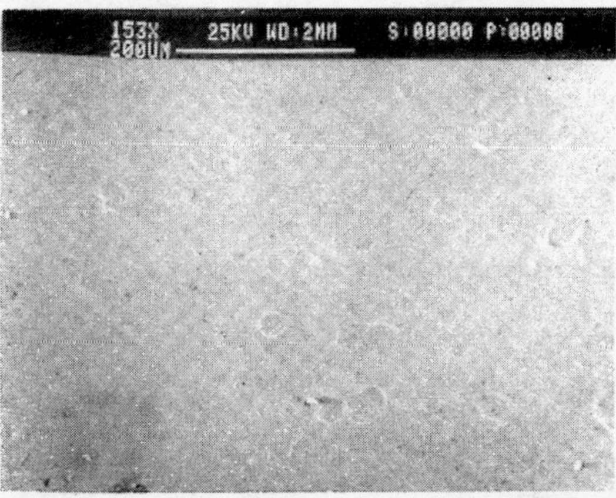

Figure 6.10a Osmium-tetroxide stained non-binding film peeled away from cooked ground chicken meat. (From Clardy, C. B. and Dawson, P. L., *Poult. Sci.*, 74, 1053, 1995. With permission.)

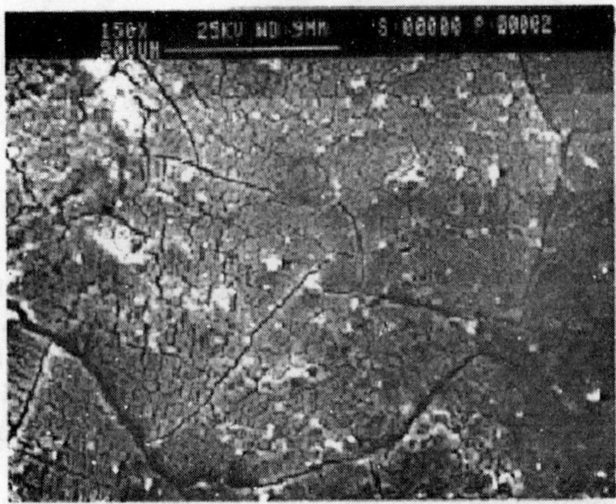

Figure 6.10b Osmium-tetroxide stained binding film peeled away from cooked ground chicken meat and showing adhering meat residue. (Clardy, C. B. and Dawson, P. L., *Poult. Sci.*, 74, 1053, 1995. With permission.)

Emerging technologies

Active packaging systems can be described as systems that interact with the environment and/or the food itself. Active packaging systems include those that scavenge oxygen, absorb moisture, and have selective gas permeability or change permeability with a change in temperature.

Oxygen scavengers

The use of oxygen scavengers is a novel approach that may have merit in selected poultry products. The addition of an oxygen scavenger within the package along with a physical barrier package such as PVdC or EVOH can maintain nearly a 0% oxygen level inside the package. Chemical oxidizing systems such as metaxylene adiamide plus a cobalt salt cata-

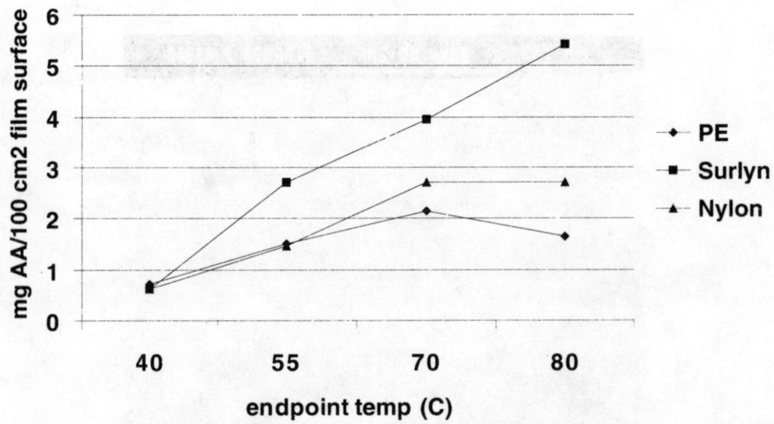

Figure 6.11 Bound amino acid concentrations (mg amino acid/100 cm^2 of film surface) from three film surfaces exposed to a weak chicken protein solution (12 mg protein/ml) at different endpoint temperatures. Samples were heated at 1°C/min.

lyst or enzyme reacting systems using glucose oxidase and catalase can actually remove oxygen from the package environment.[23] A system exists using mixed iron powder and calcium hydroxide that scavenges both oxygen and carbon dioxide.[24] These oxygen-reducing packages will inhibit the growth of aerobic spoilage bacteria but may create a favorable environment for pathogenic anaerobes.

Moisture absorbers

Because purge can facilitate bacterial growth, a moisture absorber placed in the package or as part of the film will slow the growth of bacteria. Absorbent pads placed beneath fresh poultry reduce the buildup of purge in the package. Films with entrapped propylene glycol will absorb moisture from the surface of meat when contacting its surface[24] and may have applications in extending the shelf-life of fresh poultry.

Temperature-compensating

There are films available that can switch permeability properties abruptly at specific temperatures. The change in permeability is accomplished by the use of long chain fatty alcohol-based side chains that will orient in a linear pattern changing from a random alignment which allows the change in permeability. While originally designed for use with respiring plant materials, there may be applications in poultry to maintain quality yet restrict microbial growth in products that are frozen in transit then thawed for retail display.

Antimicrobial packaging

Antimicrobial compounds have been added to the package to inhibit the growth of spoilage and pathogenic bacteria. Edible films and coatings can act as carriers of antimicrobial compounds as well as barriers to microorganisms. Most of the reported work with antimicrobial films and coatings has utilized acids carried in a variety of materials.[25–31] Sorbic acid has been incorporated into corn zein,[25] and methyl cellulose and hydroxy propyl methylcellulose[26] as coating to inhibit bacterial growth on food surfaces. Calcium alginate was used as a carrier for acetic acid and lactic acid to reduce *Listeria monocytogenes* populations on beef surfaces.

There are commercial films made using proprietary processes that incorporate a chlorinated phenoxy compound in the interstitial spaces of the polymer matrix. Nearly all commercial overwrap and vacuum-skin films are produced by a heat-extrusion method. The exceptions are some meat casings produced from collagen. Films using soy and corn protein have been formed by heat extrusion to carry antimicrobials within their structure.[32] Creating films from proteins by the heat extrusion method is a new technology that will enable the protein films to act as a carrier to deliver the antimicrobial to the food product.[33] Nisin and lysozyme in combination with EDTA when incorporated into the film structure of soy and corn protein films inhibit the growth of selected strains of Gram positive and Gram negative bacteria.[32] Nisin has also been incorporated into protein films and PE films and found to retain its antimicrobial activity (Figure 6.12).[34]

Further testing of these films has evaluated their effectiveness against *L. monocytogenes* (Figure 6.13) and *E. coli*. When the bacteria were exposed directly to the films, three to four log reductions in *L. monocytogenes*[33] and two to three log reductions in *E. coli*[32] were found. Nisin formulations have also been delivered to the surface of fresh poultry meat using agar and calcium alginate.[35] Average log reduction of *Salmonella typhimurium* populations exceeded three and four log cycles after 72 and 96 h of exposure at 4°C. These

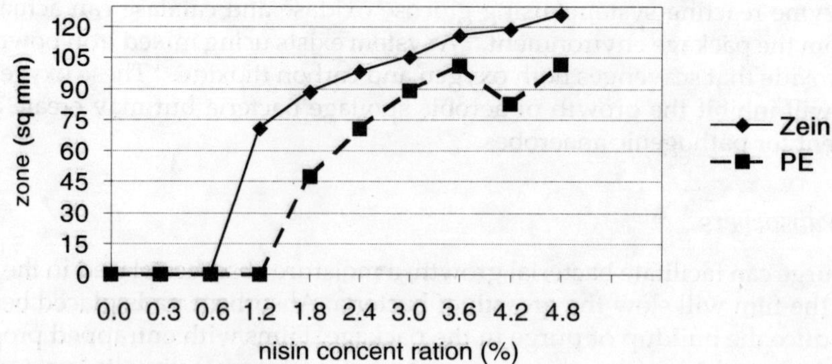

Figure 6.12 Bacterial zone of inhibition size for corn protein (zein) and polyethylene (PE) films containing different levels of nisin. (From Hoffman, K., Han, I.Y., and Dawson, P.L., *Int J. Microbiol.*, submitted for publication.)

nisin formulations have also been added to absorbent meat pads reducing *S. typhimurium* (Figure 6.14) populations up to four to five logs, and in some cases resulting in no recoverable cells.[36]

Another meat coating under development is chitosan, a carbohydrate derived from the skeleton of shellfish. This is a waste product of commercial shellfishing and can be processed to form a coating that has antifungal and antibacterial properties. Chitosan coatings reduced the total bacterial population on chicken drumsticks by one log (90%) compared to non-coated meat.[33]

The addition of combinations of antimicrobial compounds to packaging films has resulted in inhibition of both *Salmonella* and *E. coli* species (Figure 6.15).[34] The combinations of EDTA with nisin or with lauric acid or EDTA/lauric acid/nisin inhibited the growth of *E. coli* while EDTA with lauric acid or EDTA/lauric acid/nisin effectively inhibited *S. enteriditis.*

Aseptic packaging

Poultry meat products that would be contained in aseptic packages would be small meat particulates found in sauces, soups, and stews. These meat particulates are from both intact

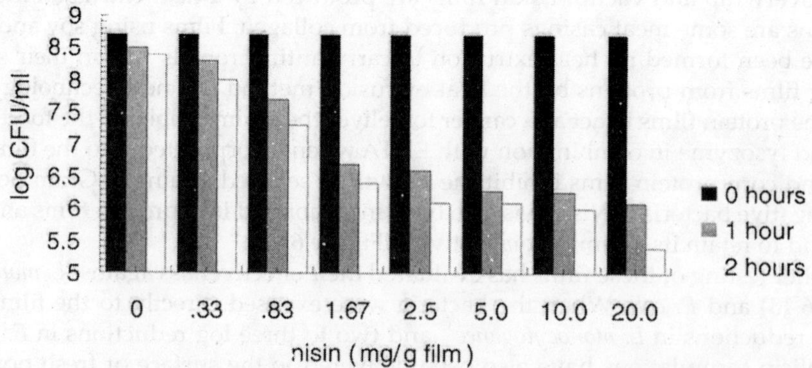

Figure 6.13 Log reductions of *Listeria monocytogenes* after exposure to films containing different levels of nisin.

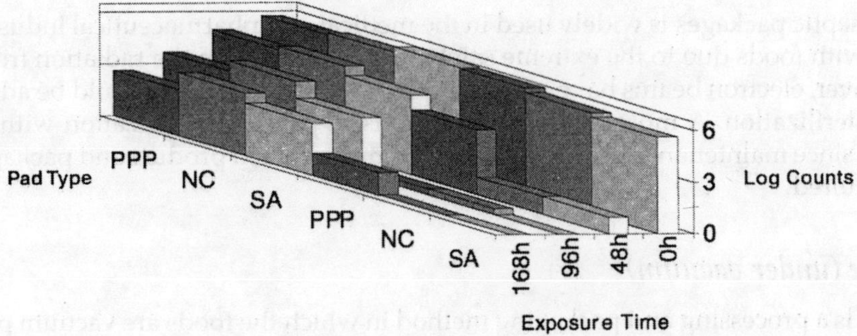

Figure 6.14 Inhibitory effect of nisin-containing pads on *Salmonella typhimurium.*

and restructured sources. Aseptic packaging and aseptic processing are inseparable by virtue of the interaction between the two in producing the final product. The major advantage of aseptic packaging is the reduction of the initial microbial load in the food and maintenance of package integrity after sterilization. The total process can be described as presterilization of the food before filling the sterile food into a presterilized package within a sterile environment followed by closing of the package in a sterile manner. While most packaging materials are sterile immediately after their production, they are easily contaminated by dust and handling during storage and prior to use. Therefore, sterilization for the aseptic process/package system must occur just before filling. Sterilization of the food in the aseptic system is most often accomplished using high-temperature, short-time processing. Other methods such as ohmic or microwave heating are also used to thermally process foods containing poultry meat particulates. These food sterilization processes follow traditional thermal death time methodology for assuring commercial sterility.

The methods of package sterilization range from steam and high heat for metal containers to non-heating methods for flexible containers such as hydrogen peroxide, ultraviolet (UV) radiation, or ionizing radiation. To ensure complete sterilization of the entire package surface, hydrogen peroxide treatment can be coupled with hot-air drying, ultrasonic energy, UV radiation, or copper ions. There are major drawbacks with UV radiation including limited penetration into liquids, no sterilization in surface shaded by package geometry or dust and the presence of rare microbial species that survive UV radiation damage and eventually can repair damaged DNA. Ionizing gamma ray radiation to

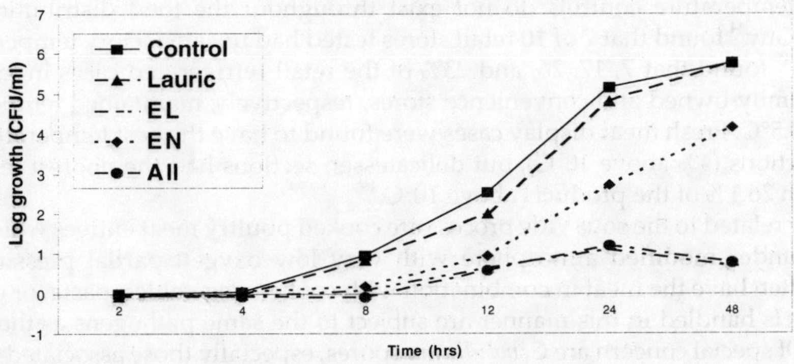

Figure 6.15 Effects of lauric acid, EDTA/lauric acid (EL), EDTA/nisin (EN), and EDTA/lauric acid/nisin (All) in corn zein films against 10^4 cfu/ml *Salmonella enteritidis.*

sterilize aseptic packages is widely used in the medical and pharmaceutical industry, however, not with foods due to the extreme safety measure to screen the radiation from workers. However, electron beams have gained approval for food use and could be adapted for package sterilization. A more likely application is in-package sterilization with ionizing radiation, since maintenance of a sterile zone and presterilized product and package would not be required.

Sous vide (under vacuum)

Sous vide is a processing and packaging method in which the foods are vacuum packaged, then cooked and stored under refrigeration after cooling. The product is usually reheated prior to consumption. The advantages of the sous vide process include cooking the meat in its own juices, sealing volatile flavor compounds in the package, and minimal loss of moisture or nutrients, resulting in a more flavorful, tender, and nutritionally complete product. The sous vide products are touted as retaining their "just-cooked" flavor for several weeks under refrigerated storage.[37] Concern about the safety of meats and foods packaged using the sous vide method has arisen, since the process is designed to produce the desired organoleptic properties without attention to proper commercial sterility guidelines.[38] The relatively mild heat treatment associated with cooking may not kill all vegetative cells and will certainly not inactivate spores. Sous vide products are formulated with few or no preservatives, are minimally heat-processed and thus are not shelf-stable, and are packaged under vacuum, which inhibits spoilage organisms but is an ideal environment for the growth of some pathogens.[38] The mild heat treatment accompanied with vacuum packaging tends to select for *Clostridium botulinum.* The outgrowth of *C. botulinum* spores and subsequent toxin production is likely to occur if spores are present since commercial sterility has not been assured.[39] While refrigeration will prevent the outgrowth of *C. botulinum,* this alone does not guarantee the safety of the food.[40, 41] The meat and poultry group of the National Advisory Committee on Microbiological Criteria for Foods recommended that refrigerated foods containing cooked, uncured meat should receive a heat treatment sufficient to achieve a 4 log reduction for *L. monocytogenes.*[42] Smith et al.[43] recommended a more intense heat treatment for these products, sufficient to achieve a 12–13 log reduction of *Streptococcus faecalis.* Other psychrotrophic pathogens of concern are *Yersinia enterocolitica* and *E. coli.* Sous vide products subjected to mild temperature abuse during storage, distribution, or preparation would add the risk of food poisoning from proteolytic strains of *C. botulinum, Staphylococcus aureus, Vibrio parahaemolyticus, Bacillus cereus,* and *Salmonella* spp. Adequate temperature controls do not exist throughout the food distribution system. Wyatt and Guy[44] found that 7 of 10 retail stores tested had unsatisfactory temperature control. Harris[45] found that 7, 17, 26, and 23% of the retail refrigerated cases in major, independent, family-owned and convenience stores, respectively, maintained temperatures at or above 10.5°C. Fresh meat display cases were found to have the best temperature control of other sections (4% above 10°C), but delicatessen sections had the poorest temperature control with 26.1% of the products above 10°C.[46]

Closely related to the sous vide process are cooked poultry meat entrees which are then packaged under modified atmosphere with very low oxygen partial pressures. These products often have the meat in combination with cooked vegetables, pasta, or on a bed of rice. Products handled in this manner are subject to the same pathogens as the sous vide products. Of special concern are *C. botulinum* spores, especially those associated with poultry since they are capable of outgrowth at and above 5°C. While these products have a good record to date, the danger for serious food-borne illness is always present.

Summary

Poultry packaging has many functions in addition to the obvious containment of the product. These functions are dependent on the properties of the packaging materials and how they interact with the food and the environment around them. The latest packaging system developments involve active packaging that can improve the product once contained and more efficient systems that facilitate distribution by reduced bulk or refrigeration.

References

1. Barron, F. B., *Food Packaging and Shelf Life: Practical Guidelines for Food Processors*, South Carolina Cooperative Extension Service and Clemson University. EC 686, 1995.
2. Miltz, J., Food Packaging, in *Handbook of Food Engineering*, Heldman, D. R. and Lund, D. B., Eds., Marcel Dekker, New York, 1992.
3. Timmons, D., "Dryer fryer"—is CVP the ultimate bulk pack?, *Broiler Bus.*, Dec. 10, 1976.
4. Shrimpton, D. H. and Barnes, E. M., A comparison of oxygen permeable and impermeable wrapping materials for the storage of chilled eviscerated poultry, *Chem. Ind.*, 1492, 1960.
5. Rey, C. R. and Kraft, A. A., Effect of freezing and packaging methods on survival and biochemical activity of spoilage organisms on chicken, *J. Food Sci.*, 36, 454, 1971.
6. Stollman, U., Johansson, F., and Leufven, A., Packaging and Food Quality, in *Shelf-life Evaluation of Food*, Man, C. M. D. and Jones, A. A., Eds., Blackie Academic, New York, 1994.
7. Dawson, P. L., Han, I. Y., Vollor, L. M., Clardy, C. B., Martinez, R. M., and Acton, J. C., Film oxygen transmission rate effects in ground chicken meat quality, *Poult. Sci.* 74, 1381, 1995.
8. Lawlis, T. L. and Fuller, S. L., Modified-atmosphere packaging incorporating and oxygen-barrier shrink film, *Food Technol.*, 44(6), 124, 1990.
9. Haines, R. B., The influence of carbon dioxide on the rate of multiplication of certain bacteria as judged by viable counts, *J. Soc. Chem. Ind.*, 52, 13, 1933.
10. Barnes, E. M., Impey, C. S., and Griffith, N. M., The spoilage flora and shelf life of duck carcasses stored at 2 or −1°C in oxygen-permeable or oxygen-impermeable film, *Br. Poult. Sci.*, 20, 491, 1979.
11. Ogilvy, W. S. and Ayres, J. C., Post-mortem changes in stored meats. II. The effect of atmospheres containing carbon dioxide in prolonging the storage life of cut-up chicken, *Food Technol.*, 5, 97, 1951.
12. Shaw, R., MAP of meats and poultry, in *Conference Proceedings, Modified Atmosphere Packaging (MAP) and Related Technologies*, September 6–7, Campden & Chorleywood Food Research Association, Campden, U.K., 1995.
13. Greengrass, J., Films for MAP foods, in *Principles and Applications of Modified Atmosphere Packaging of Foods*, Parry, R. T., Ed., Blackie Academic and Professional, Glasgow, G. B., 63, 1993.
14. Sander, E. H. and Soo, H. M., Increasing shelf-life by carbon dioxide treatment and low temperature storage of bulk pack fresh chickens packaged in nylon surlyn film, *J. Food Sci.*, 43, 1519, 1978.
15. Thomson, J. E., Microbial counts and rancidity of fresh fryer chickens as affected by packaging materials, storage atmosphere, and temperature, *Poult. Sci.*, 49, 1104, 1970.
16. Baker, R. C., Darfler, J., and Vadehra, D. V., Effect of storage on the quality of chicken frankfurters, *Poultry Sci.*, 51, 1620, 1972.
17. Seigel, D. G., Technical aspects of producing cook-in-hams, *Meat Process.*, 11, 57, 1982.
18. Terlizzi, F. M., Perdue, R. R., and Young, L. L., Processing and distributing cooked meats in flexible films, *Food Technol.*, 38(3), 67, 1984.
19. Rosinski, M. J., Barmore, C. R., Dick, R. L., and Acton, J. C., Research note: Film-to-meat-adhesion strength for a cook-in-the-film packaging system for a poultry meat product, *Poult. Sci.*, 69, 360, 1990.

20. Rosinski, M. J., Barmore, C. R., Dick, R. L., and Acton, J. C., Film sealant and vacuum effects on two measures of adhesion at the sealant-meat interface in a cook-in package system for processed meat, *J. Food Sci.*, 54, 863, 1989.
21. Clardy, C. B. and Dawson, P. L., Film type effects on meat-to-film adhesion examined by scanning electron microscopy, *Poult. Sci.*, 74, 1053, 1995.
22. Clardy, C. B., Han, I. Y., Acton, J. C., Wardlaw, F. B., Bridges, W. B., and Dawson, P. L., Protein-to-film adhesion as examined by amino acid analysis of protein binding to three different packaging films, *Poult. Sci.*, 77, 745, 1998.
23. Yoshii, J., Recent trends in food packaging development in consideration of environment, *Packag. Jpn.*, 13(67), 74, 1992.
24. Labuza, T. P. and Breene, W. M., Applications of "active packaging" for improvement of shelf-life and nutritional quality of fresh and extended shelf-life food, *J. Food Process. Preserv.*, 13(1), 31, 1989.
25. Torres, J. A. and Karel, M., Microbial stabilization of intermediate moisture food surfaces. III. Effects of surface preservative concentration and surface pH control on microbial stability of an intermediate moisture cheese analog, *J. Food Process. Preserv.*, 9, 107 1985.
26. Vojdani, F. and Torres, J. A., Potassium sorbate permeability of methylcellulose and hydroxy-propyl methylcellulose coatings: effects of fatty acids, *J. Food Sci.*, 55, 941, 1990.
27. Rico-Pena, D. C. and Torres, J. A., Oxygen transmission rate of edible methylcellulose-palmitic acid film, *J. Food Process. Eng.*, 13, 125, 1990.
28. Siragusa, G. R. and Dickson, J. S., Inhibition of *L. monocytogenes* on beef tissue by application of organic acids immobilized in a calcium alginate gel, *J. Food Sci.*, 46, 1010, 1992.
29. Davidson, P. M. and Juneja, V. K., Antimicrobial agents, in *Food Additives*, Branen, A. L., Davidson, P. M., and Salminen, S., Eds., Marcel Dekker, New York, 1990, 83.
30. Robach, M. C. and Sofos, J. N., Use of sorbates in meat products, fresh poultry and poultry products: a review, *J. Food Prot.*, 55, 1468, 1982.
31. Maas, M. R., Glass, K. A., and Doyle, M. P., Sodium lactate delays toxin production by *Clostridium botulinum* in cook-in-bag turkey products, *Appl. Environ. Microbiol.*, 55(9), 2226, 1989.
32. Padgett, T., Incorporation of food-grade antimicrobial compounds into biodegradable packaging films, *J. Food Prot.*, 61, 1330, 1998.
33. Dawson, P. L., Developments in antimicrobial packaging, in *Proceedings of the 33rd National Meeting in Poultry Health and Processing*, 1998, 94.
34. Hoffman, K., Han, I. Y., and Dawson, P. L., Efficacy of antimicrobial agents in corn zein films, *Int J. Microbiol.*, submitted.
35. Natrajan, N. and Sheldon, B. W., Evaluation of bacteriocin-based packaging and edible film delivery systems to reduce *Salmonella* in fresh poultry, *Poult. Sci.*, 74 (Suppl.1), 31, 1995.
36. Sheldon, B. W., Efficacy of nisin impregnated pad for the inhibition of bacterial growth in raw packaged poultry, *Poult. Sci.*, 75 (Suppl.1), 97, 1996
37. Baird, B., Sous Vide: What's all the excitement about? *Food Technol.*, 44(11), 92, 1990.
38. Rhodehamel, E. J., FDA's concerns with sous vide processing, *Food Technol.*, 46(12),73, 1992.
39. Conner, D. E., Scott, V. N., Bernard, D. T., and Kautter, D. A., Potential *Clostridium botulinum* hazards associated with extended shelf-life refrigerated foods: a review, *J. Food Safety*, 10, 131, 1989.
40. Palumbo, S. A., Is refrigeration enough to restrain foodborne pathogens, *J. Food Prot.*, 49, 1003, 1986.
41. Moberg, L., Good manufacturing practices for refrigerated foods, *J. Food Prot.*, 52, 363, 1989.
42. U.S. National Advisory Committee on Microbiological Criteria for Foods (USNACMCF), Recommendations of the U.S. National Advisory Committee on Microbiological Criteria for Foods:I HACCP principles, II meat and poultry, III seafood, *Food Control*, 2(4), 202, 1991.
43. Smith, J. P., Toupi, C., Gagnon, B., Voyer, R., Fiset, P. P., and Simpson, M. V., Hazard analysis and critical control point (HACCP) to ensure the microbiological safety of sous vide processed meat/pasta product, *Food Microbiol.*, 7, 177, 1990.

44. Wyatt, L. D. and Guy, V., Relationship of microbial quality of retail meat samples and sanitary conditions, *Food Prot.*, 43, 385, 1980.
45. Harris, R. D., Kraft builds safety into next generation refrigerated foods, *Food Process.*, 50(13), 111, 1989.
46. Daniels, R. W., Applying HACCP to new-generation refrigerated foods at retail and beyond, *Food Technol.*, 45(6), 122, 1991.

Selected reading

Principles and Applications of Modified Atmosphere Packaging of Foods, 2nd ed., Blackstone, B. A., Ed., Aspen Publishers, Gaithersburg, MD, 1999.

Controlled/Modified Atmosphere Packaging of Foods, Brody, A. L., Ed., Food and Nutrition Press, Trumbull, CT, 1989.

The Microbiology of Poultry Meat Products, Cunningham, F. E. and Cox, N. A., Eds., Academic Press, San Diego, CA, 1987.

Packaging Foods with Plastics, Jenkins, W. and Harrington, J. P., Eds., Technomic Publishing, Lancaster, PA, 1991.

Food Packaging, Robertson, G. L., Ed., Marcel Dekker, New York, 1998.

chapter seven

Meat quality: sensory and instrumental evaluations

Brenda G. Lyon and Clyde E. Lyon

Contents

0-8493-0120-3/01/$0.00+$.50

Introduction

There are several dimensions to quality. Quality products are those that meet some need or expectation of consumers and are safe and wholesome as well. Products that can be produced and sold to meet a demand at a profit for producers are quality products. Products that meet processing and handling guidelines set by agencies charged with protecting the commercial food supply are quality products. Quality has several dimensions, depending on whose viewpoint is needed: regulatory personnel, producer, and ultimately consumer.

Consumers are interested in appearance, aroma/odor, taste, texture, and sound, which are all quality characteristics measured by use of the senses. Human testers measure these characteristics (sensory attributes) by evaluating products and marking their responses on paper or electronic scoresheets. Instruments can measure characteristics that are directly related to the physical or chemical components of the product. These two types of measurements are used together to draw conclusions and make assumptions about quality. This chapter deals with quality factors perceived and measured by consumers (appearance, aroma/odor, taste, texture, and sound) and how these factors relate to chemical or physical component characteristics that can also be measured.

Sensory quality attributes

Sensory evaluation is analysis of product attributes perceived by the human senses of smell, taste, touch, sight, and hearing. People (consumers or users of the product) are used to assessing the sensory characteristics and providing a response. Instruments are used to measure some physical or chemical characteristic that influences the sensory stimulus perceived and responded to by the human. Instruments do not measure sensory characteristics. However, instruments are sought that provide a corollary measurement that can predict or relate to the anticipated sensory experience. Both human and instrumental methods are critical when assessing sensory quality. Human assessment is more complicated. People differ in their innate ability to sense stimuli. They differ in the experiences with foods that allow a base for the neurological categorizing of a stimulus and the subsequent varieties of responses that can be given. Instruments on the other hand can be calibrated and programmed to respond consistently in a given way, but the meaning of the responses has to be interpreted by humans and validated by the human sensory experience.

Evaluating food with the five senses

The five senses are taste, smell, sight, touch, and hearing. The responses to food are shown in Figure 7.1.

Aroma and taste

The senses of smell and taste are interrelated and assess the quality attribute known as flavor. Volatiles are small molecules released from the food (during heating, chewing, etc.) that react with receptors in the oral or nasal cavities. Signals are sent to the brain where they are processed. This processing results in responses that indicate whether the sensation was sweet, sour, salt, or bitter (four basic tastes) and whether the sensation can be identified more specifically (e.g., brothy, chickeny, fruity, etc.). Primary receptors for the four basic tastes are on the tongue and other surfaces of the oral cavity. Receptors for volatiles are located in the various sections of the nasal cavity. Sniffing is a technique used to collect a concentration of the volatiles and force them to the receptors in the nasal cavity for processing and identification.

Sight, touch, and hearing

The senses of sight, touch, and hearing are related to the structure and state of product components. With the sense of sight, the sensory attributes of color and appearance are evaluated. Receptors in the eyes are stimulated by light waves, causing signals to be sent to the brain for processing. Therefore, appearance and color of foods involve the eyes as the sense organ of the body and the components of the object (food) that reflect or transmit light. Instrumentally, color is measured with instruments that determine the amount of light reflected by the object at each wavelength. Color is very complicated. Humans measure color as a composite, whereas instruments break the color into individual wavelengths.

Examples of texture characteristics perceived by sight are smoothness and bumpiness. The physical characteristics of texture are the mechanical and geometrical characteristics that are related to structure. These include strength, size, shape, and type of components perceived as the product breaks down due to some force applied. The force could be the

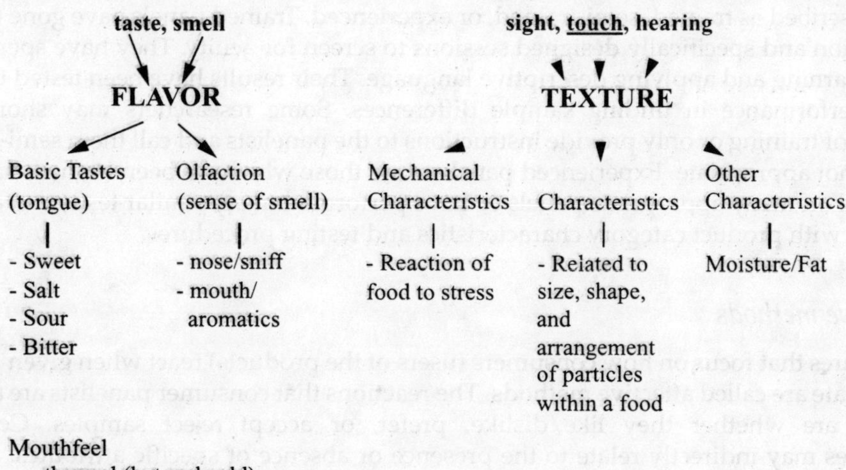

Figure 7.1 Components of the basic senses.

teeth or it could come from an instrument. Other characteristics such as oily, greasy, wet, and dry relate to mouthfeel and the sense of touch. The sense of hearing can also be used to evaluate texture. For example, crunchiness may be an important quality in the batter and breading of poultry products.

Other characteristics

Chemical and thermal mouthfeels such as cool, warm, hot, and cold are the other characteristics perceived by the senses. These are called the trigeminal sensations and are related to responses to stimuli on the cells of the linings of the mouth, tongue, and throat.

Sensory methods to evaluate poultry quality

There are two general types of sensory methods. Laboratory/analytical methods use a small number of panelists to determine if a difference exists between samples and the nature, direction, and intensity of the difference. Consumer affective methods involve a larger number of panelists and include tests that measure how consumers feel or react to the product to provide a measure of preference, acceptance, and like/dislike. There are different panel criteria for laboratory and affective methods.

Laboratory/analytical methods

Methods that focus on detecting whether differences exist in products and how those differences might be described are called laboratory/analytical methods. Small panels (6 to 12 assessors) of people who have been screened for their sensory acuity and ability to describe products are used. Laboratory panels may be composed of staff or of outside persons paid to attend sensory training and testing sessions. The key factor is that the panelists have been screened and trained to evaluate products for specific characteristics, not for whether they like or dislike the product. Therefore, the focus in these tests is the product attributes using panelists as the measuring tools or instruments. Performance of the panel must be measured to determine if their responses are reliable and consistent. Some panels have been described as trained, semi-trained, or experienced. Trained panels have gone through orientation and specifically designed sessions to screen for acuity. They have spent many hours learning and applying descriptive language. Their results have been tested to determine performance in finding sample differences. Some researchers may shorten the process of training or only provide instructions to the panelists and call them semi-trained. That is not appropriate. Experienced panelists are those who have been trained, have participated on many appropriate panels or have performed many similar tests, and are very familiar with product category characteristics and testing procedures.

Affective methods

Procedures that focus on how consumers (users of the products) react when given samples to evaluate are called affective methods. The reactions that consumer panelists are asked to convey are whether they like/dislike, prefer, or accept/reject samples. Consumer responses may indirectly relate to the presence or absence of specific attributes. Do consumers like the product? How well do they like it? Which sample is more spicy? Or more tender? Do they prefer the product well enough to always purchase this brand over another? Do they *accept* this product, even though they would *prefer* one less spicy? The panelists used in these studies must be users of the product categories. Consumer panels require larger numbers of people than do trained panels in order to sample or test responses from a user population and then extrapolate the conclusions to a general population. The consumer respondents are not trained or screened, except to determine demographic profiles for relating to larger populations. The focus in consumer/affective testing is the *behavior* of the panelist in relation to the product as the *stimulus* presented to the consumer.

Determining which type of test

There are six fundamental questions to determine whether to use a difference/discriminative test method or an affective test method.

1. Do the samples differ?
2. If so, on what sensory parameters do the samples differ?
3. Can the difference be quantified?
4. What is the direction of the difference? (i.e., more salty, less hard?)
5. How does this compare to similar products?
6. Does this have importance at the consumer level?

Generally, from the order of questions, that difference/discriminative or descriptive tests come first, so that the characteristics of the product are known. Then consumers are asked for acceptance, preference, or like/dislike in order to assess whether the known differences are important to the consumer.

In product development and marketing research, another approach is emerging. Consumer research determines how the concept of new products might be accepted and determines what characteristics consumers want. Products are then designed with those characteristics, using trained panels to screen and evaluate prototypes.

In any case, the purpose and function of the panel type remains the same. Consumer panels are large to represent the feelings or purchase behavior of people toward the test product. Trained panels are small numbers of people who have been screened to have good acuity of the senses and whose task is to pinpoint discernible differences in samples.

Considerations in conducting sensory tests

This section will focus more on the smaller panels used for difference testing. In-house panels can be made up of staff or students within the company or department. However, screening and training are important and in order to screen and train panelists, they must first be selected for their ability to detect small differences in aroma, taste, or texture. Panelists must also be able to describe the characteristics. Although taste and smell are of great importance, so is good health, a positive attitude, and motivation to perform the tests without bias. Willingness and reliability to attend and participate in training and testing sessions are equally important. A very important point to remember is that training panelists involves more than explaining a scoresheet. Trained panelists function as sensitive instruments, making responses to specific tasks that are totally separate from their personal opinions of like/dislike.

Sample presentations and preparation

Samples presented for evaluation must come from a common and uniform source. This is a difficult aspect of sensory testing when dealing with muscle food products, because they are not as homogeneous as some other samples, such as grains or liquids. Choices for sampling poultry meat depend on the test question, one being how many samples are needed at one time. Another factor is how samples are to be cooked, sectioned, and presented so that each panelist receives nearly identical samples under identical conditions.

The actual sample presented to the panelist should be uniform in size. Serving temperature should be uniform throughout the sample piece. Appropriate implements should be provided for evaluation (fork, toothpick). Filtered water should be provided between samples for mouth cleansing to prevent taste carryover. Sometimes, unsalted crackers or apples or other products are needed as well.

Methods of preparing the product are also determined by the test objective. Some studies have been conducted where roasting in an oven was appropriate. Questions to be considered were placement of pieces on the pan, placement of pan in oven, how to check internal temperature without disrupting the cooking cycle, whether to roast covered or uncovered, what oven temperature to use, and what internal temperature to use.

An example of a test preparation and sampling scheme used by Lyon and Lyon[1] involved cooking broiler breasts in heat-and-seal bags immersed in water. This procedure provided the best control for sample identification by labeling the bags and handling a large number of samples during cooking. It was also appropriate to record individual breast weights before and after cooking to determine cook yield and to conduct further analysis on the cooked meat and fluid/solids liberated during heating.[2–4] The effects of cooking method on subsequent quality attributes of broiler breast meat have been reported.[5–7]

Testing room

The area or room where panelists are presented samples and perform the tests requires specific environmental controls, such as constant, comfortable temperature and humidity, and freedom from extraneous odors, noise, and other distractions. This control is necessary because human testers are designed to perceive and process many stimuli constantly and unconsciously. In order that panelists might concentrate on smaller numbers of specific stimuli (i.e., the test sample), they must be given an area that minimizes any stimuli other than those of the test. In addition to the environmental controls, individual booths are needed so that samples are presented to the panelist in isolation from other panelists to avoid distraction and to avoid any collaboration on the part of panelists. A floor plan for a self-contained sensory laboratory is shown in Figure 7.2.

Lighting must also be controlled. If appearance of the sample is an important task of the test, the lighting must not provide shadows and the spectrum of the light must be appropriate for the use of the sample. On the other hand, if taste or mouth texture are key aspects of the test, then special lighting might be needed to mask differences that the panelists would use as cues to selecting different samples based on appearance rather than the taste or texture under investigation. Some labs use red, green, or even blue lighting. A monochromatic light often used is sodium vapor lighting that imparts an even spectrum of orange, brown.

Test areas can range from portable partitions set up at a table to large testing facilities housing a complete sensory evaluation laboratory including a waiting area, a training room, testing booths with computerized data input systems, serving areas, and kitchen/preparation areas. The key point is that the more control there is over the environment where the test is performed, the more confidence the evaluator has that the panelists are responding to stimuli in the product rather than stimuli to their surroundings.

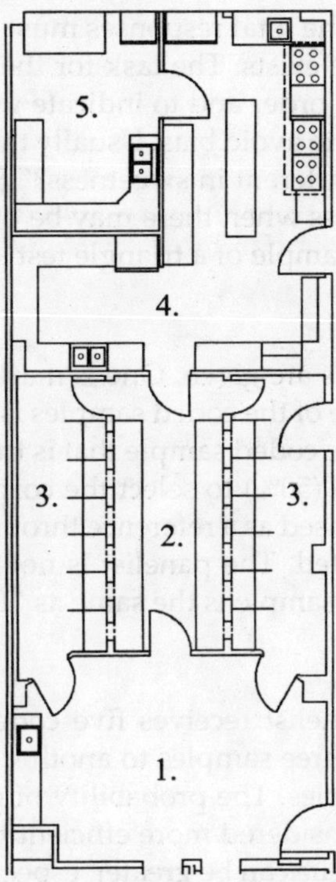

Figure 7.2 Sensory evaluation laboratory floor plan. (1) Training area includes tables, chairs, writing board, other visual aids. (2) Serving area to provide sample presentation to the individual booths in testing areas on each side. (3) Testing areas in the laboratory includes six individual booths on each side of the serving area. The individual booths are also equipped with computer components for electronic data input. (4) Preparation area for sinks, cabinets, counterspace, ovens. (5) Laboratory area for sample analysis, including color, electronic nose, instrumental texture, hood for aroma reference preparation.

Specific sensory test formats

Difference/discriminative tests

Difference/discriminative tests are conducted under the premises that the panelist evaluates a set of samples and determines whether any samples differ from another. If a significant number of panelists detect a difference, then a true difference exists. The treatments are known to the experimenter who scores the test responses as correct or not correct and determines significance from tables based on the number of samples, number of panelists, and the statistical probabilities of chance in selecting the correct sample. Details on test features and the statistical tables for data interpretation can be found in several popular textbooks.[8–9]

Difference tests include triangle, duo-trio, paired comparison, A not A, two of five, and three of five. These tests usually involve determining if two treatments differ. Multiples of either are presented and the panelists must select one or two based on stated criteria. Responses are recorded for whether the answer is correct or incorrect.

Triangle test

In the triangle test, panelists are presented with three coded samples, two the same, one different (odd). Each panelist has a one in three (33.3%) chance of choosing the correct sample by random selection. Therefore, the total responses must be higher than one third in order to conclude that a true difference exists. The task for the panelist is to taste or smell, etc., the three coded samples in given order and to indicate which is different. Sample order is randomized by the experimenter to avoid bias. Usually there is no qualifier to the test question such as, "Which sample is different in sweetness?" Such a question tends to lead the panelist to look only for sweetness when there may be other cues that determine the true differences in the samples. An example of a triangle test scoresheet is shown in Figure 7.3.

Duo-trio test

In the duo-trio test, three samples are given. One is marked as "Reference" and the other two samples are given codes. One of the coded samples is the same as the "Reference." The task for the panelist is to select the coded sample that is the same as the reference. The panelist has a one chance out of two (50%) to select the correct sample by random selection. Either of the two samples can be used as a reference throughout the whole test, or the selection for reference can be alternated. The panelist is not given a specific characteristic to focus on, but must decide which sample is the same as "Reference."

Two-out-of-five test

In the two-out-of-five test, a panelist receives five coded samples. Two of the samples belong to one set and the other three samples to another set. The task of the panelist is to identify the set of two alike samples. The probability of guessing the right answer in this test is 1 in 10, and is therefore considered more efficient than the triangle test. However, a disadvantage is that sensory fatigue can be greater, especially if the test is used for taste or oral texture. This test is used successfully with tasks involving visual, auditory, or tactile senses.

Paired-comparison test

In a paired-comparison test, the respondent receives two coded samples (a pair) and is asked to evaluate both, comparing the intensity of some specific characteristic. The specific response is to record which of the two has the greater (or lesser) intensity of that attribute being studied. In this test, a specific attribute may be given to the panelist to focus on in the evaluation, such as which is sweeter.

Usually in difference tests, the task is to determine whether or not a difference exists. If there is a difference, further tests might be presented to determine on what basis the samples might differ or in what direction the samples might differ.

Name:____________________ Date:____________

Directions: Two of the samples are the same. One is different.
Place a check by the sample that is different.

Sample #	Check the ODD sample
526	
344	
879	

Figure 7.3 Triangle test score sheet.

Ranking tests

Ranking tests are similar to directional-difference paired-comparison tests, except that more than one sample is presented and panelists are asked to place samples in the set in some sort of order. For example, rank from most tender to least tender. Rank samples from most sweet to least sweet. These are examples of evaluating samples on a specific criteria, i.e., tenderness or sweetness, and of indicating the direction of difference in that characteristic. A consumer test (large number of untrained panelists) could also be a ranking test if the task requested is that he/she place the samples in order based on least acceptable to most acceptable or vice versa.

Category scaling

Difference tests can also involve category scales in which products are tested for specific attributes and the panelists are asked to rate the amount that the characteristic is present. Category scales can also be used for consumer testing in which the specific attribute rated is degree of like/dislike, acceptance, or preference. The format of the scales can be numbers (i.e., 1 to 5) anchored with a specific term, such as very tender, moderately tender, etc. The scale can be unstructured, anchored only at ends and middle with either adjectives or faces (i.e., frown to smile). The panelist marks on a line from left to right to indicate the point their response to the product attribute is on the continuum. The response on the line is measured with a ruler or automatically if a computerized system is used. The values of the response, whether as the structured category scales or unstructured line scales, are analyzed for their distribution variances by analysis of variance.

Descriptive analysis

Descriptive analysis is a form of sensory testing in which trained panelists determine the perceptible attributes in a product set and score the intensity of the attributes that are present. Flavor or texture may be profiled, or a profile can be developed for all the major important attributes of a product from its initial appearance to the feeling left in the mouth after the sample is swallowed.

Flavor profile

The first flavor profile method was introduced in 1949 by the Arthur D Little Company.[8] Flavor characteristics are described and quantified in a consensus manner by trained sensory panelists. Much of the work of the panelists is done around a table where they first analyze products individually and then discuss their responses as a group. The order that aroma, flavor, and mouthfeeling characteristics appear is important. A simplified intensity scale is used to indicate where the sample is in an attribute range from detectable to very strong. Because the final result is usually a group decision, statistics are not used to analyze the data.

Texture profile

A method of evaluating sensory texture characteristics of products and relating these to instrumental rheology principles was developed at General Foods Research in the early 1960s.[10–12] Attributes were classified and defined to describe texture from the first bite to after swallowing. Terminology and references were developed to illustrate various classifications of characteristics. Mechanical characteristics dealt with resistance to breakdown (hardness, cohesiveness, springiness). Geometrical characteristics dealt with the size, shape, and orientation of the individual components or particles that form the structure and how they behave when that structure is disturbed by force, such as chewing. Finally, the moisture and fat properties were also considered to be part of the texture modality. Evaluating these characteristics required trained panels. A scaling system that allowed food references to be ranked or scored with the intensity of a predominant characteristic that the food displayed was also developed. With numbers to indicate intensity, the results of individual panelists could be statistically analyzed.

Other profiling methods

Building on the work of the original flavor profile and texture profile, variations of the descriptive profiling methods have emerged, some now trademarked by their creators, including Quantitative Descriptive Analysis (QDA)[9] and Sensory Spectrum.[8] Both of these involve development of descriptive language by the panels and providing intensity values that can be statistically analyzed. There are some differences in the way that terminology is developed. Also, QDA uses an intensity scale that the panel selects based on the range of products to be evaluated. Sensory Spectrum developed a universal scale to measure the intensity of any identified character note in comparison to another. For example, a 0 to 15 scale could be used to rate sweetness of beverages comparing them to the sweetness of sucrose solutions ranging from 2% (score of 2) to 10% (score of 10). Another example is the grape character note scored as a 4 in grape Kool-Aid and a score of 12 for grape-note intensity in Welch's grape drink (Table 7.1). Against this background of intensive training by the panel, the intensity of brothy notes in chicken soup or stewed chicken could be scored by one panel and understood by another sensory panel trained in the same descriptive method. Language or terms to describe the individual attributes are developed by the panel members with a panel leader to guide them and provide references.

Variations of these methods have also been reported and used successfully. Free-choice profiling lets panelists develop their own terms and score the intensities. Advanced statistical procedures are needed to interpret the results.

Rating scales

The rating scales that are used with these methods take the form of a continuous line that represents a low or no level of intensity to a very high level. Sometimes intensity terms

Table 7.1 **Example of the Universal Scale[a] to Provide Scores of Intensity of a Character Note in Evaluating Aroma or Taste**

Scale value (i.e., score)	Character note (descriptive term)	Ref. (food example)
2	Soda	Saltines
4	Grape	Grape Kool-Aid
7	Orange	Orange Juice concentrate
9.5	Orange	Tang
10	Grape	Welch's grape juice
12	Cinnamon	Big Red chewing gum

[a] Intensity of any specific taste or aroma character note fits on a common scale (like a ruler).

Source: Adapted from Meilgaard, M. C., Civille, G. V., and Carr, B. T., *Sensory Evaluation Techniques*, 3rd ed., CRC Press, Boca Raton, FL, 1999.

are presented as anchors along the line at distinct intervals or at the ends. When the intervals are clearly marked, the scale is said to be structured. When there are no marked points between the lines, the scale is unstructured and the panelist uses a mental cue for intensity.

Instrumental methods of analysis

Texture is considered the most important characteristic of poultry meat and is the attribute most affected by age of the bird and processing procedures. Because of the importance of texture, a great deal of emphasis has been placed on instrumental procedures to evaluate the structure of muscle fibers. Bourne[13] noted several important truths about instrumental texture measurements. Many tests are applicable to more than one type of food, so it is more useful to classify the texture measurements by type of test rather than by commodity. He noted that the basic process of chewing food to break down the food for swallowing occurs, regardless of what kind of food is in the mouth. Another truth was that the fundamental instrumental tests were developed by scientists and engineers interested in the theory and practice of materials or construction to measure well-defined rheological properties. Those theories may not be as useful in measuring what is happening in the mouth during mastication. As a matter of fact, the expectations of the tests are opposite for the two groups of scientists. The engineer wants to measure the strength of material in order to design a structure that will withstand forces applied to it without breaking, while the food scientist wants to measure the strength of food, and frequently weakens the structure so that it will break easier. In this situation, food texture measurement might be considered more of a study of the weakness of materials rather than strength of materials.

Instrumental procedures to estimate tenderness of meat have been studied and widely accepted by researchers and quality control (QC) personnel since the 1950s. These procedures offer repeatability to obtain numerical values that should relate to tenderness. The danger of using instrumental procedures is putting too much value in the "number" without understanding what it really represents. Texture has historically been viewed in an overly simplistic manner, so the research was geared toward finding a single measurement or number to encompass the entire mastication process and arrive at an either/or decision: tender or tough. An accurate description of poultry meat tenderness involves more than a single instrumental value, since most treatments alter postmortem biochemical events and affect not only tenderness, but also moisture-binding characteristics such as juiciness and moisture release.

Selected texture instrumental methods

Unless noted otherwise, the focus of this section will be on breast muscle/meat because this economically significant part of the carcass has received the vast majority of the research attention. The breast has received this attention because of its postmortem biochemistry (see Chapter 4) and subsequent fiber characteristics that impact finished product quality. It should be noted that any of these methods can be used to evaluate leg/thigh meat and ground/comminuted products as well as intact meat products. One simply needs to determine the objective of the analysis and choose the appropriate method. Shearing may be most important for whole-muscle while compression may be best for frankfurters or cohesiveness for restructured products like nuggets and patties.

The majority of the instrumental data used to determine tenderness in cooked poultry meat have been generated on the Warner-Bratzler (W-B) or the Kramer Shear Press (KSP). These procedures are designed to shear or cut through fibers of muscle. Another technique, instrumental Texture Profile Analysis (TPA) data, has been used to generate texture information for poultry meat products. An in-depth discussion of the concept and measurement of food texture was published by Bourne[13] and will only be briefly summarized here.

Shear test

Shear tests have been used for many years. Samples are positioned so that a single blade or multiple blades cut perpendicular to the fibers. The basic principle of the test is that the total force to cut through the sample is related to the tenderness/toughness of the cooked sample. The force has historically been recorded in weight measurements (i.e., lb, kg), but these can be converted to the force unit of Newtons, if appropriate.

Warner-Bratzler shear device. The W-B shear device has been used to shear or cut red meat and poultry samples for the last 50+ years.[14] The device is small and portable, consisting of a rectangular blade with a triangular hole cut from the center. This blade is attached to a circular fan scale. The sample of known dimensions, usually a circular core for red meat or a rectangular strip for poultry, is placed in the triangular notch of the single blade. Two bars are lowered by a hydraulic motor and the sample is pushed across the apex of the triangular notch. As the bars are lowered across the sample, the peak force to shear across the fibers is recorded in lb or kg on the circular fan scale. The benefits of this device are its reliability, ruggedness, ease of use, portability, and low cost (less than $1200). The device lends itself to on-site quality control work. The limiting factor is that only peak load or peak shear force is generated during the test, so the researcher or QC personnel must have sufficient background sensory panel data to add validity to the shear values (Figure 7.4).

Kramer Shear Press (KSP). The other shear test that has been extensively used for red meat and poultry texture research is performed with a shear cell based on the KSP.[15] The shear test cell is composed of two main parts, a metal box with slots which holds the sample and a top part with 5 or 10 blades spaced to fit into the slots. This device is attached to

Figure 7.4 Warner-Bratzler shearing device.

Figure 7.5 Warner-Bratzler shear cell for an Instron UTM.

a system designed to move the multiple blades down and through a rectangular sample placed in the cell. The multiple blades are lowered across the sample. They initially compress and then shear across the fibers forcing the resulting strips out the bottom of the slotted cell. Results are recorded as kg/g of sample weight. The KSP is rugged, but it is much heavier, less portable, and more expensive. It has been modified from its original design to predict quality of lima beans and used to measure textural properties of a variety foods including fruits and other vegetables.

Both of the blade designs of the original W-B and KSP systems have been reproduced on other instruments such as the Instron Universal Testing Machine™ (UTM); (Instron Corp., Canton, MA) and the Texture Technologies Texture Analyzer™ (Texture Technologies Corp., Scarsdale, NY). The multiple blade cell is also referred to as the Allo-Kramer shear cell. The W-B blade and an Allo-Kramer shear cell are pictured in Figures 7.5, 7.6, respectively. The newer systems are accompanied by software to program the machines and to record more dimensions of the force/distance or force/time curves.

Texture profile analysis

The instrumental TPA was introduced as a way to generate multiple textural attributes for food.[10–12] The need for a multiple-point test was reinforced by Breene[16] who noted that texture is complex and multiple point procedures would be more useful than single point procedures. The TPA was recently updated by Meullenet et al.[17]

A typical two-curve TPA for chicken meat is shown in Figure 7.7. The significant attributes are noted and defined. Significant attributes such as hardness, springiness, cohesiveness, and chewiness can be separated and analyzed. A TPA sample is usually a circular core taken from the cooked meat. A decision must be made by the researcher on percent of compression during the test. In the literature, ranges reported for percent compression range from 60 to 80% of the original height of the core. Compressing less than 60% usually does not compress the sample enough to result in measurable changes, while compressing more than 80% usually destroys the sample matrix so much during the first compression

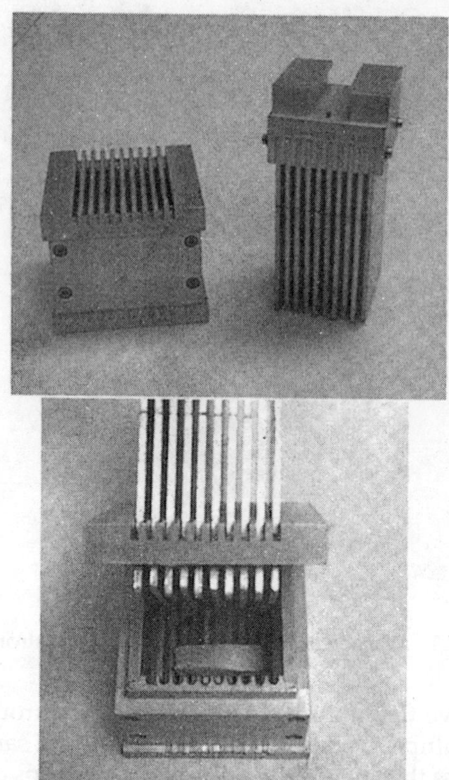

Figure 7.6 Allo-Kramer shear cell for an Instron UTM.

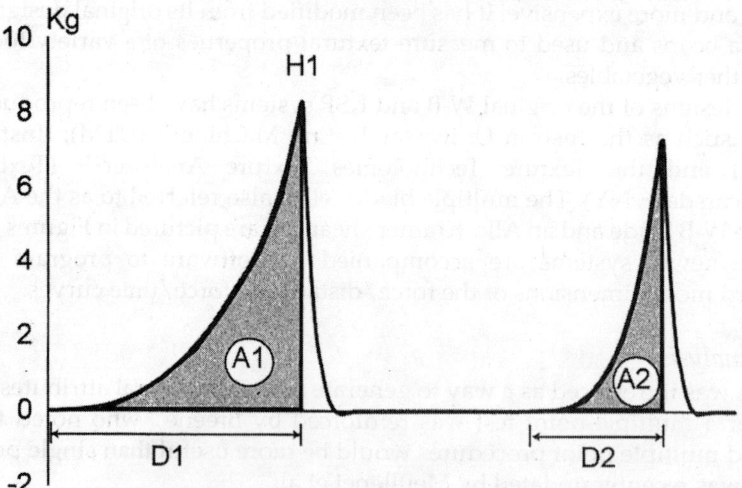

Hardness = H1= max force in kg of first peak
Cohesiveness = A2/A1= area of peak 2 in square mm / area of peak 1 in square mm
Springiness = D2/D1= distance 2 in mm / distance 1 in mm
Chewiness = Hardness x Cohesiveness x Springiness

Figure 7.7 A typical texture profile analysis (TPA) curve pattern for chicken meat. From Lyon et al., *J. Applied Poultry Res.*, 1, 27, 1992. With permission.

that the second compression curve yields little or no information. The core to be evaluated is placed on a flat metal plate, and the top metal plate attached to the load cell is positioned to contact the sample (initial point). The percent compression is converted to cross head travel from the initial point. After the first compression and return of the cross head to the initial point, the cross head is immediately engaged for the second compression. The TPA is more of a research tool than the shear tests. The TPA is more sensitive and versatile than the W-B or KSP shears. However, the purchase and maintenance costs for instruments such as the Instron UTM or the Texture Analyzer are much higher, and they are not as portable.

Sample considerations for shear or profile tests

Regardless of the type of instrumental test, sample dimensions play a major role in the results and should always be described or referenced.[18, 19] Physical characteristics of the product as used by the consumer should be taken into account when evaluating the meat sample. For example, if the treatment imposed has a direct effect on meat thickness due to muscle contraction (postmortem/postchill deboning time), then the difference in thickness should be part of the test. However, if the research goal is to evaluate the sensitivity of instrumental procedures, then uniform sample dimensions (height and width) would be required. A sampling scheme for both sensory and instrumental tests from a broiler breast muscle used in this lab is illustrated in Figure 7.8.

Relationships between instrumental procedures and sensory panels for texture

As noted earlier, there is a danger in reducing the complex continuum of texture to a single objective number. A number of studies have been conducted to help determine the relationship between instrumental and sensory data related to texture.[20–22]

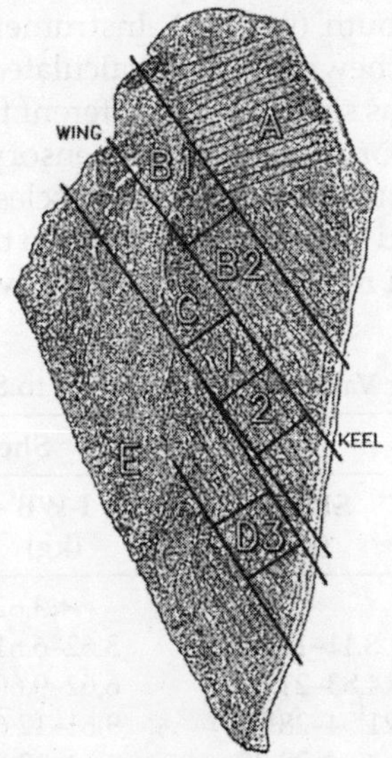

Figure 7.8 Diagram of scheme to sample individual cooked breast muscle for sensory and instrumental tests. Section B = 1.9 cm wide strip used for Warner-Bratzler shears. Section C strip was cut 1.9 cm wide. Ends were trimmed and pieces 1 and 2 (approximately 1.9 cm^2) used for panelists. Sections A, D, and E used in studies involving diced samples. From Lyon, B.G., and Lyon, C.E., *Poultry Sci.*, 75, 812, 1996. With permission.

Intact muscle

Intact muscle samples were used in a study[20] to correlate sensory scores from a five-member untrained panel to KSP values. Processing treatments were used to simulate a wide range of texture in the cooked meat. The authors noted that the KSP and sensory scores were correlated, that KSP values greater than 8 kg/g of sample weight were tender to very tender, and that due to a wide 95% confidence interval the five-member untrained panel was too small to measure sensory reaction. Two studies by Lyon and Lyon[21, 22] increased the number of untrained panelists to 24 to determine the texture relationship of broiler breast meat to 4 instrumental tests. Then 4 breast muscle deboning times ranging from after feather removal (0 hour postmortem) to after whole carcass aging for 24 hours were used to provide the texture spectrum from tough to tender. The four instrumental tests were the bench-top W-B, Allo-Kramer (i.e., KSP), W-B attached to an Instron (I-WB), and a single blade version of the multi-bladed KSP (SB-AK). All shearing apparati, except for the bench-top W-B were attached to an Instron UTM.

Results are summarized in tabular form in Table 7.2. The significance of the results is that instead of a single number for each shear test, a range of values corresponding to the sensory panel perception of tenderness was established for each test. The sensory scale is not an either/or (tough or tender), but a gradation of values from very tough to very tender. These data are used by quality control personnel to verify process control and ensure optimum tenderness for customers.

Lyon and Lyon[23] reported on the relationship between the TPA and a trained panel's response to intact broiler breast meat by using 4 postmortem deboning times (<5 min, 2, 6, 24 hours) and two cook methods (heat-seal bags in water and microwave) as variables. In a series of sessions, the 8-member trained panel developed 17 attributes and rating scales to evaluate texture (Table 7.3). The attributes developed by the panel to evaluate the samples represented a 4-stage profile ranging from the first compression with molar teeth without biting through the sample (stage 1) to impressions at the point of swallowing and the "afterfeel" properties in the mouth (stage 4). Instrumental TPA attributes of hardness, springiness, cohesiveness, and chewiness were calculated. Meat from muscles deboned 5 min and 2 hours postmortem was significantly different from those deboned 6 or 24 hours postmortem for 16 of the 17 sensory attributes. No sensory differences were noted for meat from muscles deboned 6 or 24 hours postmortem. Muscles removed 5 min postmortem had significantly higher hardness and chewiness values than those deboned 2, 6, or 24 h. Within deboning time, the panel scored meat cooked via microwaves as more juicy and wet and

Table 7.2 **Instrumental Shear Values Corresponding to Sensory Tenderness Categories**

	Shear apparatus[1]			
Sensory tenderness	**SB-AK (kg)**	**I-WB (kg)**	**MB-AK (kg/g)**	**B-WB (kg/g)**
Very tender	<8.11	<3.62	<5.99	<3.46
Moderately-slightly tender	8.11–14.82	3.62–6.61	6.00–8.73	3.47–6.40
Slightly tender-slightly tough	14.83–21.53	6.62–9.60	8.74–11.48	6.41–9.35
Slightly-moderately tough	21.54–28.24	9.61–12.60	11.49–14.24	9.36–12.30
Very tough	>28.25	>12.60	>14.25	>12.40

[1] Devices used and attached to Instron Universal Testing Machine were single blade Allo-Kramer (SB-AK), Warner-Bratzler blade (I-WB), and multi-blade Allo-Kramer (MB-AK). The fourth device was bench-top Warner-Bratzler (B-WB).

Source: Adapted from Lyon, C. E. and Lyon, B. G., *Poult. Sci.*, 69, 1420, 1990 and Lyon, B. G. and Lyon, C. E., *Poult. Sci.*, 70, 188, 1991.

Table 7.3 **Descriptive Texture Attributes and Definitions Used to Evaluate Intact Broiler Muscle**

Term	Definition
Stage I. Place sample between molars. Compress slowly (3 cycles) without biting through the sample	
1. Springiness	Degree to which sample returns to original shape after partial compression. (Scale: low to high.)
Stage II. Place the sample between molars. Bite through the sample (no more than 6 cycles) using the rate of 1 chew per second	
2. Initial cohesiveness	Amount of deformation before rupture. (Scale: low = very little deformation before rupture to high = high degree of deformation before rupture.)
3. Hardness	Force required to bite through the sample to rupture it. (Scale: low to high.)
4. Initial juiciness	Amount of moisture in the meat. (Scale: low, dry to high, juicy.)
Stage III. Place the sample between the molars. Chew at the rate of 1 chew per second. At 15 to 25 chews, begin evaluation of the attributes below	
5. Hardness	Force necessary to continue biting through the sample. (Scale: low to high.)
6. Cohesiveness of mass	How the sample holds together during chewing. (Low = fibers break easily, wad dissipates; grows high = wad in size, resists break down.)
7. Saliva produced	Amount of saliva produced in the mouth during sample manipulation to mix with sample to ready it for swallowing. (Scale: none to much.)
8. Particle size and shape	Description of size and shape of the particles as sample breakdown continues on chewing. (Scale: fine small particles to coarse, large particles.)
9. Fibrousness	Degree of fibrousness or stringiness. (Scale: small to large.)
10. Chewiness	(Scale: tender, chewy, tough.)
11. Chew count	Number of chews to get sample ready to swallow.
12. Bolus size	Size of wad at point of swallowing. (Scale: small to large.)
13. Bolus wetness	Amount of moisture in or moisture feel of wad at point ready swallowing.
Stage IV. Evaluate the following at the point the sample is swallowed	
14. Ease of swallowing	(Scale: easy to hard.)
15. Residual particles	Amount of loose particles left in mouth after swallowing.
16. Toothpack	Amount of material packed in and around teeth. (Scale: none to much.)
17. Mouth-coating	Amount of moisture and fat coating the oral cavity after swallowing. (Scale: low to high.)

From Lyon, B. G. and Lyon C. E., *Poult. Sci.*, 69, 329, 1990. With permission.

as having less residual particles and toothpack compared to the meat cooked in water. By TPA, the microwave cooked meat was more cohesive and chewy than the meat cooked in water.

The panel results significantly correlated to the instrumental TPA. For example, the muscles removed 5 min postmortem were more springy, cohesive, harder, produced more saliva on chewing, had a larger bolus size, were harder to swallow, and had more toothpack. Needless to say, this is significantly more than a single force shear value and adds to the broad spectrum of attributes that we term "texture." The impact and complexity of juiciness are evident in the panel results.

Ground poultry meat texture studies

A series of studies published in the late 1970s and 1980s[24–26] characterized the texture of poultry products made from ground and comminuted meat with various ingredients. All three studies utilized both sensory panel methods and instrumental texture measurements. A wide range of quality attributes was evaluated (proximate composition, water-holding capacity, color, rancidity, and cook loss). In one of the studies,[24] use of mechanically deboned poultry meat as the meat source (with and without skin) in conjunction with two levels of structured protein fiber (15 and 25%) was evaluated by a 5-member trained panel using the QDA technique. In addition, a scale to reflect overall impression of the products was included.

In another study,[25] six patty formulations containing different amounts of mechanically deboned broiler meat (MDBM), hand deboned fowl meat (HDFM), and structured protein fiber (SPF) were characterized for proximate composition, rancidity (measured as thiobarbituric acid or TBA values), color (Hunter L, a, b values), force to shear (W-B), and sensory properties. Sensory properties were evaluated using QDA. As the level of MDBM decreased, moisture and protein contents, lightness (L values), and shear values increased correspondingly; fat content, redness (a values), and TBA values decreased. Sensorially, as the level of MDBM decreased, the products were perceived as being lighter, more chewy and elastic, and less juicy. Based on the instrumental and sensory data, the authors noted that interchangeable ratios of 40:60/60:40 MDBM and HDFM could be incorporated with SPF to yield products of good quality. These multiple point results illustrate the benefits of integrating instrumental and sensory analysis to arrive at decisions involving finished product quality. A spider-web diagram illustrating part of the results is shown in Figure 7.9.

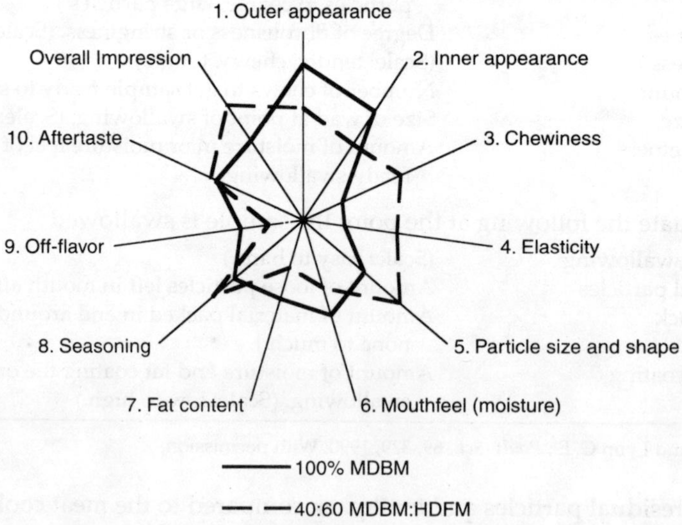

Figure 7.9 A spider-web diagram of sensory evaluation of mechancally deboned broiler meat (MDBM) and hand deboned fowl meat (HDFM). (Adapted from Lyon, B.G., Lyon, C.E., and Townsend, W.E., *J. Food Sci.*, 43, 1656, 1978.)

Data points are placed on the various lines representing each attribute. The center represents a value of "0" and the values increase away from the center point. The differences in attributes such as outer appearance, chewiness, elasticity, particle size/shape, and overall impression are easily noted. In this example, the combination of 40% MDBM: 60% HDFM was superimposed on the 100% MDBM patty product for visual comparison of attributes.

In yet another study, Lyon et al.[26] used TPA to determine differences between mixed and flake-cut MDPM in patties containing either 15 or 25% SPF. The six-member trained panel also evaluated juiciness using a seven-point intensity scale. Positive, significant correlation coefficients between instrumental and sensory measures of hardness, springiness, and chewiness indicated that the Instron and the panel were in good agreement.

Color

Color is very complex and is a major component of appearance in poultry meat or products. Instrumental methods to measure color of an object are based on a light source and a detector. Objects absorb and reflect light wavelengths that are detected by an instrument or an observer. Results of instrumental detectors have little meaning unless validated by the human observer. Therefore, numerical values provided by colorimeters are almost always associated with a color/appearance term in order to understand the meaning. For example, "lightness" is associated with "L values," "redness" with "a values" and "yellowness" with "b values" when an "Lab" color coordinate system is used. A typical colorimeter used in research and quality assurance is shown in Figure 7.10.

Fletcher[27] reviewed poultry meat color, color measurements, methods used to measure color, and summarized color defects associated with poultry. The review of meat color covered raw meat and many of the factors that affect meat color such as sex, age, strain, processing procedures, cooking temperature, and freezing. Of particular significance at the present time are the factors that influence "pinking" of breast meat. The significance from both quality and safety standpoints is the assumption of insufficient cooking time/temperature. This is a problem with immediate economic ramifications (returned shipments of

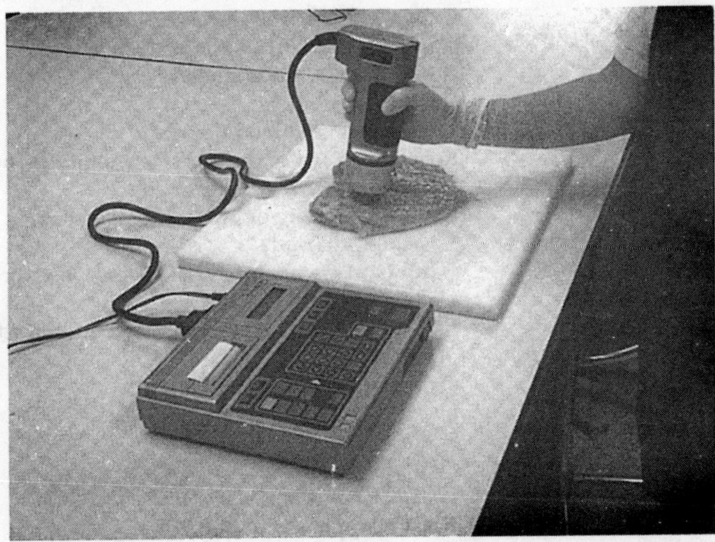

Figure 7.10 A colorimeter being used to determine the numeric color value of a broiler breast fillet.

cooked product). Other specific color defects include the relationship between lightness (paleness) and poor protein functionality,[28] and the consumer objection to color variation between meat pieces in a retail package.[29]

Flavor

Flavor analyses of poultry or poultry meat involves methods to extract compounds that are assumed to contribute to aroma. Taste is usually associated with the basic solutions of salt, sweet, sour, and bitter, while aroma is associated with stimulation of receptors in the nasal cavity by volatiles released by foods. Instruments that separate compounds and indicate their concentrations include gas chromatography (GC), high pressure liquid chromatography (HPLC), and sensing devices referred to as "electronic noses."

GC and HPLC are methods that separate extracts of the food into individual compounds. Although individual compounds and classes of compounds have been identified, they must be related to sensory response by descriptors. Sensory descriptors determined by a trained panel for evaluation of cooked chicken were developed by Lyon[30] (Table 7.4). Farmer[31] listed as many as 34 main compounds that are considered to be key components of cooked chicken flavor. Taken alone, the individual compounds do not always exhibit the aroma of individual perceived aromas from the samples. For example, 2-acetyl-pyrroline, a key odor compound in cooked poultry meat, is described as "popcorn." Re-combining certain chemicals to create an aroma-specific character note is not always successful. Sensory panels can tell the difference.

The electronic nose is a name given to instruments comprised of arrays of materials (metal oxides, conducting polymers) that record an electrical charge or resistance response when a stream of volatiles is passed over them. Several sensors of varying materials give varying responses so that a pattern emerges for a given sample. Key to the instruments is

Table 7.4 **Terms Used for Profiling the Taste and Aroma of Fresh and Reheated Chicken Meat**

Descriptive term	Definition
	Aromatic, taste sensation associated with:
Chickeny	Cooked white chicken muscle
Meaty	Cooked dark chicken muscle
Brothy	Chicken stock
Liver, organy	Liver, serum or blood vessels
Browned	Roasted, grilled or broiled chicken patties (not seared, blackened or burned)
Burned	Excessive heating or browning (scorched, seared, charred)
Cardboard, musty	Cardboard, paper, mold or mildew; described as nutty, stale
Warmed-over	Reheated meat; not newly cooked nor rancid, painty
Rancid, painty	Oxidized fat and linseed oil
	Primary taste associated with:
Sweet	Sucrose, sugar
Bitter	Quinine or caffeine
	Feeling factor on tongue associated with:
Metallic	Iron or copper ions

From Lyon, B. G., *J. Sensory Studies*, 2, 55, 1987. With permission.

analysis of data by multivariate statistical programs that can detect pattern differences and also develop algorithms that can later recognize this pattern as belonging to a certain sample. Supposedly, the technique is based on how a stream of volatiles will pass across the receptors in a human's nose, detect the differences, and attach a recognition to the pattern for future identification.

Factors that influence or contribute to meat quality

Many factors influence poultry meat quality. Some factors are more significant than others (see Chapters 2–4). Rigor condition of the breast muscle at the time of removal from its skeletal restraints (deboning time) significantly affects the texture of this economically important part of the carcass (Chapter 4). Time of breast muscle removal involves postmortem muscle biochemistry (pH decline, lactic acid increase, ATP depletion) as well as physiology (gross and microscopic muscle fiber contraction, i.e., sarcomere lengths). Time of breast muscle removal is really a "double-edged sword" depending on the condition of the meat in the finished product. On one side, higher pH noted in prerigor muscle equates to increased water-holding and emulsifying capacity which are important for ground and comminuted products. On the other side, the same high pH equates to objectionable toughness in intact cooked meat. Froning and Neelakantan[32] reported that a pH of 5.9 or higher could be used to indicate a prerigor condition in broiler and turkey breast meat, and that the pH was below this value within 30 min of death. Lyon et al.[33] reported prerigor pH values of 6.1 and 6.3 for broilers and mature hens, respectively, within 20 min of death, but values lower than 5.9 after 1.5 hours.

The relationship between postmortem time, muscle biochemistry, and ultimate texture has been illustrated by many researchers. Lyon et al.[34] noted that broiler breast muscles deboned immediately postchill had significantly higher pH values, 6.22, and the cooked meat required greater force to shear, 15.19 kg compared to muscles deboned at 1, 2, 4, 6, 8, or 24 hours postchill. The most rapid pH decline was noted during the first hour postchill, and no significant differences in pH or shear values were noted after 4 hours postchill time prior to deboning. Sams and Janky[35] evaluated the effects of water and brine chilling on broiler breast meat pH and tenderness. They added a "hot boned" group of muscles which were removed from the carcasses immediately after feather removal (picking). The other 2 treatments were breast removal after 1 hour chilling, and after 24 hours aging. For the muscles chilled in water, the highest pH and KSP values were noted for the "hot boned" group, 6.4 and 10.2 kg/g, respectively. The muscles removed after chilling were intermediate, and the lowest pH and KSP values were noted for the 24-hour aged group. Dawson et al.[36] using similar conditions noted the same postmortem deboning time and shear force (KSP, kg/g of sample weight) pattern for broiler breast meat, with highest KSP values noted at 0.17 hours postmortem holding time and the lowest at 24.33 hours, 17.8 and 4.1 kg/g, respectively.

There are contradictions in the literature as to the effect of sex of the broiler on tenderness of the cooked meat. Simpson and Goodwin [37] and Farr et al.[38] reported that shear values for male broilers were significantly lower than those for females. Other researchers, including Goodwin et al.[39] reported that sex of the bird did not influence tenderness. Lyon et al.[2] reported on the effects of postchill broiler breast muscle deboning time, fillet holding time, and sex of the bird on tenderness. The two sexes of birds were raised under commercial conditions, processed on separate days, and the data analyzed within each sex. Mean raw breast weights were 163 and 122 grams for males and females, respectively.

Under the conditions of the Lyon et al.[2] study in which weight of the breast samples was not controlled or adjusted, it would appear that meat from female broilers was more tender for all treatments. The larger size and weight of the male breasts probably contributed to the increase in force to shear the samples. For the larger muscles removed from the males immediately postchill (0 hour of aging), there would have been an accompanying loss in area and an increase in thickness due to muscle shortening facilitated by elevated ATP levels. This shortening pattern of the *pectoralis* muscle was reported by Papa

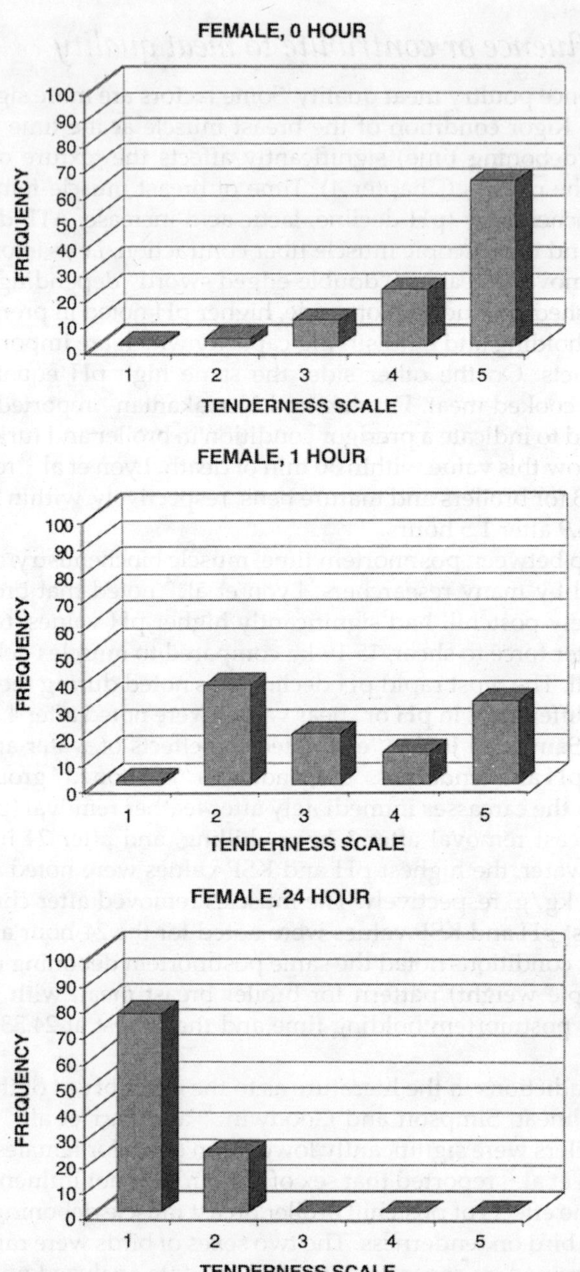

Figure 7.11 Subjective tenderness distribution for female broiler breast muscle over postmortem time. (Adapted from Lyon, C.E. and Lyon, B.G., *Poult. Sci.*, 69, 1420, 1990.)

and Lyon.[40] As noted earlier, since the condition of the sample is part of the variable being studied, the height and/or width of the sample was not standardized.

To determine the practical importance of the numerical decrease in W-B values, Lyon and Lyon[21] superimposed the shear values on a tenderness scale established by a 24-member untrained sensory panel. Frequency distribution of W-B values for female samples which correspond to the panel's perception of tenderness are illustrated in Figure 7.11. Since fillet holding time prior to freezing was not significant, the data were combined into a single value for postchill deboning time. Eighty-five percent of muscles removed from female broilers immediately postchill (0 hour) would be classified as "moderately to very tough" (categories 4 and 5). This percentage decreased to 43% if the muscles were left on the skeleton for 1 hour. Muscles removed 24 hours postchill were all in the "moderately to very tender" portion of the scale (categories 1 and 2) for both sexes. Without the sensory panel perception data, the W-B shear values have less meaning and while the numbers can be statistically analyzed, their practical importance would be limited.

Conclusions

Poultry meat quality is a complex issue which will become increasing important as more new products are introduced to consumers. Students, researchers, quality control, and management personnel must appreciate this complexity and all work together to provide the appropriate and complete information. The "marriage" of sensory and instrumental methodology is critical to providing the correct answers and making the best decisions about product quality.

References

1. Lyon, B. G. and Lyon, C. E., Research Note: shear value ranges by Instron Warner-Bratzler and single blade Allo-Kramer devices that correlate to sensory tenderness, *Poult. Sci.*, 70, 188, 1991.
2. Lyon, C. E., Lyon, B. G., Papa, C. M., and Robach, M. C., Broiler tenderness: effects of postchill deboning time and fillet holding time, *J. Appl. Poult. Res.*, 1, 27, 1992.
3. Lyon, C. E., Bilgili, S. F., and Dickens, J. A., Effects of chilling time and belt flattening on physical characteristics, yield, and tenderness of broiler breasts, *J. Appl. Poult. Res.*, 6, 39,1997.
4. Lyon, B. G. and Lyon, C. E., Assessment of three devices used in shear tests of cooked breast meat, *Poult. Sci.*, 77, 1585, 1998.
5. Lyon, B. G. and Lyon, C. E., Effects of water-cooking in heat-sealed bags versus conveyor-belt grilling on yield, moisture, and texture of broiler breast meat, *Poult. Sci.*, 72, 2157, 1993.
6. Dunn, N. A. and Heath, J. L., Effect of microwave energy on poultry tenderness, *J. Food Sci.*, 44, 339, 1979.
7. Lyon, C. E. and Wilson, R. L., Effects of sex, rigor condition, and heating method on yield and objective texture of broiler breast meat, *Poult. Sci.*, 65, 907, 1986.
8. Meilgaard, M. C., Civille, G. V., and Carr, B. T., *Sensory Evaluation Techniques*, 3rd ed., CRC Press, Boca Raton, FL, 1999.
9. Stone, H. and Sidel, J. L., *Sensory Evaluation Practices*, 2nd ed., Academic Press, San Diego, CA, 1993.
10. Szczesniak, A. S., Classification of textural characteristics, *J. Food Sci.*, 28, 385, 1963.
11. Szczesniak, A. S., Brandt, M. A., and Friedman, H. H., Development of standard rating scales for mechanical parameters of texture and correlation between the objective and the sensory methods of texture evaluation, *J. Food Sci.*, 28, 397, 1963.
12. Friedman, H. H., Whitney, J. E., and Szczesniak, A. S., The texturometer— a new instrument for objective texture measurement, *J. Food Sci.*, 28, 390, 1963.
13. Bourne, M. C., *Food Texture and Viscosity: Concept and Measurement*, Academic Press, New York, 1982.

14. Bratzler, L. J., Determining the tenderness of meat by using the Warner-Bratzler method, *Proc. Second Ann. Reciprocal Meat Conf. National Livestock and Meat Board*, 1949, 117.
15. Kramer, A. K., Guyer, R. B., and Rogers, H., New shear press predicts quality of canned limas, *Food Eng.* (NY), 23, 112, 1951.
16. Breene, W. M., Application of texture profile analysis to instrumental food texture evaluation, *J. Texture Stud.*, 6, 53, 1975.
17. Meullenet, J. F., Lyon, B. G., Carpenter, J. A., and Lyon, C. E., Relationship between sensory and instrumental texture profile attributes, *J. Sensory Stud.*, 13(1), 77, 1997.
18. Lyon, B. G. and Lyon, C. E., Assessment of three devices used in shear tests of cooked breast meat, *Poult. Sci.*, 77, 1585, 1998.
19. Smith, D. P., Lyon, C. E., and Fletcher, D. L., Comparison of the Allo-Kramer shear and texture profile methods of broiler breast meat texture analysis, *Poult. Sci.*, 67, 1549, 1988.
20. Simpson, M. D. and Goodwin, T. L., Comparison between shear values and taste panel scores for predicting tenderness of broilers, *Poult. Sci.*, 53, 2042, 1974.
21. Lyon, C. E. and Lyon, B. G., The relationship of objective shear values and sensory tests to changes in tenderness of broiler breast meat, *Poult. Sci.*, 69, 1420, 1990.
22. Lyon, B. G. and Lyon, C. E., Research Note: shear value ranges by Instron Warner-Bratzler and single-blade Allo-Kramer devices that correspond to sensory tenderness, *Poult. Sci.*, 70, 188, 1991.
23. Lyon, B. G. and Lyon, C. E., Texture profile of broiler *Pectoralis major* as influenced by post-mortem deboning time and heat method, *Poult. Sci.*, 69, 329, 1990.
24. Lyon, C. E., Lyon, B. G., Townsend, W. E., and Wilson, R. L., Effect of level of structured protein fiber on quality of mechanically deboned chicken meat patties, *J. Food Sci.*, 43, 1524, 1978.
25. Lyon, B. G., Lyon, C. E., and Townsend, W. E., Characteristics of six patty formulas containing different amounts of mechanically deboned broiler meat and hand deboned fowl meat, *J. Food Sci.*, 43, 1656, 1978.
26. Lyon, C. E., Lyon, B. G., Davis, C. E., and Townsend, W. E., Texture profile analysis of patties made from mixed and flake-cut mechanically deboned poultry meat, *Poult. Sci.*, 59, 69, 1980.
27. Fletcher, D. L., Poultry meat colour, *Poultry Meat Science*, CABI Publishing, New York, 1999, 159.
28. Owens, C. M., Hirschler, E. M., McKee, S. R., Martinez-Dawson, R., and Sams, A. R., The characterization and incidence of pale, soft, exudative turkey meat in a commercial plant, *Poult. Sci.*, 79, 553, 2000.
29. Fletcher, D. L., Color variation in commercially packaged broiler breast fillets, *J. Appl. Poult. Res.*, 8, 67, 1999.
30. Lyon, B. G., Development of chicken flavor descriptive attribute terms aided by multivariate statistical procedures, *J. Sensory Stud.*, 2, 55, 1987.
31. Farmer, L. J., Poultry meat flavour, *Poultry Meat Science*, CABI Publishing, New York, 1999, 127.
32. Froning, G. W. and Neelakantan, S., Emulsifying characteristics of pre-rigor and post-rigor poultry muscle, *Poult. Sci.*, 50, 839, 1971.
33. Lyon, C. E., Hamm, D., Thomson, J. E., and Hudspeth, J. P., The effects of holding time and added salt on pH and functional properties of chicken meat, *Poult. Sci.*, 63, 1952, 1984.
34. Lyon, C. E., Hamm, D., and Thomson, J. E., pH and tenderness of broiler breast meat deboned various times after chilling, *Poult. Sci.*, 64, 307, 1985.
35. Sams, A. R. and Janky, D. M., The influence of brine chilling on tenderness of hot-boned, chill-boned, and age-boned broiler breast fillets, *Poult. Sci.*, 65, 1316, 1986.
36. Dawson, P. L., Janky, D. M., Dukes, M. G., Thompson, L. D., and Woodward, S. A., Effect of post-mortem boning time during simulated commercial processing on the tenderness of broiler breast meat, *Poult. Sci.*, 66, 1331, 1987.
37. Simpson, M. D. and Goodwin, T. L., Tenderness of broilers affected by processing plants and seasons of the year, *Poult. Sci.*, 54, 275, 1975.
38. Farr, A. J., Atkins, E. H., Stewart, L. J., and Loe, L. C., The effects of withdrawal periods on tenderness of cooked broiler breast and thigh meats, *Poult. Sci.*, 62, 1419, 1983.
39. Goodwin, T. L., Andrews, L. D., and Webb, J. E., The influence of age, sex, and energy level on tenderness of broil rs, *Poult. Sci.*, 48, 548, 1969.
40. Papa, C. M. and Lyon, C. E., Shortening of the *Pectoralis* muscle and meat tenderness of broiler chickens, *Poult. Sci.*, 68, 663, 1989.

chapter eight

Microbiological pathogens: live poultry considerations

Billy M. Hargis, David J. Caldwell, and J. Allen Byrd

Contents

Introduction — significance of the problem

Food-borne illness is a significant worldwide public health problem. The Council for Agricultural Science and Technology, in a 1994 report entitled "Foodborne Pathogens: Risks and Consequences" estimated that as many as 9000 deaths and 6.5–33 million illnesses in the U.S. each year are caused by ingestion of contaminated foods. In 1996, the Foodborne Diseases Active Surveillance Network (FoodNet) collected data on nine food-borne diseases in several sites within the U.S.[1] Since the start of this program, *Campylobacter* and *Salmonella* have been the leading causes of laboratory-confirmed food-borne illness. In 1997, *Campylobacter* (3966 cases) and *Salmonella* (2204 cases) accounted for over 76% of the

0-8493-0120-3/01/$0.00+$.50

confirmed foodborne-related diseases.[2] In direct comparisons between *Campylobacter* and *Salmonella, Campylobacter* outnumbered *Salmonella* detection 10 to 1 in college students and 2 to 1 in the general population in the U.S.[3]

Salmonella (non-typhoid) nevertheless continues to be a predominate food-borne pathogen worldwide, and poultry and poultry products are, reportedly, a prevailing vehicle for salmonellosis.[4, 5] One study calculated an annual $1.4 billion *Salmonella*-related loss in human productivity, medical expenses, and increased animal production costs in the U.S. alone.[6] For these reasons, control of *Salmonella, Campylobacter,* and other food-borne pathogens continues to gain recognition as a serious research priority by many regulatory agencies.[1,7] Methods to control infections in poultry flocks prior to slaughter are just beginning to be elucidated, and considerable progress will undoubtedly be made in this area in the near future. Nevertheless, it is clear that the origin of these pathogens in poultry processing plants is in the flocks of product origin. As such, antemortem food-borne pathogen control can have a major impact in reducing contamination of fresh product with these agents of human food-borne illness as intervention strategies are elucidated, understood, and implemented. Current knowledge of antemortem *Salmonella* and *Campylobacter* intervention is discussed below.

Relative importance of specific poultry-derived pathogens

Salmonella *and* Campylobacter

As described above, *Campylobacter* and *Salmonella* are by far the principle pathogens derived from poultry which infect humans through food. As discussed below, most *Salmonella* and *Campylobacter* contamination of poultry originates from antemortem poultry infections. That these organisms infrequently cause apparent clinical disease in poultry flocks compounds the problem of antemortem identification and intervention.

Escherichia coli

Although *E. coli* is a commonly monitored organism in poultry processing plants, the principle reason for concern with *E. coli* is as an indirect indicator of fecal contamination. By far, the majority of *E. coli* isolates from poultry are relatively host adapted for birds and are not considered potential human pathogens.[8] However, poultry are highly susceptible to infection with *E. coli* 0157:H7, a highly pathogenic organism causing hemorrhagic enteritis in humans.[9, 10] Of greater concern is the single documented isolation of *E. coli* 0157:H7 from poultry meat[11] and the actual association of a human food-borne outbreak of diarrheal disease associated with contaminated turkey product.[12] These reports indicate that poultry are susceptible to this important human pathogen and that precautions should be taken to avoid the possible introduction of *E. coli* 0157:H7 into poultry flocks by strictly limiting contact with other animal species (particularly cattle and cattle feces) which are more commonly infected. This is especially true considering that poultry flocks infected with *E. coli* 0157:H7 could present a very real food safety threat and new regulatory interests. Nevertheless, *E. coli* contamination of poultry carcasses is not now generally considered as a significant direct food safety issue, although regulatory interest in this organism as an indicator of fecal contamination will likely continue.

Staphylococcus *species*

Staphylococcosis is an important disease problem for poultry, with *Staphylococcus* species contributing to a variety of antemortem disease problems. More importantly, typical and atypical *Staphylococcus aureus* isolates of poultry are capable of producing enterotoxins which can cause *Staphylococcus* food-borne illness in humans.[13–15] While *Staphylococcus* isolates recovered from live poultry have been principally phage typed to host-adapted biotypes common to poultry and are not believed to be infectious for humans,[16] these isolates can potentially serve as a source for enterotoxin production in mishandled products post-processing.[17] Contamination of poultry carcasses with enterotoxigenic *Staphylococcus* isolates of human origin often occurs at processing.[18,19] While staphylococcosis often represents an important economic problem for growing poultry, most cases are believed to be secondary to other diseases or immunosuppressive conditions in live birds,[17] indicating a role for general health management in reducing the incidence of *Staphylococcus*-related diseases in commercial poultry. It is important to note that *Staphylococcus* species are ubiquitous in vertebrate animals (including humans) and are considered normal and sometimes beneficial flora of the skin and mucous membranes. Considering the ubiquitous nature and the documented role of processing plants and plant workers in contributing to staphylococcal contamination of poultry carcasses, it seems unlikely that preslaughter intervention is a critical control point for these potential pathogens.

Listeria monocytogenes

As discussed in Chapter 9, *L. monocytogenes* is an important food-borne pathogen which is sometimes associated with poultry products. *L. monocytogenes* can be commonly isolated from soil and feces and is capable of causing infections in many vertebrate animals, occasionally causing clinical disease in poultry (see Barnes[20] for review). While poultry are a potential source of *L. monocytogenes* contamination, most human outbreaks have occurred from contaminated cooked ready-to-eat products that are held at refrigeration temperatures, allowing for amplification of numbers of this psychrophillic pathogen to levels infectious for humans.[21]

Potential for Salmonella *and* Campylobacter *antemortem intervention*

Research has clearly demonstrated that the reduction of microbial contamination of processed poultry requires the identification of both pre- and postharvest critical control points where contamination may occur, and the implementation of integrated control programs.[22–27] While implementation of intervention strategies to reduce the incidence of *Salmonella*- and *Campylobacter*-infected broiler flocks is an important and necessary goal, the large number of potential sources of *Salmonella* may limit our ability to always prevent infections with these pathogens in live poultry flocks. For example, wild birds, pets, rodents, and people have been implicated as fomites for transmission of *Salmonella* into broiler flocks.[28] Because of the high animal density in modern broiler production, *Salmonella* is likely to amplify in an infected flock and persist through slaughter and processing. According to Lillard,[29] the Food Safety and Inspection Service conducted a survey of poultry processing plants which showed that only 3–4% of broilers coming into processing plants were *Salmonella* positive, whereas 35% of processed broilers leaving the plant were *Salmonella* positive. While considerable progress has been made in recent years toward decreasing processing plant cross-contamination with *Salmonella*, the source of origin clearly involves antemortem infection. Sarlin and co-workers[30] demonstrated that flocks that were determined to have only low level *Salmonella* infections, or where *Salmonella* was undetectable prior to slaughter, could enter processing plants and remain essentially *Salmonella* free through the sequential stages of processing, prior to processing

Table 8.1 **Salmonella-Positive Samples Obtained by the Skin or Carcass Rinse Procedure from the First Three Consecutive Broiler Flocks Processed During a Single Day at a Commercial Processing Facility**

	Salmonella-positive/total (%)			
Flock	**Post-feather picker[1]**	**Post-evisceration[1]**	**Prechill[2]**	**Postchill[2]**
1	0/50 (0.00%)[a]	1/50 (2.00%)[a]	2/50 (4.00%)[a]	3/50 (6.00%)[a]
2	5/25 (20.00%)[b]	2/25 (8.00%)[b]	17/25(68.00%)[a]	17/25 (68.00%)[a]
3	1/25 (4.00%)[b]	1/25 (4.00%)[b]	1/25 (4.00%)[b]	17/25 (68.00%)[a]

[1] Excision and culture of skin (approximately 2 × 6 cm) from the ventral aspect of the thoracic inlet.
[2] Culture of rinse from entire carcass.
[3] Excised skin samples were not obtained from flock 1 at the prechill sampling point.
[a,b] Values differ significantly ($p < 0.05$) within rows using the chi square test of independence.
Source: Adapted from Sarlin, L. L., Barnhart, E. T., Caldwell, D. J., Moore, R. W., Byrd, J. A., Caldwell, D. Y., Corrier, D. E., DeLoach, J. R., and Hargis, B. M., *Poult. Sci.*, 77, 1253, 1998.

of the first contaminated/infected flock of the day. However, after the first *Salmonella* contaminated/infected flock was processed, carcasses from a subsequent non-contaminated/infected flock were cross-contaminated during processing (Table 8.1). These data clearly indicate the importance of antemortem contamination/infection. When live poultry infections with food-borne pathogens are limited or not-present in the antemortem flock, there is no potential for this flock to bring contamination into a clean processing plant. Alternatively, these data clearly demonstrate that contaminated/infected flocks can certainly serve as a pathogen source for contamination of subsequently processed clean flocks. Thus, antemortem prevention/intervention has considerable merit in reducing the incidence of contamination of flocks with food-borne pathogens.

The work of Sarlin et al.[30] also suggests that some benefits (reduction in total carcass contamination incidence) could be immediately achieved in problem complexes by antemortem identification of contaminated/infected flocks and processing negative flocks as the first flocks of the day. Anecdotal evidence from a commercial processing plant suggests that this approach can be useful under emergency conditions where mandated reductions are essential.

The upper gastrointestinal tract and carcass contamination

It has long been known that the ceca and large intestine are the primary sites of *Salmonella* colonization. Thus, since the early 1970s intestinal contents have traditionally been regarded as the major focal point for controlling *Salmonella* contamination in processing plants.[31–33] Much attention is directed toward the limitation and identification of intestinal rupturing with regard to mandatory feed withdrawals, types of processing equipment used, and the visual evaluation of viscera required by federal inspectors.

Until recently, little emphasis has been given to the chicken crop as a source of *Salmonella* contamination. Indeed, the ceca have been identified as the primary site of *Salmonella* colonization in poultry.[31,32] For this reason, cecal and intestinal contents have traditionally been considered to be the primary source of *Salmonella* contamination of rearing house floor litter, the skin and feathers of broilers, and processed carcasses after rupture of the intestinal tract during evisceration in processing facilities. Although the recoverable number of *Salmonella* from cecal contents is often much greater from colonized cecal contents than from contaminated crop contents, recent research has suggested that *Salmonella*-

and/or *Campylobacter*-contaminated crops may be relatively more important as a source of carcass contamination at commercial processing. Leakage of crop contents onto carcasses during processing was found to occur 86 times more frequently than cecal contents during processing.[34] Of equal importance, crop contents were far more likely to be contaminated with *Salmonella*[34, 35] and *Campylobacter*[36] than ceca at processing, providing a strong suggestion that crops and upper gastrointestinal contents may provide an important, and perhaps major, source of carcass contamination at processing. While these data clearly implicate upper gastrointestinal contents as a critical control point for pathogen contamination of poultry carcasses, it is important to mention that, to date, there is no suggestion that visible ingesta is predictive for carcass contamination with *Salmonella*. Indeed, unpublished studies from our laboratories confirm that there is no relationship between the identification visible ingesta or feces on processed broiler carcasses and *Salmonella* recovery incidence. Proponents of rules based on visible contamination should be reminded that, unlike visible ingesta or feces, microbial pathogen contamination is not visibly detectable. Furthermore, as *Salmonella* are motile and able to specifically bind to attachment sites on the carcass with great affinity,[37] it is very difficult to mechanically remove this pathogen with washing or brushing. Thus, it may not be surprising that visible indicators on the product during processing are not inherently useful for improving microbial food safety.

Antemortem contamination of the upper gastrointestinal tract

Humphrey et al.[38] found an increased incidence of recoverable *Salmonella enteritidis* phage type 4 within the crop during increased feed withdrawal time. Withdrawal of feed for 24 h had a marked and significant impact on recoverable *S. enteritidis* in the crop. In fed broilers, only 2 of 16 crop samples were *Salmonella* positive, whereas crops of broilers subjected to feed withdrawal were *Salmonella* positive in 11 of 16 samples. Confirming these results, our laboratories also found an increased incidence of *Salmonella* positive crops (two- to three-fold) following feed withdrawal as compared to crops obtained from full-fed broilers under both experimental and commercial field conditions.[39] In a more recent study,[35] the incidence of *Salmonella* recovery from crop contents increased significantly ($p < 0.05$) in 5 of 9 commercial flocks during feed withdrawal with 7/360 (1.9%) being *Salmonella* positive before feed withdrawal vs. 36/359 (10%) being *Salmonella* positive after feed withdrawal. In contrast to the observed three- to five-fold increase in crop contamination frequency due to feed withdrawal, little effect on *Salmonella* recovery from ceca was reported in two studies.[35, 39]

Unpublished results from our laboratory have also demonstrated that environmental photointensity greatly influenced the number of broilers that were observed to peck at the litter during simulated feed withdrawal. In this experiment, the number of broilers that were observed to peck at the litter during a 30-min observation period was recorded by 4 or 8 independent observers near the same pens containing 100 broiler chickens placed at approximately 1 ft^2/broiler. Observations were conducted prior to feed removal (full feed) and at 4 and 8 h after feed removal under conditions of high light intensity (44–46 footcandles; fc) or low light intensity (0.3–0.5 fc). Reducing the intensity of light resulted in 6.8-fold reduction in this behavior at 6 h of feed withdrawal. If our previously published hypothesis[39] that ingestion of litter and feces during feed withdrawal is responsible for the observed marked increase in *Salmonella* contamination of crops is correct, reducing or eliminating light intensity during the withdrawal period may reduce the frequency of crop contamination, possibly translating to reduced carcass contamination at processing.

Although it now seems clear that increased consumption of litter and feces during feed withdrawal probably contributes to the increase in *Salmonella* contamination of crop contents prior to harvest, Corrier et al.[40] have demonstrated that crop pH increased significantly during an 8-h feed withdrawal under commercial and experimental conditions, a phenomenon attributed to decreased fermentation activity and decreased lactic acid production in the crop. In this study, the decreased crop lactic acid levels and increased crop pH was associated with significantly increased *Salmonella* recovery from crops following 4 or 8 h of feed withdrawal. This study indicates that, in addition to increased exposure to *Salmonella* resulting from the ingestion of contaminated litter, the ingested bacteria are exposed to a crop environment that contains a reduced concentration of lactic acid, is less acidic, and is therefore more compatible with *Salmonella* survival.

More recently, Byrd and co-workers[36] examined the effect of feed withdrawal on *Campylobacter* isolation from crops of market-age commercial broiler chickens prior to capture and transport to the processing plant. In this study, the incidence of *Campylobacter* isolation from the crop was determined immediately before and after feed withdrawal in 40 7-week-old broiler chickens obtained from each of 9 separate broiler houses. Ceca were collected from broilers in six of the same flocks for comparison with the crop samples. Feed withdrawal caused a significant increase in *Campylobacter*-positive crop samples in seven of the nine houses sampled. Furthermore, the total number of *Campylobacter*-positive crops increased significantly from 25% before feed withdrawal to 62.4% after the feed withdrawal period. Similar to the limited effect of feed withdrawal on *Salmonella* recovery from ceca, feed withdrawal was not found to affect *Campylobacter* recovery frequency from ceca in this study.

When the potential high frequency of crop rupture and leakage at commercial processing is considered,[34] the high frequency of *Campylobacter* recovery (62.4%) from crops following feed withdrawal[36] may suggest that the crop may serve as a potential critical control point for *Campylobacter* as well as *Salmonella.* Ongoing research is directed toward investigating this potential source of carcass contamination with *Campylobacter.*

Antemortem crop contamination intervention

Feed deprivation has been shown to change the microenvironment in the crop by reducing the number of lactobacilli, decreasing the concentration of volatile fatty acids, and increasing crop pH.[38, 40] Furthermore, changes in the crop microenvironment during feed deprivation have the potential to increase the expression of invasion genes of pathogenic bacteria required for intestinal invasion. One way to reverse the increasing crop pH due to feed withdrawal would be to re-acidify the crop. Recently, we evaluated the use of 0.5% lactic acid in the drinking water during a simulated 8-h pre-transport feed withdrawal. All broilers were challenged with 10^6 cfu *S. typhimurium* (ST) by oral gavage 24 to 48 h prior to feed withdrawal (FW) in a total of 5 experiments. ST was recovered from 53% of the control crops compared to 31% of the lactic acid-treated crops. Reductions in recovery incidence were also associated with reduced numbers of ST recovered (e.g., control: log 1.45 cfu/crop; lactic acid: 0.79 cfu/crop). In an additional on-farm commercial study, broilers were provided a 0.44% lactic acid drinking water solution during a 10-h FW and pre-FW crop, post-FW crop, and prechill carcass wash samples were collected for *Campylobacter* and *Salmonella* determination.[35, 36] Crop contamination with *Salmonella* was significantly reduced by lactic acid treatment (2/50; 4%) as compared to controls (23/50; 46%). Importantly, *Salmonella* isolation incidence from prechill carcass rinses was significantly reduced by almost ten-fold, but *Campylobacter* isolation incidence was only reduced by 25% by pre-harvest lactic acid treatment. These studies suggest that incorporation of some organic acids in the drinking water during pre-transport FW may reduce *Salmonella* contamination of crops and broiler carcasses at processing. Other disinfectants with more potential efficacy, which can be administered in the drinking water during preslaughter feed withdrawal, are under investigation. Limitations of this approach include palatability of the product, as reduced preslaughter water consumption will likely cause product shrinkage, and acceptability of candidate compounds for use in a food-producing animal immediately prior to slaughter.

Chemical litter treatments

If litter acidity is reduced below about pH 5, conditions are unfavorable for *Salmonella* and other potential pathogens.[40] To achieve this, chemical treatment can be added to the litter to lower the pH and reduce ammonia production. Such treatments must be cost effective and safe for farm workers. Several chemical additives have been used to decrease the pH of poultry litter. Examples of these chemicals include aluminum sulfate,[41, 42] ferrous sulfate,[43] phosphoric acid,[44] sodium bisulfate,[45, 46] and acetic acid.[47]

Moore and co-workers[45] evaluated several chemical treatments for ammonia utilization and phosphorus solubility and found that aluminum sulfate was best at reducing ammonia volatilization, followed by phosphoric acid, ferrous sulfate, sodium bisulfate, and calcium-ferrous-sulfate. All treatments significantly reduced litter pH when compared to the control litter. Aluminum sulfate was most effective in controlling both ammonia volatilization and phosphorus solubility. These data suggest that aluminum sulfate has some possible environmental benefits by reducing phosphorous runoff into ground water; however, the initial cost per treatment of the house was higher compared to the other treatments. In another study, sodium bisulfate was shown to be effective in controlling *Salmonella, Clostridium*, and *Pasturella* in the litter.[48] Furthermore, the application of this product was effective in litter acidification and extended the life of insecticides for the control of darkling beetles.

Very recently our laboratories have evaluated the effect of hydrated lime on *Salmonella* and *Campylobacter* survival in used poultry litter. These studies indicate that concentrations of lime as low as 2% (wt/v) markedly reduced recovery of *Salmonella* from artificially innoculated litter 8 h after treatment. Early data from growth trials suggest that incorporation of hydrated lime (2%) in new poultry litter did not negatively affect performance of turkey poults during a six-week pen trial.[49] While the commercial applicability of adding hydrated lime to poultry litter has not been proven, these results are encouraging, particularly when the relatively low cost and low environmental impact of hydrated lime is considered.

Role of biosecurity

Because paratyphoid *Salmonella* serovars that can infect poultry and humans are not host adapted, they can and do infect a variety of vertebrate species. A wide variety of animals can serve as potential reservoirs for *Salmonella* infection of poultry, including rodents, wild birds, domestic animals, and humans.[28] *Salmonella* can be frequently isolated from feed sources, particularly when high levels of meat, fish, blood and/or bone meal are included in the diet.[50–52] Heat treatment of the finished feed[25, 52] can greatly reduce feed contamination with a number of potential pathogens, including *Salmonella.* Nevertheless, while there are very good reasons for producing and maintaining high quality pathogen-free feed, the actual role of feed contamination in causing poultry infections has not been clearly established. Similarly, strict and comprehensive biosecurity procedures are critical for maintaining the health of poultry flocks and limiting the introduction of numerous disease problems. As such, biosecurity is an important component of poultry health management.

Current understanding of the epizootiology of *Salmonella* infection in broiler and turkey flocks suggests that the predominant serovars of *Salmonella* that arise in poultry flocks can be traced to the hatchery, and are likely related to infections and shedding from parent breeder flocks,[53] although occasional apparent transmission through environmental contamination from a previous flock has been documented.[54] More recently, some evidence of vertical transmission of *Campylobacter* infections in chickens has been discussed, suggesting that infections of parent flocks may be an important source of *Campylobacter* infection of commercial broilers. While biosecurity and feed quality control are important components of a poultry health program, the authors believe that reliance on these factors commonly leads to a false sense of security with regard to *Salmonella* and *Campylobacter* control and may not prove to be major critical control points for antemortem intervention. Serious consideration of the role of vertical transmission and the role of breeder flocks as sources of *Salmonella* and *Campylobacter* infections in commercial poultry flocks is clearly warranted for the development of a successful antemortem control program.

Live haul/transport considerations

Broilers that have undergone feed deprivation for 4 h or greater are caught and transported to the processing plant. During transportation to the processing plant, broilers are exposed to transport coops that may be contaminated with *Campylobacter* and *Salmonella*.[22,55,56] Stern and co-workers[55] found that transport increased the total incidence of *Campylobacter* on post-transport broilers (56% positive) when compared to pre-transport broilers (12.1%). Furthermore, the mean total number of *Campylobacter* detected on each carcass increased from 2.71 to 5.15 log10 cfu.[55] Similarly, Hoop and Ehrsam[56] reported that 32% of the unwashed transport coops were contaminated with *C. jejuni* from a single processing plant in Switzerland. However, transport does not seem to uniformly increase the frequency of *Campylobacter* contamination in all plants as exceptions have been noted.[24]

Salmonella-positive cecal carriers were found to increase during experimental shipping conditions from 23.5% (control) to 61.5% (shipped).[22] Similarly, Jones and co-workers[25] found that 33% of unwashed transport coops were contaminated with *Salmonella* although the broilers transported in this study were cloacal-negative for *Salmonella*. That effective cleaning and disinfection of transport coops and equipment is an important component of any flock biosecurity program should be considered when evaluating the total costs of these procedures.

The presence of food-borne pathogens on transport materials is indicative of a potential source of external and internal contamination of broilers. Mead and co-workers[57] evaluated transport coops using an *E. coli* marker organism for contamination after a normal system of cleaning coops using chlorine water. These researchers found that 50% of the crates remained positive after steam cleaning.[57] The importance of providing clean transport crates has been illustrated as one of the last steps to maintain or reduce external carcass contamination prior to entering the processing plant. Typically, birds entering the processing plant have been contaminated with *Salmonella* (up to 60–100%) and *Campylobacter* (80–100%).[58–60] Salmonellae have been shown to firmly attach to the skin of broilers entering the processing plant,[29] and all indications are that avoidance of contamination is much preferable to remediation efforts. *Campylobacter* is even more ubiquitous within poultry processing plants as this genus has been recovered from scald tanks, feathers, chill tanks, and processing equipment.[59] Furthermore, processing shackles and tanks were contaminated with *Campylobacter* in duck, turkey, layer, and broiler processing plants.[61]

Medications

Therapeutic and prophylactic use of antibiotics to control *Salmonella* has brought about many debates. Antimicrobial agents are regularly used in controlling hatchery problems and to attempt to prevent and clear *S. enteritidis* infections.[62, 63] Several antibiotics have been reported to increase the incidence of *Salmonella* colonization, possibly by suppressing the growth of beneficial bacteria that may produce pathogen-inhibiting compounds.[64, 65] Consistent use of subtheraputic levels of antibiotics to promote growth may also lead to drug-resistant bacterial isolates.[66–68]

Competitive exclusion

The gastrointestinal tract of newly hatched chicks is essentially sterile and highly susceptible to colonization/infection with pathogenic bacteria.[69] One approach to prevent the colonization of pathogenic bacteria is to accelerate establishment of normal intestinal flora in chicks as early as possible, thus providing a source of competition for subsequent pathogens to which the host bird may be exposed. Competitive exclusion (CE) is the delivery of a suspension of healthy adult cecal microflora, and was first described in 1973.[70] The benefits of competitive exclusion treatment on reducing *Salmonella* and *Campylobacter* shedding and environmental contamination are now well documented.[71–74]

The protective effect of CE has been explained by competition for attachment sites (Figure 8.1), production of volatile fatty acids, decreased oxidation-reduction potential, and competition for nutrients.[70, 75] CE products may consist of cultures in which the bacterial composition is known (defined) or unknown (undefined). Defined cultures offer some additional safety since there is decreased likelihood of introduction of unintended, potentially pathogenic organisms. Presently, one competitive exclusion product is licensed by the U.S. Food and Drug Administration for use in poultry for the prevention/reduction of *Salmonella* infections (Preempt™, M.S. BioScience, Madison, WI) and an additional undefined culture which is presently undergoing testing (Mucosal Starter Culture, Continental Grains, Chicago, IL).

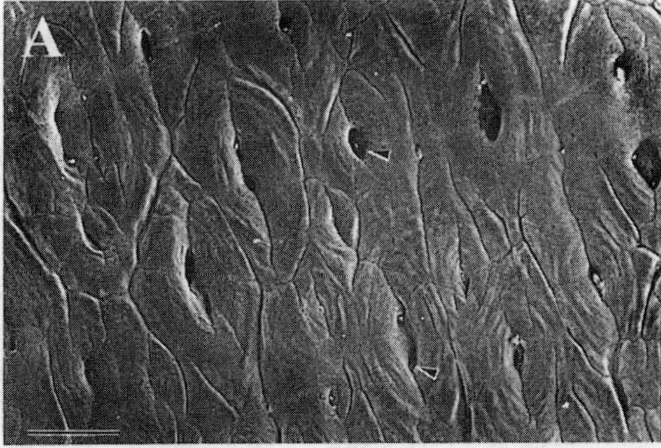

Figure 8.1 Scanning electron micrograph of cecal mucosa (mid-cecum) from broiler chicks 48 h after hatch. Bars represent 100 μm. Courtesy of Dr. Robert E. Droleskey, USDA-ARS, College Station, TX. Panel A: Normal untreated chick cecal mucosa. Cecal crypts are present (arrowheads) without the large clumps of bacteria seen in similar segments following treatment with a competitive exclusion product.

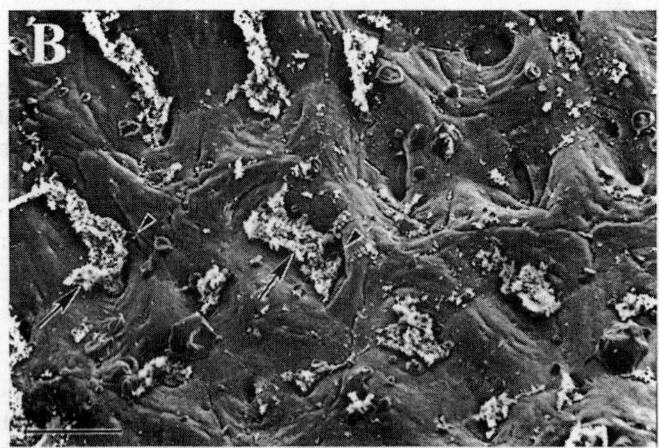

Figure 8.1 (continued) Panel B: Similar section of mucosa from a chick treated at hatch with a commercial competitive exclusion product. The majority of crypts (arrowheads) contain large clumps of bacteria (large arrows). A few crypts devoid of bacteria are also present (small arrows).

In commercial field trials, both CE cultures were effective in controlling *Salmonella* cecal colonization in market-age broilers.[27, 75] Day-of-hatch chicks provided a single dose of a defined CE culture had a significantly ($p < 0.05$) lower incidence of *Salmonella* recovery (0%) compared to the non-treated controls (7%). Similarly, market-age broilers treated with an undefined CE culture in a two-step procedure (spray at hatchery and via drinking water at placement) had significantly fewer *Salmonella*-positive carcasses prior to chilling (6.7%) compared to non-treated controls (12.8%). One defined CE product has been shown to be effective in protecting chicks against colonization under experimental conditions by *S. enteritidis* (PT13, PT 4), *S. gallinarum*, *S. typhimurium*, *Clostridium perfringens*, and *E. coli* O157:H7.[74, 76–79]

The use of CE products can be an effective component in an integrated control program. The efficacy of CE products is dependent on the volume and concentration of the dose as well as the delivery method. CE cultures often consist of bacterial organisms that are sensitive to antimicrobial components which, therefore, may diminish the effectiveness of the culture. One should consider that CE products will not eliminate the food-borne pathogens but should be included in a complete integrated control program. As effective CE programs can reduce intestinal colonization by a number of pathogens, these programs may offer alternatives to low level antimicrobial use, which should also be considered with regard to cost of an integrated antemortem food-borne pathogen control program.

Vaccination

Vaccination programs are used to prevent or reduce the spread of pathogenic viruses and bacteria and generally depend on recognition of specific antigens (epitopes) by the immune system by the host. Because there are a large number of *Salmonella* serovars, each with individual epitopes which do not elicit cross protection against other serovars, there has been little traditional emphasis on development of generic *Salmonella* vaccines. *Campylobacter* vaccines for poultry are not faced with the problem of large serovar diversity. While one study suggested that an orally administered inactivated *Campylobacter* vaccine could sometimes reduce shedding in vaccinated chickens,[80] effective commercial vaccines have yet to be developed for commercial meat poultry use.

In contrast to generic *Salmonella* vaccination, vaccination for specific *Salmonella* serovar strains (e.g., *S. enteritidis, S. gallinarum*) has gained considerable acceptance in countries with endemic problems with these more devastating serovars, particularly in breeders and table egg production chickens (see Shivaprasad[81] for review). Such vaccines should be considered in the event of emergence of a high level problem related to a specific *Salmonella* serovar.

A live-type commercial vaccine with a double gene deletion that is avirulent and immunogenic has been reported[82] and other specific deletion mutants have been proposed.[83, 84] Day-of-hatch chicks vaccinated with this live-type vaccine have been shown to have serological protection to homologous and heterologous *Salmonella* serotypes, possibly through a mechanism similar to competitive exclusion.[85, 86] Furthermore, maternal antibodies can be demonstrated in eggs and chicks from breeders vaccinated with this vaccine. These antibodies are reported to reduce salmonellae colonization and to provide protection to laying hens up to 11 months post-inoculation.[86] However, susceptibility to antimicrobial agents commonly used in poultry production can reduce or eliminate the efficacy of live vaccines. While much of the published research appears encouraging, live *Salmonella* vaccines have not gained widespread commercial acceptance within the U. S. for paratyphoid *Salmonella* control to date.

Summary

Clearly, successful antemortem intervention programs for food-borne pathogens must be integrated and must approach multiple critical control points. To date, there is no single identified critical control point that will assure reductions of food-borne pathogens, but integration of multiple approaches, focused on known critical control points, has been partially effective. Present evidence indicates that a major factor for ultimate success of antemortem intervention will be the production and maintenance of food-borne pathogen-free breeder flocks, a problem compounded by the necessity of feed restriction during growth and stress associated with production. Emerging areas of antemortem food-borne pathogen control include the use of effective competitive exclusion products, treatment of drinking water with organic acids during the preslaughter feed withdrawal, and the treatment of litter with acidification or alkalinization products. The use of organic acids during feed withdrawal and the effects of environmental treatments have been recently reviewed.[87] Similarly, the benefits of competitive exclusion treatment on reducing *Salmonella* and *Campylobacter* shedding and environmental contamination are now well documented.[71–74] As the future of conventional use of some antimicrobial compounds in commercial poultry production is in question, competitive exclusion may find a new role in poultry health and production, with the additional benefits of food-borne pathogen control.

References

1. United States Department of Agriculture, Food Safety Inspection Service, FSIS/CDC/FDA Sentinel Site Study: The Establishment and Implementation of an Active Surveillance System for Bacterial Foodborne Diseases in the United States, Report to Congress, February, 1997.
2. United States Department of Agriculture, Food Safety Inspection Service, Foodnet: An Active Surveillance System Bacterial Foodborne Diseases in the United States, Report to Congress, April, 1998.
3. Tauxe, R. R. V., *Salmonella:* A postmodern pathogen, *J. Food Prot.*, 54, 563, 1991.
4. Bean, N. H. and Griffin, P. M., Food-borne disease outbreaks in the United States, 1973–1987: Pathogens and trends, *J. Food Prot.*, 53, 804, 1990.

5. Persson, U. and Jendteg, S. I., The economic impact of poultry-borne salmonellosis: how much should be spent on prophylaxis?, *Int. J. Food Microbiol.*, 15, 207, 1992.
6. Madie, P., *Salmonella* and *Campylobacter* infections in poultry, in *Proc. Solvay Chicken Health Course*, Grunner U. Peterson Massey University, Palmerston North, New Zealand, 1992.
7. National Research Initiative Competitive Grants Program, Program Descriptions, Cooperative State Research Service; United States Department of Agriculture, Washington, D.C., Section 32.0: Food Safety, 1991.
8. Barnes, H. J. and Gross, W. B., Colibacillosis, in *Diseases of Poultry*, 10th Ed., Calnek et al. Eds. Iowa State University Press, Ames, IA, 1997, 131.
9. Beery, J. T., Doyle, M. P., and Schoeni, J. L., Colonization of chicken cecae by *Escherichia coli* associated with hemorrhagic colitis, *Appl. Environ. Microbiol.*, 49, 310, 1985.
10. Stavric, S., Buchanan, B., and Gleeson, T. M., Intestinal colonization of young chicks with *Escherichia coli* 0157:H7 and other verotoxin-producing serotypes, *J. Appl. Bacteriol.*, 74, 557, 1993.
11. Doyle, M. O. and Schoeni, J. L., Isolation of *Escherichia coli* 0157:H7 from retail fresh meats and poultry, *Appl. Environ. Microbiol.*, 53, 2394, 1987.
12. Griffin, P. M. and Tauxe, R. V., The epidemiology of infections caused by *Escherichia coli* 0157:H7, other enterohemorrhagic *E. coli*, and the associated hemolytic uremic syndrome, *Epidemiol. Rev.*, 13, 60, 1991.
13. Evans, J. B., Ananaba, G. A., Pate, C. A., and Bergdoll, M. S., Enterotoxin production by atypical *Staphylococcus aureus* from poultry, *J. Appl. Bacteriol.*, 54, 257, 1983.
14. Gibbs, P. A., Patterson, J. T., and Harvey, J., Biochemical characteristics and enterotoxigenicity of *Staphylococcus aureus* strains isolated from poultry, *J. Appl. Bacteriol.*, 44, 57, 1978.
15. Harvey, J., Patterson, J. T., and Gibbs, P. A., Enterotoxigenicity of *Staphylococcus aureus* strains isolated from poultry: raw poultry carcasses as a potential food-poisoning hazard, *J. Appl. Bacteriol.*, 52, 251, 1982–258.
16. Deveriese, L. A., Devos, A. H., Beumer, J., and Moes, R., Characterization of staphylococci isolated from poultry, *Poult. Sci.*, 51, 389, 1972.
17. Skeeles, J. K., Staphylococcosis, in *Diseases of Poultry*, 10th Ed., Calnek et al., Eds., Iowa State University Press, Ames, IA, 1997, 247.
18. Adams, B. W. and Mead, G. C., Incidence and properties of *Staphylococcus aureus* associated with turkeys during processing and further-processing operations, *J. Hyg.*, 91, 479, 1983.
19. Notermans, S., Dufrenne, J., and van Leeuwen, W. J., Contamination of broiler chickens by *Staphylococcus aureus* during processing: incidence and origin, *J. Appl. Bacteriol.*, 52, 275, 1982.
20. Barnes, H. J., Other bacterial diseases, in *Diseases of Poultry*, 10th Ed., Calnek et al. Eds., Iowa State University Press, Ames, IA, 1997, 289.
21. Marsden, J. L., Industry Perspectives on *Listeria monocytogenes* in foods: raw meat and poultry, dairy, *Food Environ. Sanit.*, 14, 83, 1994.
22. Rigby, C. E. and Pettit, J. R., Changes in the *Salmonella* status of broiler chickens subjected to simulated shipping conditions, *Can. J. Comp. Med.*, 44, 374, 1980.
23. Goren, E., de Jong, W. A., Doornenbal, P., Bolder, N. M., Mulder, R. W. A. W., and Jansen, A., Reduction of *Salmonella* infection of broilers by spray application of intestinal microflora: a longitudinal study, *Vet. Q.*, 10, 249, 1988.
24. Jones, F., Axtell, R. C., Rives, D. V., Scheideler, S. E., Tarver, F. R., Walker, R. L., and Wineland, M. J., A survey of *Campylobacter jejuni* contamination in modern broiler production and processing systems, *J. Food Prot.*, 54, 259, 1991.
25. Jones, F., Axtell, R. C., Rives, D. V., Scheideler, S. E., Tarver, F. R., Walker, R. L., and Wineland, M. J., A survey of *Salmonella* contamination in modern broiler production, *J. Food Prot.*, 54, 502, 1991.
26. Stavric, S. and D'Aoust, J. Y., Undefined and defined bacterial preparations for the competitive exclusion of *Salmonella* in poultry—a review, *J. Food Prot.*, 56, 173, 1993.
27. Blankenship, L. C., Bailey, J. S., Cox, N. A., Stern, N. J., Brewer, R., and Williams, O., Two-step mucosal competitive exclusion flora treatment to diminish salmonellae in commercial broiler chickens, *Poult. Sci.*, 72, 1667, 1993.

28. Krabisch, P. and Dorn, P., The importance of living vectors for the dissemination of *Salmonella* in broiler flock rodents, cats and insects as vectors, *Berl. Muench. Tieraerztl. Wochenschr.*, 93, 232, 1980.
29. Lillard, H. S., Factors affecting persistence of *Salmonella* during the processing of poultry, *J. Food Prot.*, 52, 829, 1989.
30. Sarlin, L. L., Barnhart, E. T., Caldwell, D. J., Moore, R. W., Byrd, J. A., Caldwell, D. Y., Corrier, D. E., DeLoach, J. R., and Hargis, B. M., Evaluation of alternative sampling methods for *Salmonella* critical control point determination at broiler processing, *Poult. Sci.*, 77, 1253, 1998.
31. Fanelli, M. J., Sadler, W. W., Franti, C. E., and Brownell., J. R., Localization of salmonellae within the intestinal tract of chickens, *Avian Dis.*, 15, 366, 1971.
32. Snoeyenbos, G. H., Soerjadi, A. S., and Weinack, O. M., Gastrointestinal colonization by *Salmonella* and pathogenic *Escherichia coli* in monozenic and holoxenic chicks and poults, *Avian Dis.*, 26, 566, 1982.
33. Corrier, D. E., Hargis, B. M., Hinton, A., Lindsey, D., Caldwell, D. J., Manning, J., and DeLoach, J. R., Effect of cecal colonization resistance of layer chicks to invasive *Salmonella enteritidis*, *Avian Dis.*, 35, 337, 1991.
34. Hargis, B. M., Caldwell, D. J., Brewer, R. L., Corrier, D. E., and DeLoach, J. R., Evaluation of the chicken crop as a source of *Salmonella* contamination for broiler carcasses, *Poult. Sci.*, 74, 1548, 1995.
35. Corrier, D. E., Byrd, J. A., Hargis, B. M., Hume, M. E., Bailey, R. H., and Stanker, L. H., Presence of *Salmonella* in the crop and ceca of broiler chickens before and after preslaughter feed withdrawal, *Poult. Sci.*, 78, 45, 1999.
36. Byrd, J. A., Corrier, D. E., Hume, M. E., Bailey, R. H., Stanker, L. H., and Hargis, B. M., Effect of feed withdrawal on the incidence of *Campylobacter* in crops of preharvest broiler chickens, *Avian Dis.*, 42, 802, 1998.
37. Lillard, H. S., Effect of surfactant or changes in ionic strength on the attachment of *Salmonella typhimurium* to poultry skin and muscle, *J. Food Sci.*, 53, 727, 1988.
38. Humphrey, T. J., Baskerville, A., Whitehead, A., Rowe, B., and Henley, A., Influence of feeding patterns on the artificial infection of laying hens with *Salmonella enteritidis* phage type 4, *Vet. Rec.*, 132, 407, 1993.
39. Ramirez, G. A., Sarlin, L. L., Caldwell, D. J., Yezak, C. R., Jr., Hume, M. E., Corrier, D. E., Deloach, J. R., and Hargis, B. M., Effect of feed withdrawal on the incidence of *Salmonella* in the crops and ceca of market age broiler chickens, *Poult. Sci.*, 76, 654, 1997.
40. Corrier, D. E., Byrd, J. A., Hargis, B. M., Hume, M. E., Bailey, R. H., and Stanker, L. H., Survival of *Salmonella* in the crop contents of market-age broilers during feed withdrawal, *Avian Dis.*, 43, 453, 1999.
41. Huff, W. E., Moore, P. A., Balog, J. M., Bayyari, G. R., and Rath, N. C., Evaluation of toxicity of alum (aluminum sulfate) in young broiler chickens, *Poult. Sci.*, 75, 1359, 1996.
42. Moore, P. A. and Miller, D. A., Decreasing phosphorus solubility in poultry litter with aluminum, calcium and iron amendments, *J. Environ. Qual.*, 23, 325, 1994.
43. Huff, W. E., Malone, G. W., and Chaloupka, G. W., Effect of litter treatment on broiler performance and certain litter quality parameters, *Poult. Sci.*, 63, 2167, 1984.
44. Reece, F. N., Bate, B. J., and Lott, B. D., Ammonia control in broiler houses, *Poult. Sci.*, 58, 754, 1979.
45. Moore, P. A., Jr., Daniel, T. C., Edwards, D. R., and Miller, D. M., Evaluation of chemical amendments to reduce ammonia volatilization from poultry litter, *Poult. Sci.*, 75, 315, 1996.
46. Terzich, M., Quarles, C., Goodwin, M. A., and Brown, J., Effect of Poultry Litter Treatment (PLT) on death due to ascites in broilers, *Avian Dis.*, 42, 385, 1998.
47. Parkhurst, C. R., Hamilton, P. B., and Baughman, G. R., The use of volatile fatty acids for the control of microorganisms in pine sawdust litter, *Poult. Sci.*, 58, 801, 1974.
48. Terzich, M., The effects of sodium bisulfate on bacteria load of poultry litter and bird performance, in *Proc. 68th Northeastern Conference on Avian Diseases*, Penn State University, June 10–12, 1996.
49. Hargis, B. M., Caldwell, D. J., and Byrd, J. A., unpublished data, 2000.

50. Morris, G. K., McMurray, B. L., Galton, M. M., and Wells, J. G., A study of the dissemination of salmonellosis in a commercial broiler chicken operation, *Am. J. Vet. Res.*, 30, 1413, 1969.
51. MacKenzie, M. A. and Bains, B. S., Dissemination of *Salmonella* serotypes from raw feed ingredients to chicken carcasses, *Poult. Sci.*, 55, 957, 1996.
52. Shrimpton, D. H., The *Salmonella* problem of Britian, *Milling Flour Feed*, Jan:16–17, 1989.
53. Caldwell, D. J., Hargis, B. M., Corrier, D. E., Williams, J. D., Vidal, L., and DeLoach, J. R., Evaluation of persistence and distribution of *Salmonella* serotype isolation from poultry farms using drag-swab sampling, *Avian Dis.*, 39, 617, 1995.
54. Byrd J. A., Origin and relationship of *Campylobacter* and *Salmonella* contamination of poultry during processing, *Poult. Sci.*, 78 (Suppl. 1), 4, 1999.
55. Stern, N. J., Clavero, M. R. S., Bailey, J. S., Cox, N. A., and Robach, M. C., *Campylobacter* spp. in broilers on the farm and after transport, *Poult. Sci.*, 74, 937, 1995.
56. Hoop, R. and Ehrsam. H., Ein beitrag zur epidemiologie von *Campylobacter jejuni* and *Campylobacter* coli in der Hünnermast, *Schweiz. Arch. Tierheilknd*, 129, 193, 1987.
57. Mead, G. C., Hudson, W. R., and Hinton, M. H., Use of a marker organism in poultry processing to identify sites of cross-contamination and evaluate possible control measures, *Br. Poult. Sci.*, 35, 345, 1994.
58. Acuff, G. R., Vanderzant, C., Hanna, M. O., Ehlers, J. G., Golan, F. A., and Gardner, F. A., Prevalence of *Campylobacter jejuni* in turkey carcass processing and further processing of turkey products, *J. Food Prot.*, 49, 712, 1986.
59. Wempe, J. M., Genigeorgis, C. A., Farver, T. B., and Yusufu, H. I., Prevalence of *Campylobacter jejuni* in two California chicken processing plants, *Appl. Environ. Microbiol.*, 45, 355, 1983.
60. Kotula, K. L. and Pandya, Y., Bacterial contamination of broiler chickens before scalding, *J. Food Prot.*, 58, 1326, 1995.
61. Baker, R. C., Paredes, M. D. C. Paredes, and Qureshi, R. A., Prevalence of *Campylobacter* in poultry meat in New York State, *Poult. Sci.*, 66, 1766, 1987.
62. Goodnough, M. C. and Johnson, E. A., Control of *Salmonella enteritidis* infections in poultry by polymyxin B and trimethoprim, *Appl. Environ. Microbiol.*, 57, 785, 1991.
63. Muirhead, S., *Feed Additive Compendium*, Miller Publishing, Minneapolis, MN, 1994.
64. Manning, J. G., Hargis, B. M., Hinton, A., Corries, D. E., DeLoach, J. R., and Creger, C. R., Effect of nitrofurazone or novobiocin on *Salmonella enteritidis* cecal colonization and organ invasion in leghorn hens, *Avian Dis.*, 36, 334, 1992.
65. Manning, J. G., Hargis, B. M., Hinton, A., Corrier, D. E., DeLoach, J. R., and Creger, C. R., Effect of selected antibiotics and anticoccidials on *Salmonella enteritidis* cecal colonization and organ invasion in Leghorn chicks, *Avian Dis.*, 38, 256, 1994.
66. Kobland, J. D., Gale, G. O., Gutafson, R. H., and Simkins, K. L., Comparison of therapeutic versus subtherapeutic levels of chlortetracycline in the diet for selection of resistant *Salmonella* in experimentally challenged chickens, *Poult. Sci.*, 66, 1129, 1987.
67. Gast, R. K. and Stephens, J. F., Effect of kanamycin administration to poultry on the proliferation of drug-resistant *Salmonella*, *Poult. Sci.*, 67, 689, 1988.
68. Gast, R. K., Stephens, J. F., and Foster, D. N., Effect of kanamycin administration to poultry on the proliferation of drug-resistant *Salmonella*, *Poult. Sci.*, 67, 699, 1988.
69. Jayne-Williams, and Fuller, D. J. R., The influence of intestinal microflora on nutrition, in *Physiology and Biochemistry of Domestic Food*, Bell, D. J. and Freeman, B. M., Eds., Academic Press, London, 1971, 74.
70. Nurmi, E. and Rantala, M., New aspects of *Salmonella* infection in broiler production, *Nature*, 241, 210, 1973.
71. Wierup M., Wahlstrom, H., and Engstrom, B., Experience of a 10-year use of competitive exclusion treatment as part of the *Salmonella* control programme in Sweden, *Int. J. Food Microbiol.*, 15, 287, 1992.
72. Schoeni J. L. and Wong, C. L., Inhibition of *Campylobacter jejuni* colonization in chicks by defined competitive exclusion bacteria, *Appl. Environ. Microbiol.*, 60(4), 1191, 1994.
73. Corrier D. E., Nisbet, D. J., Byrd, J. A., II, Hargis, B. M., Keith, N. K., Peterson, M., and DeLoach, J. R., Dosage titration of a characterized competitive exclusion culture to inhibit *Salmonella* colonization in broiler chickens during growout, *J. Food Prot.*, 61, 796, 1998.

74. Nisbet, D. J., Tellez, G. I., Lowery, V. K., Anderson, R.C., Garcia, G., Nava, G., Kogut, M. H., Corrier, D. E., and Stanker, L. H., Effect of a commercial competitive exclusion culture (PREEMPT) on mortality and horizontal transmission of *Salmonella gallinarium* in broiler chickens, *Avian Dis.*, 42, 651, 1998.
75. Corrier, D. E., Nisbet, D. J., Scanlan, C. M., Hollister, A. G., Caldwell, D. J., Thomas, L. A., Hargis, B. M., Tomkins, T., and DeLoach, J. R., Treatment of commercial broiler chickens with a characterized culture of cecal bacteria to reduce salmonellae colonization, *Poult. Sci.*, 74, 1093, 1995.
76. Corrier, D. E., Hinton, A., Jr., Ziprin, R. L., Beier, R. C., and DeLoach, J. R., Effect of dietary lactose on cecal pH, bacteriostatic volatile fatty acids, and *Salmonella typhimurium* colonization of broiler chicks, *Avian Dis.*, 34, 617, 1990.
77. Corrier, D. E., Nisbet, D. J., Scanlan, C. M., Tellez, G., Hargis, B. M., and DeLoach, J. R., Inhibition of *Salmonella enteritidis* cecal and organ colonization in leghorn chicks by a defined culture of cecal bacteria and dietary lactose, *J. Food Prot.*, 56, 377, 1994.
78. Nisbet, D. J., Corrier, D. E., Ricke, S. C., Hume, M. E., Byrd, J. A., and DeLoach, J. R., Cecal propionic acid as a biological indicator of the early establishment of a microbial ecosystem inhibitory to *Salmonella* in chicks, *Anaerobes*, 2, 345, 1996.
79. Byrd, J. A., Nisbet, D. J., Corrier, D. E., and Stanker, L. H., Use of continuous-flow culture system to study the interaction between *Clostridium perfringens* and a mixed microbial competitive exclusion culture (CF3), *Biosci. Microflora*, 16 (Suppl.), 15, 1997.
80. Rice, B. E., Rollins, D. M., Mallinson, E. T., Carr, L. J., and Sam, W., *Campylobacter jejuni* in broiler chickens: Colonization and humoral immunity following oral vaccination and experimental infection, *Vaccine*, 15, 1922, 1997.
81. Shivaprasad, H. L., Pullorum disease and fowl typhoid, in *Diseases of Poultry*, 10th Ed., Calnek et al., Eds., Iowa State University Press, Ames, IA, 1997, 82.
82. Curtiss, R. S. and Kelly, M., *Salmonella typhimurium* deletion mutants lacking adenylate cyclase and cyclic AMP receptor protein are avirulent and immunogenic, *Infect. Immun.*, 55, 3035, 1987.
83. Zhang-Barber, L., Turner, A. K., and Barrow, P. A., Vaccination for control of *Salmonella* in poultry, *Vaccine*, 17, 2538, 1999.
84. Sydenham, M., Gillian, D., Bowe, F., Ahmed, S., Chatfield, S., and Dougan, G., *Salmonella enterica* serovar typhimurium *surA* mutants are attenuated and effective live oral vaccines, *Infect. Immun.*, 68, 1109, 2000.
85. Hassan, J. O. and Curtiss, R., Development and evaluation of an experimental vaccination program using a live-avirulent *Salmonella typhimurium* strain to protect immunized chickens against challenge with homologous and heterologous *Salmonella* serotypes, *Infect. Immun.*, 62, 5519, 1994.
83. Hassan, J. O. and Curtiss, R., Efficacy of live avirulent *Salmonella typhimurium* vaccine in preventing colonization and invasion of laying hens by *Salmonella typhimurium* and *Salmonella enteritidis*, *Avian Dis.*, 41, 783, 1997.
87. Byrd J. A., Corries, D. E., Caldwell, D. J., Bailey, R. H., Brewer, R. L., Stanker, L. H., and Hargis, B. M., Effect of selected organic acids on the control of Salmonella in market-age broilers during feed withdrawal, *Poult. Sci.*, 78 (Suppl. 1), 85, 1999.

chapter nine

Poultry-borne pathogens: plant considerations

Donald E. Conner, Michael A. Davis, and Lei Zhang

Contents

0-8493-0120-3/01/$0.00+$.50

Introduction

Food safety concerns

Microbial contamination of carcasses is a natural result of procedures necessary to produce retail products from live animals.[1,2] Contamination of poultry meat products can occur throughout initial processing, packaging, and storage until the product is sufficiently cooked and consumed. Heavy loads of bacteria enter the processing plant with the live bird, and these bacteria can be disseminated throughout the plant during processing. Most of the bacterial contaminants are non-pathogenic, and are associated with meat spoilage. However, poultry serve as reservoirs for a number of pathogens including, *Salmonella* serotypes, *Campylobacter jejuni, Listeria monocytogenes, Clostridium perfringens,* and *Staphylococcus aureus.*

Role of poultry in food-borne disease

In terms of food safety, poultry ranks first or second in foods associated with disease in Australia, Canada, England, and Wales,[3,4] while in the U.S., total poultry is the food vehicle in 8% (ranked third) of the reported food-borne disease outbreaks.[5] Epidemiological reports indicate that more than 95% of all food-borne illnesses are the result of activities occurring after the product has left the plant;[6] that is, illness is generally the result of temperature abuse and improper handling or preparation. However, when contamination and illness occur, investigators tend to look at raw product (how it was produced, processed, and handled), and press for elimination of pathogens before the product reaches the consumer.[7] This creates a challenge to the poultry industry to improve the microbiological safety and quality of its products.

Regulatory issues

Food safety has emerged in the U.S. and worldwide as a major consumer issue, and consequently has had a major impact on food processors and regulatory policy. Since 1996, the United States Department of Agriculture-Food Safety and Inspection Service (USDA-FSIS), which is charged with the federal inspection of meat and poultry in the U.S., has implemented new regulations for the poultry processing industry. This new set of regulations, entitled "Pathogen Reduction/Hazard Analysis Critical Control Points" (PR/HACCP),[8] consists of four major requirements that processors must meet:

1. Development and implementation of Sanitation Standard Operating Procedures
2. Development and implementation of HACCP
3. *Salmonella* performance standards
4. Biotype I *Escherichia coli* performance criteria.

In addition, the USDA-FSIS also implemented a "zero fecal" performance standard in 1998. For processors, this rule mandates that carcasses entering the chiller must be free of "visible feces." In October 1999, the USDA-FSIS published a final rule, "Sanitation Requirements for Official Meat and Poultry Establishments," which requires additional facility sanitation performance standards. All of these new regulations represent a more scientific approach to maintaining and improving the microbiological safety of poultry meat products.

Primarily in response to regulatory requirements that place more responsibility on processors, poultry processors must establish well-written and documented sanitation standard operating procedures (SSOPs) and good manufacturing practices (GMPs). Likewise, processors must develop HACCP plans, and implement them for all of their products and processes. Through its inspection personnel, USDA-FSIS verifies that processors are complying with their SSOP, GMP, and HACCP plans and that the specific microbiological performance standards are met. In concert, these programs represent the necessary approach to maintaining the microbiological safety of poultry during processing.

Pathogenic microorganisms on processed poultry

Pathogens of concern

As stated above, processed raw poultry meat naturally harbors bacteria. Most of these bacteria are responsible for the spoilage of poultry meat, but are not pathogenic to humans. However, poultry products can harbor bacteria capable of causing human disease (i.e., pathogens). A number of food-borne pathogens have been isolated from processed poultry (Table 9.1).[9] Of these, *Salmonella* serotypes, *C. jejuni, L. monocytogenes, C. perfringens,* and *S. aureus* are of major concern.

Salmonella *serotypes*

Salmonella are mesophilic, faculative, Gram negative bacteria of the family Enterobacteriaceae. There are three human disease syndromes caused by *Salmonella* spp.: typhoid fever, paratyphoid fever, and gastroenteritis. Typhoid and paratyphoid fever, which are rare in developed countries, are transmitted human to human by the fecal-oral route, and humans are the only reservoir.

In contrast, gastroenteritis (non-typhoidal salmonellosis) is caused by *Salmonella enterica* serotypes, which are found in the intestinal tract of both humans and non-human animals. Poultry has been identified as a primary reservoir for these salmonellae.[10] There are >2300 serotypes. Of these, Typhimurium, Enteritidis, and Heidelberg are the most frequently isolated serotypes from human cases, and these are common poultry-borne

***Table 9.1* Food-Borne Pathogens Isolated from Processed Raw Poultry Meat**

Aeromonas ssp.	*Shigella* ssp.
Camplyobacter ssp.	*Streptococcus* ssp.
Clostridium perfringens	*Staphylococcus aureus*
Listeria ssp.	*Yersinia enterocolitica*
Salmonella serotypes	

Source: Adapted from Waldroup, A. L., Contamination of raw poultry with pathogen, *World's Poult. Sci,*. 52, 7, 1996.

serotypes, also. Human disease ranges from mild to severe and is characterized as a self-limiting infection of the lower intestinal tract. The infectious dose ranges from 10,000–1,000,000 cells. Symptoms typically appear 12–36 h after consumption of a contaminated food, and include nausea, vomiting, severe diarrhea, fever, abdominal cramps, and malaise.

While *Salmonella* serotypes ultimately originate from contaminated feces, a wide variety of environmental and food sources can harbor these pathogens. This widespread distribution demonstrates the ability of *Salmonella* to survive well in the environment. These bacteria are introduced into the processing plant with the live birds, which can harbor these pathogens in skin and feathers, as well as in the GI tract. Consequently, *Salmonella* can persist on final raw products. Disease can result when these products are handled without good hygienic practices, not properly cooked, and/or subjected to temperature abuse.

Campylobacter jejuni

Campylobacters are mesophilic, microaerophilic, Gram negative, spiral rods. *C. jejuni, C. coli,* and *C. lari* comprise the thermotolerant group of the family Campylobacteriaceae, and are food-borne human pathogens of concern. Of these, *C. jejuni* is the most prevalent food-borne pathogen. The disease caused by *C. jejuni* (*C. coli* and *C. lari,* also) is similar to that caused by *Salmonella* serotypes. According to CDC data,[11] *C. jejuni* is the leading cause of diarrheal disease in the U.S. Poultry is the primary reservoir of *C. jejuni,* and most sporadic cases of human campylobacteriosis are attributed to mishandled or improperly prepared poultry.[10] While *C. jejuni* does not survive well in the environment, it is introduced into processing facilities via the GI tract of live birds, where it then can attach to broiler skin and persist into final products.[12]

Listeria monocytogenes

L. monocytogenes is a psychrotrophic, Gram positive bacillus. This is an opportunistic pathogen, infecting primarily the immunocompromised. Pregnant women and their fetuses, AIDS patients, alcoholics, and the elderly are the most often affected. In these patients, listeriosis can progress to meningitis and therefore can be life threatening. In the immunocompetent, listeriosis is often a mild, flu-like illness. *L. monocytogenes* tends to be an environmental contaminant, and can persist in cool, damp areas of a poultry processing plant. Drains, refrigeration-freezing equipment, and other fomites are known to harbor *L. monocytogenes.* Post-process contamination of fully cooked, ready-to-eat poultry products has emerged as a major food safety issue. Therefore, processors of such products must take steps to prevent contamination.

Clostridium perfringens

C. perfringens is an anaerobic, spore-forming, Gram positive bacillus. If large numbers of *C. perfringens* spores ($>10^6$) are consumed, a toxicoinfection can occur. The toxicoinfection results when *C. perfringens* attaches and colonizes the lower GI tract where it will enter into a spore-vegetative cell cycle. An enterotoxin, which produces a profuse watery diarrhea in the host, is produced during this growth cycle. The disease is generally mild and limited to one to two days in duration.

This pathogen is found in the soil and is carried in the fowl's GI tract. Therefore, *C. perfringens* can be introduced into the processing plant with live birds. Because the organism produces spores, it can survive in harsh environments. Thus, it can spread during processing and persist into final product.

Cooked poultry products, particularly those cooked in large batches, are of greatest risk. Such products can be difficult to heat thoroughly and subsequently cool quickly. If spores are not destroyed by the cooking process, they can grow to high numbers if the product is not cooled at a sufficient rate to prevent spore germination. In meat products that provide an anaerobic environment and are held at improper temperatures, this pathogen can enter a growth cycle in which the number of spores will double every 15 min. For this reason, cooked meat products must be cooled rapidly to prevent the germination and outgrowth of *C. perfringens*. This requirement for rapid cooling of cooked products is often referred to as "product stabilization."

Staphylococcus aureus

S. aureus are aerobic, Gram positive cocci. Certain strains, typically referred to as coagulase-positive *S. aureus*, produce enterotoxins as a byproduct of their growth. These enterotoxins can cause a generally mild gastroenteritis in humans. Food-borne illness results from the ingestion of enterotoxin(s) that have been pre-formed in a food product. Therefore, conditions must exist that allow the organism to grow to high populations (>10^6cfu/g).

While *S. aureus* is part of the natural microflora of poultry, poultry-associated strains of *S. aureus* can be differentiated from human strains. The poultry-associated strains do not seem to be involved in food-borne disease.[13] In terms of coagulase-positive *S. aureus*, these typically originate from humans. Therefore, employees (food handlers) are the primary source of *S. aureus* contamination in the processing plant. Most staphylococcal intoxications involving poultry products are related to recontamination of cooked product by food handlers, followed by improper holding temperatures.[13]

Incidence of pathogens on processed poultry

Waldroup[9] provided an excellent review of the incidence of bacterial pathogens on raw poultry, as reported in the scientific literature. While the incidence of pathogens varied from report to report, Waldroup's review clearly showed that pathogens can and do occur on raw poultry. For regulatory purposes, the USDA-FSIS obtained 1297 postchill carcass rinse samples from approximately 200 broiler processing plants.[14] These samples were analyzed for the presence and populations of six prevalent food-borne pathogens and indicator bacteria. Results of this survey are shown in Table 9.2.

Table 9.2 **Results of USDA-FSIS Microbiological Baseline Study for Broiler Chickens**

Organism	Incidence (% of rinse fluids positive)	Mean population[1] (cfu/cm^2 of broiler carcass)	Mean population[2] (cfu/broiler carcass)
Campylobacter jejuni	88	4.4	5,300
Clostridium perfringens	43	1.4	1,700
E. coli O157:H7	0	NA[2]	NA
Listeria monocytogenes	15	0.02	30
Salmonella serotypes	20	0.03	38
Staphylococcus aureus	64	2.6	3,200
Biotype I *E. coli*	100	6.6	7,900
Mesophilic aerobic bacteria	100	400	480,000

[1] Level only of those positive by qualitative method.

[2] Not applicable.

[3] Based on assumed surface area of 1200 cm^2 for broiler carcasses.

Source: Adapted from United States Department of Agriculture-Food Safety and Inspection Service, 9 CFR Part 304 et al.: Pathogen Reduction; Hazard Analysis and Critical Control Point (HACCP) Systems; Final Rule, *Fed. Regis.*, 61 (no. 144), 38806, July 25, 1996.

Effects of processing on pathogen load

Live birds destined for processing represent the primary entry point into the processing plant for an exceedingly high level of bacteria. Moreover, in the absence of effective control measures during live production, birds arriving at the processing plant should be considered potential sources of the pathogens indicated above. The processing steps to which birds are subjected are designed to produce wholesome and safe final products. Thus, as birds proceed through processing, there is substantial decrease in overall bacterial load. Removal of feathers, feet, heads, and viscera serve to also remove the bulk of the bacterial load. However, given the nature of modern poultry processing, not all bacteria are eliminated. The remaining bacteria can be transferred among carcasses. The extent to which bacteria are removed from carcasses or transferred among carcasses is a function of the specific processing steps and operational conditions. Although live haul (transportation of live birds from production farms to the processing plant) may not be considered a plant process, cross contamination can be attributed to this step in the process. Transportation coops are often contaminated with *Salmonella* even after washing.[15] *Salmonella* from coops can be transferred to birds held in them and to adjacent coops.[16] *Salmonella*-contaminated coops lead to external contamination (feet, feathers, skin) and to cecal and crop carriage.[16, 17] *Salmonella* originating from live haul equipment can contribute significantly to subsequent cross contamination among carcasses during processing.[16] Factors affecting this cross contamination include close crowding, coprophagy, weather, other stressors, and time the birds are off feed, which is an additive effect of feed withdrawal at the farm, transportation distance, and time birds are kept in the holding yard. Because these factors can affect the spread of *Salmonella* during the transportation phase, they subsequently affect the level of pathogens entering the processing plant.

The process of scalding the carcass is used prior to feather removal. This process subjects the carcass to immersion in hot water, facilitating the opening of the feather pores so that the feathers may be removed more effectively. There are two types of scalding, hard and soft. A hard scald, in which the carcass is immersed in water that is greater than 55°C, removes the cuticle (or epidermis) of the skin. If the carcass is immersed in water that is ≤55°C the cuticle is not removed and the carcass is considered to be soft scalded. Scalding tends to partially remove dirt, fecal material, and other contaminants found in the feathers. However, these contaminants may be spread to other carcasses through scalder water.[18] Most plants use a countercurrent scalder, in which water for the continuous overflow is fed from the cleanest end of the scalder (that end nearest the picking machines) toward the dirtiest end. This helps to reduce the amount of cross contamination. Tests on scald water have shown that *C. perfringens* and *S. aureus* can be isolated. However, *Salmonella* spp. and *Campylobacter* spp. are usually not isolated.[13] In general, scalding has little effect on the microbiological quality and safety of raw poultry products in the retail market.[19]

While defeathering of carcasses reduces the overall bacterial load via removal of the feather, the process is of major concern to the poultry industry because the modern mechanical process of feather removal can be a major contributor to cross contamination.[18] This process usually leads to an increase in the number of non-psychotrophic organisms on the individual carcasses.[13] Defeathering has also been attributed to an increase particularly in *S. aureus* because the organism becomes embedded in the cracks of the rubber fingers. It has also been attributed to the cross contamination of carcasses by *Salmonella* spp., *Campylobacter* spp., and *E. coli.* This may be due to embedding of these microorganisms in the feather follicles after the feather is removed and before the follicle can reduce in size.[13]

Evisceration of the carcass is another area of special concern in the area of cross contamination. If the intestines of the bird are cut, fecal contamination can occur. This is especially important because of the enteric pathogens that the intestines can harbor. These pathogens can contaminate machinery and workers. Plant personnel that handle carcasses must wash their hands frequently to decrease the possibility of transferring fecal-borne pathogens among carcasses. Continuous flow of chlorinated water over machinery that frequently becomes contaminated with feces and other GI tract contents is used to prevent spread of fecal-borne pathogens via equipment contact.

An emerging area of interest concerning cross contamination of commercial broiler carcasses is crop removal. While rupture of the intestinal tract, especially the ceca, has been the major focus of cross-contamination issues, crop removal is also a major problem. In the plant, crops are more likely than the ceca to rupture by up to 86-fold.[20] This problem is further exacerbated by findings that both *Campylobacter* and *Salmonella* can be more readily extracted from the crops of market-age broilers than from the ceca. Byrd et al.[21] reported that of 359 birds sampled, 286 (62.4%) harbored *Campylobacter* in the crop, whereas only 9 of 240 (3.8%) of birds sampled harbored *Campylobacter* in the ceca. In a similar study by Hargis et al. in 1995,[20] 286 of 550 (52%) broilers studied were positive for *Salmonella* in the crop, while only 73 of 500 (14.6%) were positive for *Salmonella* in the ceca. These findings show that care must be taken in crop removal during processing to reduce cross contamination of these important pathogens.

Washing of the carcass before chilling will reduce the organic material and remove possible fecal material both inside and outside of the carcass. Poultry processors use various washing devices during processing. These can include multiple washers at various points in the line, inside/outside wash cabinets, and final wash cabinets. The water used in these devices is normally chlorinated and has a fairly high pressure. The pressure of these washing devices must be high enough to remove exterior organic material, but not so high that it will force microbes into the pores of the skin. Use of washers of this type tends to reduce levels of enteric bacteria,[22] and multiple washers from defeathering to chilling are typically more effective than a single final wash just prior to chilling.[23]

Chilling of poultry carcasses in the U.S. is usually accomplished by immersion chilling. In this procedure, carcasses are immersed in large tanks of cold water that is typically chlorinated. This process can be both beneficial and detrimental. Beneficially, most poultry carcasses are bacterially contaminated mainly on the surfaces of the carcass, both inside and out. Rapid cooling may slow the growth of mésophilic organisms. Many immersion chillers are countercurrent, meaning that clean water is pumped into the chiller close to the exit end. A counter flow of water and birds is more effective in reducing bacterial numbers on carcasses.[24] Detrimentally, large tanks that hold many carcasses can be responsible for cross contamination. Pathogens may be washed off of the skin or other surface areas of one carcass and be moved to another carcass. Waldroup et al.[25, 26] found that the incidence of *Campylobacter* spp. increased from 86.4% of prechill to 90.8% of postchill carcasses, while the incidence of salmonellae increased from prechill to postchill by 20%.[25, 26] Shackleford[7] identified scalding and immersion chilling as the major source of cross contamination in a poultry processing plant. Maintaining >25 ppm of chlorine in the chiller can reduce cross contamination by vegetative bacteria, including *Salmonella*.[27–29]

Evaporative air chilling of broiler carcasses is not common in the U. S., but this process is fairly common in the European Union. This type of chilling occurs in a large room that has an ambient air temperature of 2–4°C. Carcasses are sprayed with water at intervals to take advantage of the cooling effects of water. The carcasses are circulated around the room until the optimum deep muscle temperature (<4.4°C) is reached. Effects of air chilling on the spread of microbes have been mixed. Sanchez et al.[30] found that air-chilled carcasses contained higher APC (total) counts and higher coliform counts than carcasses that were chilled by immersion in water. They also found that both air- and immersion-chilled carcasses contained approximately the same number of psychrotrophs and generic *E. coli*. However, incidence of *Salmonella* was lower (33.3%) on air-chilled carcasses than on immersion-chilled carcasses (56.6%). These data suggest that cross contamination can still occur in air-chilling systems, especially when the carcasses are sprayed with water. Data also suggests that since air-chilled carcasses are not completely immersed in water, and some drying of the skin occurs, some microbes on the surface of the carcass may be killed during the drying process.

Through initial processing, the carcass proceeds with its skin still intact. Through immersion and washing treatments, a water film is established on the carcass surface (i.e., the skin). This water film facilitates retention of bacteria on the carcass.[31–33] Initially, bacteria in the water film can be washed away, which likely accounts for decreases in microbial load observed with carcass washing procedures. However, during the time of processing, bacteria can readily attach to or become embedded within the structure of the poultry skin.[12, 34] Once attached or embedded, bacteria are more difficult to remove or kill.[12, 35] This retention likely accounts, in part, to the persistence of pathogenic bacteria on post-chill carcasses.[9, 14]

After the carcasses are removed from the chilling system, they may be put through various other processes. Carcasses may be shipped whole, cut-up, deboned, and/or cooked. Bacteria from carcasses destined for further processing can be transferred to contact surfaces, utensils, and personnel, which then become the primary vectors of cross contamination. If the product is cooked, internal temperatures that are adequate for killing target pathogens (lethality) are required.[36] After cooking, products must be stabilized by rapid cooling to prevent outgrowth of *C. perfringens* spores.[37] Following proper cooking, products will be free of vegetative pathogens; however, recontamination of product can occur if preventive measures are absent or ineffective. Post-cooking contamination is the primary factor leading to *L. monocytogenes* contamination of ready-to-eat poultry products. *L. monocytogenes* can survive well in processing plant environments and is considered an environmental contaminant.[38] Freezing equipment has been a frequent source of this pathogen. *L. monocytogenes* can survive below −1.5°C and thus can persist in freezers.[39] Other environmental sources of *L. monocytogenes* include water, air, personnel, and all product contact surfaces.

Control of pathogens during processing

General considerations

Two overall strategies are used by poultry processors to control pathogens in the plant: GMPs and HACCP. The processing plant must provide hygienic environmental and operating conditions (i.e., follow GMPs) such that products are produced in a safe, sanitary, and wholesome manner. As referred to earlier, processing plants must also develop and implement HACCP to control pathogenic bacteria and other food safety hazards. While HACCP must be a separate program, it must also be based on solid GMPs. Therefore, these two pathogen control strategies are interrelated. As part of GMP programs or HACCP, there are specific antimicrobial treatments that can be employed to improve the microbiological safety of poultry meat.

Good manufacturing practices — HACCP prerequisites

As HACCP has evolved, it has become apparent that HACCP programs must be supported by other facility-wide sanitation and food hygiene programs. That is, certain basic requirements must be in place before HACCP can be effectively implemented. In the 1998

NACMCF document,[40] the definition of "prerequisite programs" and their relationship to HACCP was very prominent:

> "Each segment of the food industry must provide conditions to protect food while it is under their [sic] control. These conditions are prerequisite to the development and implementation of HACCP. ... Prerequisite programs provide the basic environment and operating conditions that are necessary for the production of safe, wholesome food."

Prerequisite programs, again, are not part of the formal HACCP plan and system. However, they are crucial to HACCP development, implementation, and maintenance. In contrast to HACCP, prerequisite programs typically are facility-wide in nature and not product specific; therefore, they often cross and affect many, if not all, product lines in the plant. Moreover, prerequisite programs often target objectives other than food safety, such as quality and process control. Prerequisite programs would include those programs listed in Table 9.3.

Because prerequisite programs cut across many, if not all, operations and products in the plant, it is often not possible to tie performance of a given prerequisite program to a specific product lot (or even product line). Therefore, management of prerequisite programs can differ from that needed for HACCP. Most processors typically manage their prerequisite programs within their QA/QC or other quality programs. The exception to this is when a specific prerequisite program is critical for hazard control and subsequent production of safe product. In this case, the prerequisite program or a component of the prerequisite program would be incorporated into the plant's formal HACCP plan, and would, therefore, be required to be managed as such.

The existence of prerequisite programs greatly influences the hazard analysis. Hazards identified at each step in the establishment's process are assessed for likelihood of occurrence. Adequate and reliable prerequisite programs can provide environmental and operating conditions such that potential hazards would be unlikely to occur (low risk); therefore, these potential hazards would not have to be addressed in the plant's HACCP plan. In contrast, inadequate or unreliable prerequisite programs could lead to an increased likelihood of potential hazards occurring; therefore, these hazards would have to be addressed in the plant's HACCP plan. In terms of overall impact on the plant's HACCP plan, the existence of adequate, well-implemented prerequisite programs leads to simpler, more manageable HACCP plans, while an absence of effective prerequisite programs will tend to increase the number of CCPs and complexity of the HACCP plan needed to produce safe product.

As indicated above, facility-wide hazard control is one goal of most prerequisite programs. Microorganisms can build up rapidly, and/or be transferred from one part of the plant to another if control steps are not taken. Bacteria can be introduced and spread by water, air, people, pests, and fomites (inanimate objects such as equipment, tools, utensils,

***Table* 9.3 Typical GMP or HACCP Prerequisite Programs in a Poultry Processing Plant**

Premises and facilities	Processing procedures
Cleaning and sanitation (SSOPs)	Personnel
Inbound materials	Pest control
Equipment	Product traceability and recall

etc.). In addition to these biological hazards, chemical and physical hazards could be introduced into product if proper procedures and safeguards are not in place. The proper procedures and safeguards, if not part of the plant's formal HACCP plan, will fall into prerequisite programs.

Regardless of program, there are certain requirements for all prerequisite programs. The goal of each prerequisite program should be to thoroughly address and control items that can impact food hygiene and overall wholesomeness of the product. Ideally, each prerequisite program should be based on written procedures (SOPs, see also discussion below), have assigned responsibilities, be subjected to measurement criteria and record keeping, and have prescribed corrective actions when criteria are not met. By having written programs, it will be easier to train responsible employees and implement the procedures. Moreover, the expected goals of the program will be known, which will allow for more objective assessment. Because many prerequisite programs are facility-wide, responsibility for implementing and maintaining programs often crosses departments within the plant; therefore, management of prerequisite programs must be a key consideration.

Premises and facilities

The processing plant should be located, constructed, and maintained in accordance with sound sanitary design and hygienic principles. Because pests can be vectors of food-borne pathogens, premises should minimize pest (i.e., rodents, insects, birds) harborages, such as areas of standing water, trees and shrubbery in close proximity to processing plant, bird nesting sites associated with the building, waste collection sites, etc. For this reason, the processing plant site should be well drained, landscaped with minimal shrubbery, and designed to facilitate waste management.

As overall considerations, facilities should be designed to facilitate product flow and should provide for separation of operations where appropriate. Product should flow from the area of highest microbial load to the area of lowest microbial load (e.g., raw to cooked), and not "back track." Separation of areas and of employee traffic patterns are an important consideration in preventing microorganisms from moving throughout the plant. Overall layout and design of the plant should also provide for adequate ventilation, lighting, and space for equipment and storage. Without these provisions, maintaining sanitary conditions in the plant will be more difficult.

Walls, doors, ceilings, and floors represent the interior surfaces in the plant, and, therefore, need special attention. The surfaces should be easily cleaned and sanitized, impervious to water, and minimize niches for collection or entry of microorganisms and pests. Walls should be solid, sealed for waterproofing, and be free of windows. In terms of the latter, windows are not necessary in well-ventilated and lighted plants. If windows are present, the glass should be unbreakable, the sill should be sloped to prevent collection of debris (bacteria), and windows should not be able to be opened by employees unless required by fire regulations. Doors represent another type of opening in a wall. If doors are present, it is likely that they are there to ensure separation of different areas of the plant; therefore, they should be kept closed during plant operations. In addition to meeting the same cleaning and sanitizing requirements as walls, doors should be tight fitting and well maintained. In critical areas of the plant, doors may be supplemented with other requirements such as air curtains, foot baths, etc. False or drop ceilings should be avoided because they allow for collection of bacteria and can become harborages for pests. In relation to ceilings, overhead piping, beams, etc. should be minimized. Because of their effect on heat transfer, ceiling-roof construction and insulation along with ventilation are often primary determinants (e.g., in wet processing areas) of the extent of condensation in the plant. Floors represent the interior surface most susceptible to rapid buildup of microorganisms.

Again, floors should be impervious to water, and therefore should be sealed and free of cracks. Additionally, floors should be sloped and well drained to prevent standing water. Standing water can be a breeding ground for bacteria.

In poultry and other food processing, water has always been a major premises/facility related issue. Water is used extensively in processing and cleaning operations, therefore the plant should have access to a good water source and be able to maintain water quality in the plant. Water system design and plumbing will be important in preventing water contamination in the plant. The water system should be designed to keep potable water protected from wastewater and sewage, and plumbing/maintenance activities should not compromise this protection. In recent years, other water issues, such as availability, adequacy of potability standards, conservation, reuse, and wastewater treatment, have also emerged. These issues will also impact the plant's water programs.

Cleaning and sanitation

The goal of cleaning and sanitizing is microbial control through elimination of nutrients, microbial niches, and excessive water. Microbial control is particularly important on surfaces which come in contact with product. By USDA-FSIS regulations (9 CFR Part 417), each plant must have SSOPs to which it must adhere in its day to day operations. Beyond overall plant sanitation, SSOPs are also required for all food contact surfaces, including equipment and utensils. Beyond the specific regulatory requirements, SSOPs should specify the following: item to be cleaned and sanitized, how and when the procedure is to be done (including chemical agents and other materials), and responsible personnel. A means of assessing the effectiveness of the plant's sanitation program should also be written. The USDA inspection procedures serve as verification that the processor is complying with its SSOPs. Failure to comply can result in the inspector issuing a "noncompliance report" (NR).

Inbound materials

A poultry processor will receive a number of materials into its plant that are needed in the manufacturing process. A partial list would include raw materials, ingredients, packaging materials, cleaning and sanitizing agents, processing aids, etc. All inbound materials need to be obtained from reputable suppliers. All suppliers should have verifiable food safety programs, including HACCP where appropriate, in place. Because inbound materials can affect quality and safety of product, the processor should establish written specifications for all incoming materials, and then obtain materials from suppliers who can meet the specifications. However, specifications must be realistic. For example, a specification for *Salmonella*- or *Campylobacter*-free raw poultry would be an unrealistic specification. Documentation in the form of Letters of Guarantee and Certificates of Analysis are often used in connection with supplier specifications. Once received, materials must be stored in a sanitary manner, and this manner of storage should be a written SOP for each class of inbound material. Storage should prevent against contamination that can affect product safety. For example, cleaning agents are not to be stored with food ingredients. For perishable ingredients, environmental control would be an important consideration.

Equipment

All equipment should have sanitary design, which means that it does not directly contribute to product contamination, it is constructed of nonreactive, nontoxic materials, it is easily cleaned and sanitized, etc. Initially, it is very important that equipment is installed properly and by qualified personnel. When adding new equipment, there should be adequate space for it such that cleaning and sanitation procedures can be performed properly.

All equipment requires, to some degree, preventive maintenance and repair. While repair often cannot be predicted or scheduled, preventive maintenance can and should be systematic. Therefore, preventive maintenance should follow a written SOP, and records should be kept on key pieces of equipment. Preventive maintenance helps ensure that processing steps are done as intended, and that the risk of physical and chemical hazards such as machine pieces and lubricants is minimized. Calibration of processing equipment and instruments would be a specific type of preventive maintenance program. Written preventive maintenance programs should include the following: specific equipment identification, exact procedures and frequency, records to be kept, and assignment of responsibility.

Processing procedures

Processing procedures must be strictly controlled primarily for quality control purposes. However, processing control is directly related to product safety in many instances. That is, how a procedure is carried out can have a direct bearing on product safety. A good example would be product formulation, in which restricted ingredients are used. Processing steps in which the ingredients are weighed, mixed, blended, added, etc. must be done properly to ensure that the ingredient's concentration in final product is within regulatory limits. Another example would be poultry meat cutting or portioning operations, which done incorrectly could lead to physical hazards in the form of bone or metal. For this reason, prescribed procedures, SOPs, are needed for most if not all processing steps. These written procedures serve to communicate expectations of the process and serve as the basis for training the personnel involved in the various processing operations. Beyond specific procedures, there should be certain written expectations of how product is to handled, as well as the expected flow of product during normal plant operations. These latter issues relate to the time-temperature sequence of the product, which can be a major determinant of microbial contamination.

Personnel

All employees in the plant require training. The plant's training program should include training the key personnel in their role and responsibility in producing safe product and complying with regulatory requirements. All personnel involved in processing operations must be trained in food hygiene principles, particularly personal hygiene. Training should target personal cleanliness and avoidance of product contamination, but should also emphasize the employee's responsibilities in complying with hygienic practices. A formal food safety training program consists of established training material, a training schedule, and documentation of each employee's training. An employee's training file, the documentation, should indicate education and training the employee has received related to food safety and hygiene, as well as an assessment of the employee's proficiency in appropriate topics. Employee turnover can be very high in poultry processing plants; therefore, processing plants must establish a solid training policy and be vigilant in ensuring that key personnel receive appropriate, effective, and documented training.

Pest control

The objective of pest control in the poultry processing plant is simply to prevent or eliminate pests, including insects, rodents, and birds. As stated above, these pests are vectors of pathogenic bacteria. Pest control in the processing plant is typically achieved by a multifaceted approach referred to as integrated pest management, which aids in minimizing the need for use of chemical pesticides. Of course, the use of chemical pesticide in a food environment has food safety implications; therefore, minimizing chemical pesticide use has food safety benefits. An integrated approach typically entails three practices: inspection, housekeeping, and physical/mechanical/chemical methods. Inspection must be conducted on a prescribed frequency, thus, a written SOP is needed, as are record keeping forms to document inspection results. Formal inspections should be conducted by those specifically trained in pest management, while plant personnel can conduct ongoing inspections. Housekeeping is a function of the cleaning-sanitation operations in the plant. The goal of inspection and housekeeping is to prevent pest infestation. If infestation is noted, then removal or eradication methods are needed, and there should be acceptable physical, mechanical, and chemical means available. A trained exterminator should oversee the removal or eradication procedures used in the plant. Strict records should be kept when such methods are used, particularly when chemical pesticides are involved.

Product traceability and recall

As unpleasant as it may be, recalling product once it has been produced and shipped is part of food processing. Therefore, the processor needs to develop strategies to avoid recalls in the first place, and a strategy to recall product if the need arises should be in place. Key tactics of this strategy must insure that product is effectively recovered and disposed of properly.

Hazard Analysis Critical Control Points (HACCP)

HACCP is a systematic and science-based approach to planning, controlling, and documenting the production of safe products.[41,42] In essence, HACCP is a management tool that the poultry processor can and must use to assess and manage the risks associated with its products. In contrast to GMPs, which tend to be facility-wide programs, HACCP is product specific and focuses exclusively on food safety. As covered in an earlier chapter, HACCP consists of seven principles:[40]

1. Identification and assessment of food safety hazards
2. Identification of critical control points (CCPs)
3. Establishment of critical limits for CCPs
4. Establishment of CCP monitoring procedures
5. Establishment of corrective actions
6. Establishment of verification activities and procedures
7. Establishment of record keeping procedures

In practice, the processor establishes an "HACCP team" whose responsibility is to develop and implement an effective HACCP plan for each and every product. For each product or product category, a simple diagram of all of the processing steps is prepared. At each processing step, potential food safety hazards (e.g., pathogenic bacteria and other hazards) are identified. Once identified, the hazards are assessed in terms of likelihood of occurrence and severity of illness they can cause. For hazards that are likely to occur or are of a severe nature, CCPs in the operation at which these hazards can be controlled are identified. Specific limits are assigned to each CCP. Limits must be scientifically valid for control of the identified hazard. Since limits are critical to product safety, each CCP must be systematically monitored to ensure that the hazards are under control. A "deviation" occurs when the critical limits are not met. Thus, actions to be taken in event of a deviation are prescribed in the HACCP plan. Periodically, HACCP personnel must take steps to verify that the written HACCP plan is being followed as intended in the daily plant operations.

Specific records must be kept to document that the CCPs are monitored correctly, that corrective actions are taken when needed, and that plant personnel are adhering to the written plan.

This approach emphasizes a preventive strategy rather than an inspection strategy to controlling pathogens on poultry. Every component of processing must be evaluated in the course of developing an HACCP plan, and it is the processor's responsibility to develop and implement a valid plan. USDA, through its inspection personnel, takes steps to verify that the plant's HACCP plan is being followed and that the plan is achieving pathogen control.

Antimicrobial treatments used in poultry processing

Chemical treatments

One approach to reduce the pathogens on poultry carcasses is the application of GRAS (generally regarded as safe) chemical treatments during primary processing. Many antimicrobial treatments have been researched for their efficacy against bacterial contamination of poultry. For ease of discussion, these chemicals have been grouped into four categories: chlorine and chlorine compounds, trisodium phosphate, organic acids, and others.

Chlorine compounds. Chlorine is the most common antimicrobial compound currently used in the poultry processing plant. May[22] reported that significant reductions were found with the addition of 18–25 ppm of chlorine into the chilling system. Izat et al.[43] reported that 100 ppm chlorine in chilling systems effectively reduced *Salmonella,* but a strong chlorine odor was noted. It has been determined that a level over 1200 ppm would be necessary to achieve a minimum 99% kill. The efficacy of chlorine is affected by many factors, including initial bacterial concentration, water level, organic load. temperature, pH, and trace minerals in the water. Efficacy of chlorine increases as the concentration increases, but discoloration, off-odor, and off-flavor are then associated with the carcass. One concern with the use of chlorine has been the buildup of organochlorides from the combining of chlorine with proteins. The reaction of amino acids with chlorine increases as pH increases from 3 to 9. Chlorine in high concentration can corrode metal equipment and pose a health threat to employees.

Chlorine dioxide is a more stable form of chlorine that could be used for pathogen reduction. Chlorine dioxide is more effective than chlorine in the presence of organic matter and over a wider pH range. Also, chlorine dioxide is relatively inert toward individual amino acids and will not result in off-flavors. The incidence of *Salmonella* on carcasses can be reduced from approximately 14.3 to 2.1% with 3 ppm and to 1.0% with 5 ppm chlorine dioxide. Villarreal et al.[44] added slow release chlorine dioxide (SRCD) to turkey rinse and chilling water to reduce the incidence of *Salmonella*-contaminated carcasses. Chilling carcasses in 1% SRCD and ice eliminated any recoverable *Salmonella* from turkey carcasses; meanwhile, an in-plant chlorination system reduced the incidence of *Salmonella*-contaminated carcasses from an average of 70% after evisceration to 25% after chilling.

The Sanova Food Quality System®, a microbial control application from Alcide, Inc., was approved by USDA for food safety application in poultry processing in January 1998, and for continuous on-line reprocessing in June 1998. This latter application of Sanova circumvents the current USDA rules requiring carcasses with visible fecal matter to be removed from processing lines and reprocessed manually, which is time-consuming and costly. The Sanova System is an automated system using a sprayed-on mixture of sodium chlorite and citric acid to kill *E. coli, Salmonella, Listeria, Campylobacter,* and other bacteria that can contaminate birds. Fresh, unused solution is applied via proprietary application between final bird wash and the chiller.

Trisodium phosphate (TSP). Trisodium phosphate (Na_3PO_4) is a GRAS food additive. In October 1992, USDA approved the use of TSP during poultry processing and more recently approved TSP use for continuous on-line reprocessing. Since then, poultry processors have shown a great interest in research determining the effectiveness of TSP in reducing or eliminating pathogens from poultry carcasses at the processing plant. TSP can affect the skin of carcasses, which allows bacteria to be washed from the surface of the bird more effectively. Lillard[45] reported that high pH of the whole carcass rinse and skin samples from TSP dip could account for the low counts of *Salmonella* from inoculated samples. Somers[46] studied effectiveness of TSP against planktonic (suspended) and biofilm (attached) cells of *C. jejuni, E. coli* O157:H7, *L. monocytogenes,* and *S. typhimurium* at room temperature and 10°C. At both temperatures, *E. coli* O157:H7 was the most sensitive to TSP treatment; *C. jejuni* was slightly less sensitive; followed by *S. typhimurim;* and *L. monocytogenes* was least sensitive to TSP. Hollender[47] studied effects of TSP treatment on sensory attributes of broiler carcasses and reported that appearance, flavor, and texture scores were not significantly different for treated and control samples.

Organic acids. Lactic and acetic acids are inexpensive, have GRAS status, are environmental friendly, and are naturally occurring. Izat et al.[43] found *Salmonella* incidence rates were reduced by adding 0.5–1.0% lactic acid to chilling water. Acid treatments are especially effective before bacteria are firmly attached to the meat surface. Cudjoe and Kapperud[48] reported that spraying 1 and 2% lactic acid 24 h post-inoculation did not significantly reduce *C. jejuni* populations; however, spraying 2% lactic acid 10 min post-inoculation eliminated all *C. jejuni* within 24 h. Typically, bactericidal effectiveness increases with increasing concentration or temperature; however, higher concentrations tend to bleach the carcasses. The use of acid mixtures has been studied, and Rubin[49] reported that lactic and acetic acids were slightly synergistic in their inhibitory effects on *S. typhimurium.* Adams and Hall[50] also noted an apparent synergistic interaction between acetic and lactic acids.

Acids used with surfactants have been recently investigated. These surfactants (or transdermal compounds) increase the activity of the acid by aiding in the delivery of the acid to the chicken skin, thereby loosening embedded or entrapped bacteria. Tamblyn and Conner[35] reported that when monolaurate (SPAN-20) was added to 0.5% citric, lactic, malic, and tartaric acids, and used in a simulated scalder and chiller, the activity of these acids against firmly attached *S. typhimurium* increased significantly. They also found that sodium lauryl sulfate had the same effect as SPAN-20. Applying combinations of organic acids and surfactants in a pilot processing environment effectively reduced levels of *Salmonella* serotypes, *C. jejuni,* and *L. monocytogenes* on broiler carcasses, and reduced spread of these bacteria to uncontaminated carcasses.[12, 51]

Other chemical treatments. Ozone is a powerful oxidizing and bactericidal agent. Since 1906, ozone has been used as a disinfectant to remove color, odor, and turbidity, and also to reduce the organic loads at European wastewater plants. Sheldon and Brown[52] evaluated the effects of ozone on the quality of poultry chiller water and broiler carcasses. Not only were >99% of all microorganisms washed from carcasses destroyed by the residual ozone, but a 30% reduction in chemical oxygen demand and a significant increase in light transmission of process water were also achieved. Furthermore, there were no significant carcass skin color losses, lipid oxidation, or off-flavors resulting from ozone contact. However, a number of factors still need to be considered prior to using ozone in poultry chillers or as a wastewater treatment. These factors include: equipment needs, toxicity, corrosiveness, gas containment within a unit operation, optimal ozone to water ratio, gas transfer efficiencies, government regulations, etc.

Sodium bisulfate (SBS) is a GRAS compound and approved for use in certain food applications. Yang et al.[53] studied the effectiveness of SBS on reduction of *S. typhimurium* with an inside-outside bird washer on prechilled chicken carcasses. Both total aerobes and *S. typhimurim* on the chicken carcasses were reduced by 1.66 $\log_{10}$ cfu per carcass after spraying with 5% SBS. Concentration and the spray pressure affect the bactericidal activity of SBS.[54] Visual inspection indicated that SBS spray treatments slightly discolored part of the skin of chicken carcasses.

Potassium sorbate is commonly used as a mold inhibitor, but also has antibacterial properties. Sorbic acid salts are also GRAS food additives. Using potassium sorbate dips in different concentrations can reduce both the total number of viable bacteria and the growth rate of *Salmonella* on poultry product. Dipping poultry in potassium sorbate and packing in 100% carbon dioxide has also been shown to be effective for bacterial control.[55]

Hydrogen peroxide has been investigated as a potential bactericide for pathogens on poultry and has shown potential for reducing bacterial counts. However, carcasses chilled in hydrogen peroxide tend to have a bloated skin appearance. It has also been noted that the skin of treated carcasses was rubbery and bleached, and gas and water accumulated under the skin.[43, 56]

Cetylpyridinium chloride (1-hexadecylpridinium chloride, CPC), a quaternary ammonium compound with a neutral pH, is approved for use in mouthwashes to prevent dental plaque. As a cationic surfactant, the mechanism by which CPC kills bacteria involves the interaction of basic cetylpyridium ions with the acid groups of bacteria to form weakly ionized compounds that subsequently inhibit bacterial metabolism. In poultry-processing experiments, CPC caused no carcass bloating or skin discoloration and did not corrode equipment. Upon treatment of broilers, total aerobic counts and *Salmonella* populations on carcasses were reduced by 2.16 and 2.01 $\log_{10}$ cfu per carcass, respectively after spaying with 0.5% CPC.[53]

Salmide™ is an oxy-halogen inorganic compound. A study was conducted to compare the effect of addition of 31 m*M* Salmide vs. 20 ppm chlorine to a carcass chiller. Total counts in carcass rinse samples were reduced from 8100 cfu/ml on chlorine-treated carcasses to 2700 cfu/ml on Salmide-treated carcasses. Furthermore, *C. jejuni* and *Salmonella* counts were reduced from 260 and 73 cfu/ml, respectively with chlorine treatment to <3 and <2 cfu/ml, respectively, with Salmide treatment. There was a slight tightening of the carcass skin with Salmide treatment.[57]

Physical treatments

Beyond chemical treatments, physical treatments are available to processors for pathogen control in products. In fact, temperature manipulation is the primary means by which pathogens in poultry products are controlled or eliminated. Besides temperature control, the application of ionizing radiation (irradiation) has recently been approved for raw fresh or frozen poultry, and has the potential to emerge as a means of eliminating pathogenic microorganisms from raw poultry.

Temperature control. Because most poultry-associated bacteria reproduce by binary fission, each growth cycle results in a doubling of bacterial population. Under optimal temperature and other environmental conditions, bacterial growth is characterized by a short

Table 9.4 Grouping of Bacteria Based on Temperature Effects on Growth

Bacterial group	Temperature ranges that allow growth (°C) Minimum	Optimum	Maximum
Psychrophilic	−15–5	5–30	20–40
Psychrotrophic	−5–8	20–30	30–43
Mesophilic	5–8	25–43	40–50

Source: Adapted from Ayres, J. C., Mundt, J. O., and Sandine, W. E., *Microbiology of Foods*, W. H. Freeman, San Francisco, 1980, and Banwart, G. J., *Basic Food Microbiology*, Van Nostrand Reinhold, New York, 1989.

lag phase in which bacterial numbers remain relatively constant over time, followed by a rapid growth phase in which cell numbers increase exponentially over a relatively short period of time. Under such conditions, bacteria can double in number in as little as 15 minutes. As the temperature moves below or above the optimum, bacterial growth rate will decrease. The further away from the optimum, the slower the growth rate. At some point, which is dependent on bacterial type, the organism will not be able to reproduce. Thus, a basic tenet of food safety is that the rate at which bacteria multiply is temperature dependent, and temperature can be a useful tool to control bacteria on food products.

Bacteria that occur on poultry products can be classified (Table 9.4) according to the temperature range in which they can grow (multiply in numbers).[58, 59] Therefore, the temperature at which poultry products are held can affect bacterial growth and influence the types of bacteria that will predominate. The predominant poultry spoilage bacteria are psychrotrophic or psychrophilic, while the primary food-borne pathogens associated with poultry are mesophilic. Refrigeration in addition to retarding microbial spoilage can be an effective means for preventing pathogens from increasing in number on poultry products. An exception would be *L. monocytogenes*, which has the ability to proliferate, albeit slowly, at refrigeration temperatures.

Because most of the pathogens of concern can proliferate between 5 and 50°C, this temperature range is often referred to as the danger zone. To prevent proliferation of bacteria, poultry products should be brought to 4°C or below as quickly after processing as possible according to USDA guidelines (9 CFR Part 381.66). After processing, raw products must be stored either at or below 4°C. Products should be taken through the danger range as quickly as possible when a temperature change is necessary (e.g., cut-up and deboning, cooking and chilling, etc.). During cut-up and deboning, it is impractical to keep product at or below 4°C; however, product should not exceed 10°C, and processing time during these operations should be minimized. Storage of products at temperatures below the critical zone (refrigerator or freezer) does not kill bacteria; rather, bacteria are prevented from growing. When products are subsequently held within the danger zone, bacteria can increase in number, thereby increasing the risk of disease.

In the manufacture of ready-to-eat poultry products, vegetative bacteria cells are expected to be destroyed in the cooking process. Moreover, processors of ready-to-eat poultry products are required by the USDA-FSIS to meet lethality performance standards to ensure that these products are free of vegetative pathogens. Thermal processes should be designed and validated, within a reasonable margin of safety, to eliminate vegetative pathogens. Such processes are referred to as safe harbors and should provide a 5–7 $\log_{10}$ reduction in *Salmonella* serotypes and other vegetative pathogens. Although there are many factors that can influence the rate at which bacteria are killed by heat, it is generally accepted that internal product temperatures of at least 71.1°C will provide the safe harbor lethality to ensure elimination of non-spore-forming pathogens such as the *Salmonella* serotypes, *C. jejuni, L. monocytogenes,* and *S. aureus.*[36] The spores of *C. perfringens* are typically not eliminated by such heat treatments, as spores of this organism have been demonstrated to survive, at least in part, an exposure to 80°C for 10 min.

Because spores of *C. perfringens* and other bacteria can survive typical cooking processes, products must be cooled quickly following cooking. This is referred to as product stabilization. USDA requires producers of fully cooked, ready-to-eat products to meet stabilization performance standards to ensure the spores of *C. perfringens* will not germinate and grow.[37] As general safe harbor guidelines for product stabilization, during post-cook cooling, product temperature should not remain between 54.4 and 26.6°C for more than 1.5 h nor between 26.6 and 4.4°C for more than 5 h. Other cooling cycles are acceptable if the processor can validate that they prevent the outgrowth of *C. perfringens* spores.

Irradiation. Irradiation, a process by which food is exposed to ionizing radiant energy (gamma or X-rays) to extend product shelf-life, results in the elimination of spoilage and pathogenic bacteria. In 1990, FDA approved irradiation of poultry at 1.5 to 3 KGy to eliminate pathogenic bacteria such as *Salmonella* serotypes, *E. coli* O157:H7, *C. jejuni,* and *L. monocytogenes.*[60, 61] In 1992, USDA-FSIS approved facilities to irradiate packaged fresh or frozen uncooked poultry.[62]

For irradiation of poultry, product is packaged and shipped to the irradiation facility. Low dose irradiation is applied to the packaged product at a USDA-approved facility. Product remains in its package after irradiation to reduce the risk of contamination until consumer use. The approved irradiation process greatly reduces but does not eliminate all bacteria. Refrigerated storage life is extended, but the need for cold storage is not replaced by irradiation.

The amount of energy the food absorbs is carefully controlled and monitored by the plant quality control personnel and USDA inspectors so desirable food preservation effects can be achieved while maintaining the safety, quality, and wholesomeness of the product. Irradiated food itself does not become radioactive. Facilities for irradiating food are similar to those in operation for sterilizing medical equipment and do not resemble nuclear reactors in any way. There are no explosives or materials that could cause widespread dissemination of radioactive material.

At an irradiation facility, the radiation source, cobalt-60, cesium-137, or an electron beam generator, is housed in a protective containment environment. Packaged poultry travels in pallets on a conveyor to the source, where it is exposed to gamma rays. The radiation dose is a function of the strength of the radiation source and time of exposure. Therefore the radiation dosage is typically controlled by a computerized rate of passage (e.g., conveyor speed) through the chamber.

Food irradiation is sometimes called a "cold" process because it achieves its effect with little rise in the temperature of the food. There is little if any change in the physical appearance of irradiated poultry. Irradiation causes only minor changes in the nutritional, chemical, and physical attributes of the product, and such changes are of a lesser degree than the changes caused by freezing, canning, or cooking.[60] Off-flavors and odors, which can occur with high dose irradiation, do not occur in poultry irradiated at the approved doses. Maintaining low product temperatures during the irradiation process aids in preserving quality and overall nutrient retention in irradiated poultry.

Consumer acceptance is a key issue in the adoption of irradiation as an antimicrobial treatment for poultry. Consumers are typically not knowledgeable about food irradiation. However, consumers in general show a higher level of concern for preservatives and pesticides than for food irradiation. Attitudes of conventional consumers regarding irradiation can be positively influenced by an educational effort, and the influence is most effective when the consumer can interact with someone knowledgeable about irradiation. After seeing a 10-minute video describing irradiation, interest in buying irradiated foods among California and Indiana consumers increased from 57 to 82%.[63] As irradiation becomes more widely accepted, the use of this technology for improving the microbiological safety of poultry may become a common practice.

Microbiological testing

As indicated earlier, control of pathogen requires adherence to GMPs, compliance with valid HACCP plans, and perhaps use of specific antimicrobial agents. The need for and effectiveness of these programs is assessed through microbiological information (criteria). Therefore, microbiological testing, while not a control measure per se, is an essential component of a processor's overall pathogen control strategy. Microbiological testing involves assessment of product and the processing environment. Product testing allows an assessment of the overall microbiological load, incidence of pathogens, effects of processing procedures on pathogen load, adherence to regulatory performance standards or criteria, etc. Through these types of microbiological assessments, processors can more reliably identify and assess the microbial hazards in their products and processes, as well as validate their measures used for pathogen control. Environmental testing is used primarily to assess the effectiveness of sanitation programs and other facility-wide programs designed for microbial control.

Depending on the objective of the microbiological analysis, there are specific sampling plans and methods that can be used. Sampling is a key issue. Because it is not possible to analyze 100% of a given production lot of final product or the entire processing environment, samples that are representative of the entire lot must be obtained and analyzed. Sampling plans are based on statistical probability, and therefore provide confidence when interpreting results. Furthermore, the part of the sample that is actually subjected to the analytical procedure, referred to as an analytical unit, must also be representative of the whole sample. Whole carcass rinse samples are used for broilers, swab samples are collected for turkeys, and a sample of defined volume is collected for ground or portioned product. Both qualitative and quantitative methods are utilized in poultry processing to analyze samples. Qualitative methods provide a "yes/no" answer regarding the presence of specific bacterial types in the sample, while quantitative methods provide an estimate of the number of specific bacterial types. Again, defined procedures are to be followed when microbiologically evaluating product or environmental samples.

With the establishment of USDA-FSIS performance standards and criteria for poultry products, microbiological testing has taken on more importance in the processing of safe poultry products. At present, raw whole and ground products must meet *Salmonella* performance standards, carcasses must meet Biotype I *E. coli* criteria, and cooked poultry products must meet lethality, stabilization, and *L. monocytogenes* performance standards. *Salmonella* performance standards are used by USDA-FSIS as a means to determine the validity of a processor's HACCP plan. The *E. coli* criteria are used as an indicator of control of fecal contamination (a primary source of pathogenic bacteria) in slaughter operations. Lethality, stabilization, and *L. monoctyogenes* performance standards are used to establish that the processor's cooking, cooling, and post-process handling procedures are valid for producing a safe product.

To remain in compliance with current regulatory requirements and to ensure production of safe products, poultry processors must establish an ongoing microbiological testing program. The testing program should be integrated into the plant's normal operations so that trends can be detected such that preventive measures can be taken in a timely manner. The nature of the microbiological testing program will be a function of the processor's overall food safety objectives. These objectives should reflect the demands of the processor's customers as well as regulatory compliance.

Conclusion

Live poultry arriving at the processing plant harbor a heavy load of microorganisms. Most of these microorganisms are not harmful; however, poultry are known to harbor a number of bacteria that are pathogenic to humans. Typically, these occur in low levels, and only pose a threat to the consumer if the product is not handled in a safe manner. Regardless, it is the goal of the poultry processor to produce product with as low a level of pathogens as possible, which represents the acceptable level of safety based on product type. A comprehensive approach to food safety, which encompasses adherence to GMPs, HACCP, the use of specific antimicrobial treatments, and a microbiological testing program, is required to produce final products that are safe for the consuming public.

References

1. Anderson, M. E., Marshall, R. T., Stringer, W. C., and Naumann, H. D., Efficacies of three sanitizers under six conditions of application to surfaces of beef, *J. Food Sci.*, 42, 326, 1977.
2. Dickson, J. S. and Anderson, M. E., Microbiological decontamination of food animal carcasses by washing and sanitizing systems: a review, *J. Food Prot.*, 55, 133, 1992.
3. Todd, E. C., Foodborne disease in six countries—a comparison, *J. Food Prot.*, 41, 559, 1978.
4. Todd, E. C., Poultry associated foodborne disease—its occurrence, cost, sources and prevention, *J. Food Prot.*, 43, 129, 1980.
5. Bean, N. N. and Griffin, P. M., Foodborne disease outbreaks in the United States, 1973–1987: pathogens, vehicles, and trends, *J. Food Prot.*, 53, 804, 1990.
6. Bean, N. H., Griffin, P. M., Goulding, J. S., and Ivey, C. B., Foodborne disease outbreaks, 5-year summary, 1983–1987, *J. Food Prot.*, 53, 711, 1990.
7. Shackelford, A. D., Modification of processing methods to control *Salmonella* in poultry, *Poult. Sci.*, 67, 933, 1988.
8. United States Department of Agriculture-Food Safety and Inspection Service, 9 CFR Part 304 et al.: Pathogen Reduction; Hazard Analysis and Critical Control Point (HACCP) Systems; Final Rule, *Fed. Regis.*, 61 (no. 144), 38806, July 25, 1996.
9. Waldroup, A. L., Contamination of raw poultry with pathogen, *World's Poult. Sci.*, 52, 7, 1996.
10. Bryan, F. L. and Doyle, M. P., Health risks and consequences of *Salmonella* and *Campylobacter jejuni* in raw poultry, *J. Food Prot.*, 58, 326, 1995.
11. CDC (Centers for Disease Control), *Food Net: 1998 Preliminary Data*, 1999.
12. Benefield, R. D., *Pathogen Reduction Strategies for Elimination Foodborne Pathogens on Poultry During Processing*, M.S. thesis, Auburn University, Auburn, AL, 1997.
13. National Advisory Committee on Microbiological Criteria for Foods, Generic HACCP application in broiler slaughter and processing, *J. Food Prot.*, 60, 579, 1997.
14. Food Safety and Inspection Service, *Nationwide Broiler Chicken Microbiological Baseline Data Collection Program July 1994–June 1995*, USDA, Washington, DC, 1996.
15. Rigby, C. E., Pettit, J. R., Baker, M. F., Bentley, A. H. Salomons, M. O., and Lior, H., Flock infection and transport as sources of *Salmonella* in broiler chickens and carcasses, *Can. J. Comp. Med., 44, 328, 1980.*
16. Wakefield, C. B., *Control and Consequences of Salmonella Contamination of Broiler Litter and Livehaul Equipment*, M.S. thesis, Auburn University, Auburn, AL, 1999.
17. Rigby, C. E. and Pettit, J. R., Changes in the *Salmonella* status of broiler chickens subjected to simulated shipping conditions, *Can. J. Comp. Med.*, 44, 374, 1980.

18. Mulder, R. W. A. W., Dorresteijn, L. W. J., and van der Broek, J., Cross-contamination during the scalding and plucking of broilers, *Br. Poult. Sci.*, 19, 61, 1978.
19. Bailey, J. S., Thomson, J. E., and Cox, N. A., Contamination of poultry during processing, in *The Microbiology of Poultry Meat Products*, Cunningham, F.E. and Cox, N.A., Eds., Academic Press, Orlando, FL, 1987, 193.
20. Hargis, B. M., Caldwell, D. J., Brewer, R. L., Corrier, D. E., and DeLoach, J. R., Evaluation of the chicken crop as a source of Salmonella contamination for broiler carcasses, *Poult. Sci.*, 74, 1548, 1995.
21. Byrd, J. A., Corrier, D. E., Hume, M. E., Bailey, R. H., Stanker, L. H., and Hargis, B. M., Incidence of Campylobacter in crops of preharvest market-age broiler chickens. *Poult. Sci.*, 77, 1303, 1998.
22. May, K. N., Changes in microbial numbers during final washing and chilling of commercially slaughtered broilers, *Poult. Sci.*, 53, 1282, 1974.
23. Notermans, S., Terbijhe, R. J., and van Schotghorst, M., Removing faecal contamination of boilers by spray cleaning during evisceration, *Br. Poult. Sci.*, 21, 115, 1980.
24. Brant, A. W., Gable, J. W., Hamann, J. A., Wabeck, C. J., and Walters, R. E., *USDA Agriculture Handbook 581: Guidelines for Establishing and Operating Broiler Processing Plants*, Washington, DC, 1982.
25. Waldroup, A. L., Rathgeber, B. M., and Forsythe, R. H., Effects of six modifications on the incidence and levels of spoilage and pathogenic organisms on commercially process post-chill broilers, *J. Appl. Poult. Res.*, 1, 226, 1992.
26. Waldroup, A. L., Rathgeber, B. M., Hierholzer, R. E., Smoot, L., Martin, L. M., Bilgili, S. F., Fletcher, D. L., Chen, T. C., and Wabeck, C. J., Effects of reprocessing on microbiological quality of commercial prechill broiler carcasses, *J. Appl. Poult. Res.*, 2, 111, 1993.
27. Lillard, H. S., Effect of broiler carcasses and water of treating chiller water with chlorine or chlorine dioxide, *Poult. Sci.*, 59, 1761, 1980.
28. Dye, M. and Mead, G. C., The effect of chlorine on the viability of clostridial spores, *J. Food Technol.*, 7, 173, 1972.
29. Patterson, J. T., Bacterial flora of chicken carcasses treated with high concentrations of chlorine, *J. Appl. Bacteriol.*, 31, 544, 1968.
30. Sanchez, M., Brashears, M., and McKee, S., Microbial quality comparison of commercially processed air-chilled and immersion chilled broilers, *Poult. Sci.*, 78(Suppl. 1), 68, 1999.
31. Lillard, H. S., Bacterial cell characteristics and conditions influencing their adhesion to poultry skin, *J. Food Prot.*, 48, 803, 1985.
32. Thomas, C. J. and McMeekin, T. A., Attachment of *Salmonella* spp. to chicken muscle surfaces, *Appl. Environ. Microbiol.*, 42, 130, 1981.
33. Notermans, S. and Kampelmacher, E. H., Attachment of some bacterial strains to the skin of broiler chickens, *Br. Poult. Sci.*, 15, 573, 1974.
34. Conner, D. E. and Bilgili, S. F., Skin Attachment model (SAM) for improved laboratory evaluation of potential carcass disinfectants for their efficacy against *Salmonella* attached to broiler skin, *J. Food Prot.*, 57, 684, 1994.
35. Tamblyn, K. C. and Conner, D. E., Bactericidal activity of orgain acids in combination with transdermal compounds against *Salmonella typhimurium* attached to broiler skin, *Food Microbiol.*, 14, 477, 1997.
36. Food Safety and Inspection Service, *Appendix A: Compliance Guidelines for Meeting Lethality Performance Standards for Certain Meat and Poultry Products*, USDA, Washington, DC, 1999.
37. Food Safety and Inspection Service, *Appendix B: Compliance Guidelines for Cooling Heat-Treated Meat and Poultry Products (Stabilization)*, USDA, Washington, DC, 1999.
38. Food Safety and Inspection Service, *Listeria Guidelines for Industry*, USDA, Washington, DC, 1999.
39. Ryser, E. T. and Marth, E. H, *Listeria, Listeriosis and Food Safety*, Marcel Dekker, New York, 1999.
40. National Advisory Committee for Microbiological Criteria for Food, Hazard Analysis and Critical Control Point Principles and Application Guidelines. *J. Food Prot.*, 61, 762, 1998.
41. Stevenson, K. E. and Bernard, D. T., *HACCP: A Systematic Approach to Food Safety*, The Food Processors Institute, Washington, DC, 1999.

42. Pierson, M. D. and Corlett, D. A., *HACCP Principles and Applications*, Chapman & Hall, New York, 1992.
43. Izat A. L., Colbert, M., Adams, M. H., Reiber, M. A., and Waldroup, P. W., Production and processing studies to reduce the incidence of *Salmonella* on commercial broilers, *J. Food Prot.*, 52, 670, 1989.
44. Villarreal, M. E., Baker, R. C., and Regenstein, J. M., The incidence of *Salmonella* on poultry carcasses following the use of slow release chlorine dioxide (Alcide), *J. Food Prot.*, 53, 465, 1990.
45. Lillard, H. S., Effect of trisodium phosphate on Salmonellae attached to chicken skin, *J. Food Prot.*, 57, 465, 1994.
46. Somers, E. B., Schoeni, J. L., and Wong, A. C. L., Effect of trisodium phosphate on biofilm and planktonic cells of *Campylobacter jejuni, Escherichia*, O157: H7 *Listeria, monocytogenes* and *Salmonella typhimurium*, *Int. J. Food Microbiol.*, 22, 269, 1994.
47. Hollender R., Bender, F. G., Jenkins, R. K., and Black, C. L., Consumer evaluation of chicken treated with a trisodium phosphate application during processing, *Poult. Sci.*, 72, 755, 1993.
48. Cudjoe, K. S. and Kapperud, G., The effect of lactic acid sprays on *Campylobacter jejuni* inoculated on to poultry carcass, *Acta. Vet. Scand.*, 32(4), 491, 1991.
49. Rubin, H. E., Toxicological model for a two-acid system, *Appl. Environ. Microbiol.*, 36, 623, 1978.
50. Adams, M. R. and Hall, C. J., Growth inhibition of foodborne pathogens by lactic and acetic acids and their mixtures, *Int. Food Sci. Technol.*, 23, 287, 1988.
51. Conner, D. E. and Benefield, R. D., Antibacterial activity of organic acid-surfactant treatments against foodborne pathogens on processed broiler chickens: evaluation in a pilot scale commercial processing plant, in *Proc. XIV Eur. Symp. Quality Poultry Meat*, Vol. 1, Cavalchini, L G. and Baroli, D., Eds., Bologna, Italy, Sept. 19–23, 1999.
52. Sheldon, B. W. and Brown, A. L., Efficacy of ozone as a disinfectant for poultry carcasses and chill water, *J. Food Sci.*, 51, 305, 1986.
53. Yang Z. P., Li, Y. B., and Slavik, M., Use of antimicrobial spray applied with an inside-outside birdwasher to reduce bacterial contamination on prechilled chicken carcasses, *J. Food Prot.*, 61, 829, 1998.
54. Li, Y. B., Slavik, M. F., Waker, J. T., and Xiong, H., Pre-chill spray of chicken carcasses to reduce *Salmonella typhimurium*, *J. Food Sci.*, 62, 605, 1997.
55. Robach, M. C. and Ivey, F. J., Antimicrobial efficacy of potassium sorbate dip on freshly processed poultry, *J. Food Prot.*, 41, 284, 1978.
56. Lillard, H. S. and Thomson, J. E., Efficacy of hydrogen peroxide as a bactericide in poultry chiller water, *J. Food Sci.*, 48, 125, 1983.
57. Wabeck, C. J., Methods to reduce microorganisms on poultry, *Broiler Ind.*, 34, 1994.
58. Ayres, J. C., Mundt, J. O., and Sandine, W. E., *Microbiology of Foods*, W. H. Freeman, San Francisco, 1980.
59. Banwart, G. J., *Basic Food Microbiology*, Van Nostrand Reinhold, New York, 1989.
60. Giddings, G. G. and Marcotte, M., Poultry irradiation: for hygiene/safety and market-life enhancement, *Food Rev. Int.*, 7(3), 259, 1991.
61. Thayer, D. W., Use of irradiation to kill enteric pathogens on meat and poultry, *J. Food Safety*, 15, 181, 1995.
62. USDA, *FSIS Backgrounder: Poultry Irradiation and Preventing Foodborne Illness*, Food Safety Service, USDA, Washington, DC, 1992,
63. Bruhn, C. M., Schutz, H. G., and Sommer, R., Attitude change toward food irradiation among conventional and alternative consumers, *Food Technol.*, 40(12), 86, 1986.

chapter ten

Spoilage bacteria associated with poultry

Scott M. Russell

Contents

0-8493-0120-3/01/$0.00+$.50

Introduction

Within the U. S., most poultry products are produced in the Southeast; however, a large percentage of this poultry is consumed throughout the country. Thus, part of the shelf-life of these products is eliminated during transportation of the products to their final destination. Approximately 7 billion chickens or 30 billion lb. of meat are processed in the U. S. each year, of which, 80% are marketed as fresh product. It is estimated that 2 to 4% of this meat is lost as a result of spoilage, which is equivalent to a loss of approximately $300 to $600 million per year. Thus, spoilage is of great concern to the poultry industry.

The primary causes of spoilage are as follows: (1) prolonged distribution or storage time, (2) inappropriate storage temperature, and (3) high initial bacterial counts. If fresh poultry products are held long enough at refrigerator temperatures, they will spoil as a result of the growth of bacteria that are able to multiply under cold conditions. This situation may be improved by proper rotation of stock. Product that has remained on the shelf for the longest period of time should be sold first and product that is to be sold in locations far from the processor should be transported at temperatures that are near freezing (i.e., −3.3°C), but not sufficient to freeze the muscle tissue (deep chill).

Inappropriate storage temperatures or fluctuations in storage temperature are the most avoidable causes of spoilage. Temperature abuse can occur during distribution, storage, retail display, or handling of the product by the consumer. The only means by which processors can determine whether their product has been temperature abused is to monitor temperature or evaluate bacterial populations throughout the distribution system.

Initial bacterial counts on broiler carcasses may have a direct effect on the shelf-life of fresh product as well. The initial number of bacteria on poultry is generally a function of growout procedures, production practices, and plant and processing sanitation.

Growth temperature classification

The temperature at which fresh poultry is held is of great concern to the poultry industry because it is the most important factor that affects the growth of both spoilage and pathogenic bacteria. Olsen[1] reported that, when considering the relationship of temperature to microbial life, two things must be considered: the holding temperature of the microorganism and the time the microorganism is exposed to that temperature. All living cells respond to variations in temperature in various ways and bacteria, because they are living cells, are no exception. Bacterial metabolism, physical appearance, or morphology may be altered and proliferation may be stimulated or retarded, depending upon the particular combination of temperature and time of exposure. All bacteria are only able to multiply within a defined range of temperatures. Olsen[1] reported that within this range, there is a minimum growth temperature, below which growth ceases, an optimum growth-temperature, which is the most favorable for rapid growth, and a maximum growth temperature, above which

Table 10.1 Minimum, Optimum, and Maximum Growth Temperature Ranges for Pyschrophilic and Mesophilic Bacteria

	Maximum	Minimum	Optimum
Psychrophilic	−5 to 0°C	10 to 20°C	25 to 30°C
Mesophilic	10 to 25°C	20 to 40°C	40 to 45°C

Source: Adapted from Olson, J. C., Jr., Psychrophiles, mesophiles, thermophiles, and thermodurics—What are we talking about?, *Milk Plant Mon.*, 36, 32, 1947. With permission.

growth ceases. Different species of bacteria may vary not only with regard to the temperature range within which they are able to multiply, but also in their minimum, optimum, and maximum growth temperatures.[1] The two criteria that are used to determine optimum growth conditions for a bacterial species are generation time and maximum cell population.[2] Generation time indicates the speed of cell division; whereas, maximum cell population takes into account cell death as well as cell production.

The minimum, optimum, and maximum growth temperatures for psychrotrophic and mesophilic bacteria are listed in Table 10.1. Olsen[1] placed bacteria, now considered psychrotrophic, in the psychrophilic category. Muller,[3] Zobell and Conn,[4] and Ingraham[5] objected to the term "psychrophiles" because, while many bacteria responsible for spoilage are able to survive and multiply at low temperatures, the temperatures that are optimal for growth are well above freezing. Ayres et al.[6] reported that the optimum temperature for replication of psychrophilic bacteria is between 5 and 15°C. Much earlier, Muller[3] reported that the psychrotrophic bacteria are a group of mesophiles that are able to multiply relatively slowly at a lower temperature range than most other bacteria.

A more current perspective on bacterial groupings based on growth temperatures is presented in Table 10.2. Many species of bacteria cannot be placed into any single category because their temperature range is very broad.[6] Some species of bacteria, such as *Listeria monocytogenes*, are able to grow well at both refrigerator temperatures and warm temperatures. However, these bacteria represent the exception, rather than the rule, when considering separation of bacteria based on their minimum, optimum, and maximum growth temperatures.

Factors affecting shelf-life of fresh poultry

Holding temperature

By far, the most important factor affecting psychrotrophic bacterial growth and hence, the shelf-life of fresh poultry is holding temperature. Pooni and Mead[7] reported that poultry

Table 10.2 Minimum, Optimum, and Maximum Growth Temperatures (°C)

	Minimum	Optimum	Maximum
Psychrophiles	≤0	5 to 15	±20
Low temperature mesophiles, psychrotrophic, psychroduric microorganisms	±10 to +8	20 to 27	32 to 43
Non-fastidious high temperature mesophiles	±8	35 to 43	43 to 45
Fastidious high temperature mesophiles	20 to 25	37	?

Source: Adapted from Ayres, J. C., Mundt, J. O., and Sandine, W. E., *Microbiology of Foods*, W. H. Freeman, San Francisco, CA, 1980, 55. With permission.

products may be subjected to variations in holding temperature during processing, storage, distribution, and retail sale. Ayres et al.[8] evaluated the effect of storage temperature on the shelf-life of fresh poultry. The authors reported that the average shelf-life for commercially eviscerated fresh cut-up carcasses, was 2–3, 6–8, and 15–18 days, respectively, when held at storage temperatures of 10.6, 4.4, and 0°C. Barnes[9] demonstrated that turkey carcasses that were stored at −2, 0, 2, and 50°C, developed off-odors in 38, 22.6, 13.9, and 7.2 days, respectively. Daud et al.[10] reported that broiler carcasses maintained under optimal conditions should have a shelf-life of 7 days when stored at 5°C. The rate of spoilage is twice as fast at 10°C and three times as fast at 15°C, than for carcasses stored at 5°C.[10] Hence, as storage temperatures were reduced, the shelf-life of carcasses in these studies was extended.

Moreover, Baker et al.[11] reported that the temperature and time of storage are related to shelf-life because increases in aerobic bacterial counts on ready-to-cook broiler carcasses, stored more than 7 days at 1.7 and 7.2°C, were much greater than increases in bacterial counts for corresponding carcasses stored for shorter periods of time. This study confirmed that carcasses will eventually spoil if held long enough, even if held under appropriate refrigeration, and that significantly longer shelf-life is obtained by holding carcasses at temperatures as low as possible.

Storage on ice

Studies conducted on the microbiological effects of storing chicken on ice have been somewhat conflicting. Lockhead and Landerkin[12] observed that chicken carcasses that are suspended in a refrigerator at −1.1°C do not develop spoilage odors as soon as chickens held at the same temperature surrounded by ice or ice water. In contrast to these results, Naden and Jackson[13] reported that there are significant advantages to packing poultry on ice including: (1) fresh quality is maintained longer, (2) drying out is prevented, and (3) the carcasses are more attractive in the display case. Baker et al.[11] determined that bacterial counts on ready-to-cook poultry, stored on ice for 9 days, were similar to those stored under refrigeration for 5 days at 1.7°C or 4 days at 7.2°C, indicating that storage on ice is more effective to extend shelf-life. However, others have observed that carcasses stored in crushed ice had the same shelf-life as those stored in mechanical refrigerators at −0.6°C.[14] It is interesting to note that among four separate studies, all three possible conclusions were reached (i.e., refrigeration is best, ice is best, and no difference). This may be due to the fact that different investigators used different parameters to judge spoilage, such as odor or slime production.

Evisceration

Although most poultry in the U.S. is purchased fully eviscerated or as cut-up parts, another factor purported to affect the shelf-life of fresh poultry is whether or not the carcass has been eviscerated. Lockhead and Landerkin[12] determined that eviscerated chicken carcasses developed spoilage odors sooner than New York-dressed (uneviscerated) chickens held under similar conditions. Others reported that bacterial counts on ready-to-cook poultry were much higher than on New York-dressed poultry, after four days of storage in ice at 1.7 and 7.2°C.[11] These authors attributed the increase in spoilage rate and spoilage bacteria on fully eviscerated poultry to the fact that the abdominal region of the carcass is open to contamination and the water used for washing these carcasses may be a means of spreading spoilage bacteria.[11] These results may be of interest in countries that still market New York-dressed carcasses.

Initial bacterial load

Initial bacterial load immediately after processing has also been shown to affect shelf-life. Brown[15] demonstrated that an increase in the initial bacterial load results in a concomitant dramatic decrease in shelf-life. This effect is due to the fact that much less time is required for bacterial populations to reach numbers that are high enough to produce spoilage defects when bacteria are high in number initially.

Breast meat color affects spoilage rate

A study by Allen et al.[16] demonstrated that there is a relationship between the color (as determined using C. I. E. L*a*b* measurements) of chicken breast meat fillets, the meat pH, and shelf-life of the fillets. Darker breast fillets were found to have significantly higher pH (6.08 to 6.22 for dark fillets vs. 5.76 to 5.86 for lighter fillets). Darker fillets also had significantly higher psychrotrophic plate counts and much higher subjective odor scores than the lighter breast fillets at day 7 of storage. The authors concluded that darker broiler breast meat fillets have a shorter shelf-life than lighter breast fillets, and the shorter shelf-life may be due to differences in pH.

Other factors

Spencer et al.[14] identified a number of factors that may affect shelf-life and reported that the scalder water temperature and chlorination of the chiller water were important. Under simulated commercial conditions, carcass halves scalded at 53.3°C had an average shelf-life of 1 day longer than carcass halves scalded at 60°C (both scalded for 40 s). Carcass halves scalded at 53.3°C and cooled for 2 h in ice water containing 10 ppm of residual chlorine had a shelf-life of 15.2 days, as compared to 12.8 days for control halves chilled with non-chlorinated water.

Effect of storage temperature on generation times of bacteria found on broiler carcasses

Cold storage temperatures

When fresh poultry is placed in a cold environment, conditions for replication of most species of bacteria are no longer optimal. Ayres et al.[8] reported that, the total number of bacteria on poultry stored at 0°C decreased during the first few days of storage. These authors reported that this decrease was due to the following: (1) the unsuitability of the temperature for reproduction and survival of chromogenic bacteria (pigment producers) and mesophilic bacteria, and (2) insufficient time for psychrotrophic bacteria to begin the exponential phase of growth.

Psychrotrophic bacteria are able to grow at refrigeration temperatures and spoil foods; however, the rate at which these bacteria multiply is greatly reduced. Most species of mesophilic bacteria are unable to multiply at refrigeration temperatures below 5°C.[17] Olsen and Jezeski[18] reported that generation times for mesophiles and psychrotrophs do not increase proportionally when incubation temperatures are lowered progressively from their optimum growth temperature ranges. As the lower temperature limit for *E. coli* (a mesophile commonly found on broiler carcasses) replication is approached, not only is the doubling time of the bacterium much slower, but there also is a longer lag period before it begins to multiply.[9] Barnes[9] observed that the generation time of *E.coli* at −2, 1, 5, 10, 15, 20,

25, and 30°C was 0, 0, 0, 20, 6, 2.8, 1.4, and 0.6 h, respectively. Elliott and Michener[19] reported that when mesophilic bacteria are placed at storage temperatures below 0°C, the generation time may exceed 100 h.

Elevated storage temperatures

At temperatures that are considered "mild temperature abuse" (around 10°C) the generation times of psychrotrophic bacteria are much shorter than mesophilic populations of bacteria.[20] At approximately 18°C, however, the multiplication rate of psychrotrophic and mesophilic populations of bacteria is approximately equal. When storage temperatures exceed 18°C, mesophiles proliferate much more rapidly than psychrotrophic bacteria.[20]

Bacteria involved in spoilage of poultry

Identification of bacteria responsible for spoilage of fresh chicken and other muscle foods dates back to the late 19th century. Forster[21] reported that most foods are exposed to saprophytic spoilage bacteria that are found in the air, soil, and water. The author mentioned that, when cold storage was to be used as a means of preserving foods, it was important to be able to predict the behavior of these saprophytes over a given range of temperature.[21] Glage (as reported by Ayres[22]) was one of the first researchers to isolate spoilage bacteria from the surfaces of meat that had been stored at low temperature and high humidity. This author named these bacteria *Aromobakterien.* Glage observed a total of seven species of spoilage bacteria, one of which predominated. The author reported that these bacteria were oval to rod shaped with rounded ends and that they occurred occasionally in chains. Glage (as reported by Ayres[22]) revealed that these *Aromobakterien* grew well at 2°C, but very slowly at 37°C, and their optimum growth temperature was 10 to 12°C.

In 1933, Haines[23] determined that Glage's *Aromobakterien* were similar to isolates that produced slime on meat stored at refrigeration temperatures. Haines reported that, except for some members of the *Pseudomonas* group and a few *Proteus,* microorganisms found on lean meat stored at 0–4°C mostly belong to the *Achromobacter* group. Others observed that 95% of the bacteria found on fresh beef, immediately after processing that were capable of growth at 1°C, were *Achromobacter* and some *Pseudomonas* and *Micrococcus.*[24] The authors found that during cold storage, populations of *Achromobacter* and *Pseudomonas* were able to increase, while populations of *Micrococcus* significantly decreased.

Various studies by Haines,[25] Empey and Scott,[26] and Lockhead and Landerkin[12] indicated that species of *Achromobacter* were the predominant spoilage bacteria of fresh meat and poultry. However, Ayres et al.,[8] Kirsch et al.,[27] and Wolin et al.[28] (1957) conducted studies which contradicted these earlier studies. These authors reported that species of *Pseudomonas* were more predominant than *Achromobacter.* These three groups of researchers attributed the discrepancy between their results and those of previous workers to changes in nomenclature used in the sixth edition of *Bergey's Manual of Determinative Bacteriology*[29] from that adopted in the third edition[30] that may have been used by Haines, Empey and Scott, and Lockhead and Landerkin.

In 1958, Brown and Weidemann[31] reassessed the taxonomy of the 129 psychrotrophic meat spoilage bacteria that had been isolated by Empey and Scott[26] and the authors concluded that most of these bacterial species were pseudomonads. Empey and Scott[26] previously classified meat spoilage bacteria as *Pseudomonas* largely on the basis of the production of a water soluble green pigment. Brown and Weidemann[31] determined that 21 of the strains that were originally classified as pseudomonads on the basis of pigment production, failed to produce any type of pigment. Ayres et al.,[8] using *Bergey's Manual 6th ed.*[29]

as their taxonomic guide, reported that bacterial isolates collected from spoiled, slimy carcasses were closely related to these species of *Pseudomonas: ochracea, geniculata, mephitica, putrefaciens, sinuosa, segnis, fragi, multistriata, pellucida, rathonis, desmolytica (um)* or *pictorum.* These authors revealed that, because of changes in Bergey's Manual between the 3rd[30] and 6th ed.,[29] many bacterial species that were originally reported as belonging to the *Achromobacter* genus should be reclassified as members of the genus *Pseudomonas,* because they move by means of polar flagellation.[8] Kirsch et al.[27] in separate studies achieved the same results.

In 1950, Ayres et al.[8] observed that *P. putrefaciens,* which is a common spoilage bacterium found on meat and poultry, has both lateral and polar flagella. The authors argued that this bacterium should not be placed into the genus *Pseudomonas. P. putrefaciens* is characterized by brownish colonies and is further differentiated from other pseudomonads by its highly proteolytic properties and hydrogen sulfide production.

Later, Halleck et al.[32] determined that non-pigmented *Achromobacter-Pseudomonas* type bacteria made up approximately 85% of the total bacterial populations on fresh meats during the first two weeks of storage at 1.1 to 3.3°C and during the first week of storage on meat samples held at 4.4 to 6.7°C. These authors reported that *Pseudomonas fluorescens* constituted approximately 80% of the bacterial species on meat toward the end of the storage period; however, at the beginning of the storage period, *P. fluorescens* seldom exceeded 5% of the population on fresh meats.[32]

Barnes and Impey[33] observed that the three genera of bacteria that were most commonly isolated and identified from spoiled chicken were *Pseudomonas, Acinetobacter,* and *P. putrefaciens.* The predominant pseudomonads on spoiled poultry are divided into two related categories: fluorescent or pigmented strains and non-pigmented strains.

Since the time when Barnes and Impey[33] reported that *P. putrefaciens* was determined to be a primary spoilage bacterium of fresh poultry, this bacterium has been reclassified. *P. putrefaciens* was originally classified as *Alteromonas putrefaciens.*[34] It was then changed from *Alteromonas* to *Achromobacter. Achromobacter* was transferred to the genus *Pseudomonas* in the 7th edition of Bergey's Manual.[35] MacDonell and Colwell[34] placed *P. putrefaciens* into a new genus and named it *Shewanella putrefaciens.* Thornley[36] mentioned that *Acinetobacter* was also part of the genus *Achromobacter* until the mid-1960s.

More recently, Russell et al.[37] conducted a study to identify the bacteria responsible for the production of off-odors on spoiled broiler chicken carcasses, to characterize the odors they produce, and to survey carcasses produced in different areas of the U. S to determine how consistently these spoilage organisms were found. The authors reported that the bacteria isolated from spoiled carcasses that consistently produced off-odors in chicken skin medium, regardless of the geographical location from which the chickens were obtained, were *S. putrefaciens* A, B, and D, and *Pseudomonas* (*fluorescens* A, B, and D, and *P. fragi*). These bacteria produced off-odors which resembled "sulfur," "dishrag," "ammonia," "wet dog," "skunk," "dirty socks," "rancid fish," "unspecified bad odor," or a sweet smell resembling "canned corn." However, odors produced by the spoilage bacteria were varied. Odors most associated with spoiled poultry, such as "dishraggy" or "sulfurous" odors, were produced by the bacteria that were most consistently isolated, such as *S. putrefaciens* and the pseudomonads.[37]

The origin of psychrotrophic spoilage bacteria on broiler carcasses

Psychrotrophic populations of bacteria that are found on the carcass immediately after processing originate from the feathers and feet of the live bird, the water supply in the processing plant, the chill tanks, and processing equipment.[38] These spoilage bacteria are not usually found in the intestines of the live bird. Schefferle[39] found high populations of *Acinetobacter* (10^8cfu/g) on the feathers of the bird and suggested that they may originate from the deep litter. Other psychrotrophic genera of bacteria, such as *Cytophaga* and *Flavobacterium,* are often found in chill tanks but are rarely found on carcasses.[33] The psychrotrophic species of bacteria on chicken carcasses immediately after slaughter are generally *Acinetobacter* and pigmented pseudomonads.[33] Although strains of non-pigmented *Pseudomonas* produce off-odors and off-flavors on spoiled poultry, initially, they are difficult to find on carcasses and *P. putrefaciens* (*S. putrefaciens*) is rarely found.[33]

Psychrotrophic bacteria are able to survive on processing equipment surfaces and on the floor of the processing facility because of the amount of moisture and food residue available to them. In addition, the cold temperatures in the processing plant are of little assistance in inhibiting the growth of these bacteria. Hence, proper cleaning and sanitation of processing equipment and floors is essential to reduce contamination of fresh product by populations of psychrotrophic bacteria that may become residual in the plant. Perhaps the most common culprit when investigating plants that have experienced reduced shelf-life of fresh meat and poultry products is that the cleaning crew did not clean and sanitize the equipment surfaces and floors of the plant appropriately. Proper use of high pressure, hot water, and appropriately diluted sanitizer that is effective against psychrotrophs is essential for maintaining adequate shelf-life.

Identification of spoilage flora on broilers held at elevated temperatures

Bacterial genera that are responsible for off-odors and slime on spoiled chicken are not nearly as prevalent if storage temperature is increased. Populations of spoilage bacteria on chickens held at various temperatures, as reported by Barnes and Thornley,[40] are listed in Table 10.3. Immediately after processing, the predominant bacterial species on broiler carcasses are mesophilic, such as micrococci, Gram positive rods, and flavobacteria. However,

Table 10.3 **The Spoilage Flora of Eviscerated Chickens Initially and After Storage at 1, 10, and 15°C Until Spoiled**

	Number of strains			
	Initial	**1°C**	**10°C**	**15°C**
Total strains	58	40	80	69
Gram positive rods	14	0	4	6
Enterobacteriaceae (lactose pos.)	8	0	3	10
Enterobacteriaceae (lactose neg.)	0	3	12	17
Micrococci	50	0	4	0
Streptococci	0	0	6	8
Flavobacteria	14	0	0	0
Aeromonas	0	0	4	6
Acinetobacter	7	7	26	34
Pigmented *Pseudomonas*	2	51	21	9
Non-pigmented *Pseudomonas*	0	20	12	2
Pseudomonas putrefaciens	0	19	4	4
Unidentified	5	0	4	4

Source: Adapted from Barnes, E. M. and Thornley, M. J., The spoilage flora of eviscerated chickens stored at different temperatures, *J. Food Technol.*, 1, 113, 1966. With permission.

if carcasses are held at temperature abuse temperatures, such as 10°C, *Acinetobacter*, pseudomonads, and Enterobacteriaceae are able to multiply. For carcasses held at 15°C, *Acinetobacter* and Enterobacteriaceae, whose optimum growth temperatures are higher than those of the pseudomonads, predominate.[40]

Number of bacteria needed for spoilage

High numbers (10^5 cfu/cm^2) of psychrotrophic spoilage bacteria are required on poultry surfaces before off-flavors, off-odors, and appearance defects are able to be detected organoleptically. Lockhead and Landerkin[12] were not able to detect off-odors caused by bacteria on uneviscerated broiler carcasses until bacterial concentrations reached 2.5×10^6 to 1×10^8 cfu/cm^2. Other researchers[8] observed that odor and slime were not present until bacteria exceeded 1×10^8 cfu/cm^2. Elliott and Michener[41] were able to detect odor when bacterial concentrations reached 1.6×10^5 cfu/cm^2. The authors reported that higher numbers of bacteria (3.2×10^7 to 1×10^9 cfu/cm^2) were required to produce slime.

In another study on sliced beef, Kraft and Ayres[42] found that off-odor was able to be detected when bacterial concentrations on the surface reached 2×10^6 cfu/cm^2. The authors reported that incipient spoilage was indicated by the onset of off-odor at 10^6 cfu/cm^2; however, off-odors were more easily recognized when bacterial counts on the surface of meat reached 10^7 cfu/cm^2. More recently, Dainty and Mackey[43] determined that proteolysis and slime production under aerobic conditions begins when bacterial numbers reach 10^7 to 10^8 cfu/g.

Causes of spoilage defects

Spoilage is caused by the accumulation of metabolic byproducts or the action of extracellular enzymes produced by psychrotrophic bacteria as they multiply on poultry surfaces at refrigeration temperatures. Some of these byproducts become detectable as off-odors and slime as bacteria utilize nutrients on the surface of meats. The metabolic byproducts from spoilage bacteria vary depending on the energy source available to them. When populations are low, the bacterial cells utilize glucose as their primary source of energy. The byproducts of glucose metabolism are not usually odorous and do not substantially contribute to spoilage defects. However, as bacterial populations increase and glucose availability begins to decrease, these bacteria begin utilizing other substrates, such as protein, which yields much more odorous end products.[7] Others have also reported that proteolysis of the skin and muscle tissue begins when concentrations of glucose and/or gluconate have been exhausted.[44, 45] Pooni and Mead[7] determined that initial off-odors do not result from breakdown of the protein in skin and muscle, as previously thought, but from the direct microbial utilization of low molecular weight nitrogenous compounds such as amino acids, which are present in skin and muscle.

Venugopal[46] reported that bacteria growing on the surface of muscle (meat or fish) secrete a variety of extracellular enzymes that degrade the muscle tissue, causing extensive damage. Tarrant et al.[47] and Porzio and Pearson,[48] using SDS-PAGE techniques, demonstrated that these enzymes were able to degrade myofibrillar proteins extensively into heavy meromyosin, light meromyosin, and meromysin. Schmitt and Schmidt-Lorenz[49] demonstrated that there was an increase of low molecular weight peptides (less than 50,000 Da) and free amino acids on chicken carcasses stored at 4°C.

Recently, Nychas and Tassou[50] monitored the progression of spoilage as it relates to depletion of substrates in the muscle tissue. The concentration of glucose decreased progressively and more rapidly toward the end of the storage period and occurred to a greater degree for samples held at higher temperatures. Similar observations were made for concentrations of L-lactate. Concentrations of free amino acids increased as proteolysis occurred throughout the storage period. De Castro et al.[51] demonstrated that measurement of these free amino acids, due to the production of aminopeptidases and subsequent breakdown of protein, may be used to rapidly determine the bacteriological quality of beef.

Physical development of off-odor and slime formation

Spoilage defects have been the subject of researchers for many years. Glage (as reported by Ayres[22]) found that spoilage bacteria initially produce a gray coating on the surface of meat which later turns yellow in color. As these bacteria multiply, an aromatic odor accompanies their growth. Eventually, meat surfaces become coated with tiny drop-like colonies which increase in size and coalesce to form a slimy coating. Microorganisms appear first in damp pockets on the carcass, such as folds between the foreleg and breast of a carcass, and their dispersion is promoted by condensation which occurs when a cold carcass is exposed to warm, damp air.[22]

Ayres et al.[8] identified an ester-like odor which was described as a "dirty dishrag" odor that developed on cut-up chickens. In most cases, off-odor preceded slime formation and was considered the initial sign of spoilage. Immediately after off-odors were detected, many small, translucent, moist colonies appeared on the cut surfaces and skin of the carcass. Initially, these bacterial colonies appeared similar to droplets of moisture; however, they eventually became large, white or creamy in color, and often coalesced to form a uniform sticky or slimy layer. In the final stages of spoilage, the meat began to exhibit a pungent ammoniacal odor in addition to the dirty dishrag odor,[8] which may be attributed to the breakdown of protein and the formation of ammonia or ammonia-like compounds.

Slime production has also been attributed to proteolytic activity of bacteria growing on the surface of meat and poultry. Various authors have reported that degradation of meat by pseudomonads results in the formation of slime.[44, 49, 52, 53]

Metabolic adaptation of spoilage bacteria to refrigeration temperatures

Under refrigeration (<5°C), psychrotrophic bacterial populations are able to multiply on broiler carcasses and produce spoilage defects; however, the mesophilic bacteria that initially predominate on the carcass remain the same or decrease in number.[40, 17] This phenomenon may be explained by examining some of the metabolic changes that occur in these groups of bacteria as they are exposed to refrigerator temperatures.

Effect of cold storage on cellular lipids

Specific species of bacteria cease to proliferate at a particular temperature as the temperature of their environment is lowered because, as environmental temperature decreases, so does the absorption of nutrients by bacterial cells.[54] Moreover, as environmental temperature decreases, bacteria begin to increase the content of lipids in their cell membranes. Graughran (as reported by Wells et al.[54]) observed that as primarily mesophilic species of bacteria are exposed to progressively lower temperatures, the quantity of cellular lipids increases and the degree to which these lipids are saturated increases. As lipids in the cell membrane increase, the absorption of nutrients is inhibited. Eklund[55] reported that

Brevibacterium linens contained 7.2% fat when incubated at 25°C, where it grew well; however, at 4°C, it produced 16.7% fat and multiplied poorly. In addition, the author determined that bacterial cells produce more fat at 4°C than at 9.4 or 22°C.[55] Interestingly, two typical psychrotrophic bacteria species exhibited no such temperature-induced differences when grown at 4°C.[54]

Effect of cold storage on lipase production

Research has demonstrated that the amount of lipase produced by psychrotrophic bacteria increases as a result of exposure of the bacteria to cold temperatures. Nashif and Nelson[56] observed that lipase production by *P. fragi* was high when bacteria were incubated at temperatures between 8 and 15°C; however, production of lipase was almost completely absent at temperatures of 30°C or higher. Other researchers reported that lipase production by *P. fluorescens* was the same when this organism was cultured at 5 or 20°C; however, very little lipase was produced when the bacteria were exposed to 30°C.[57]

Effect of cold storage on proteolytic activity

Changes in proteolytic activity of bacteria at low temperatures have also been studied. Peterson and Gunderson[58] reported that production of proteolytic enzymes by *P. fluorescens* was higher when this bacterium was cultured at lower temperatures. Moreover, De Castro et al.[51] reported that by measuring the production of free amino acids as a result of the production of aminopeptidases and subsequent breakdown of protein, the progression of spoilage on meat surfaces may be evaluated and the quality of the meat may be determined.

Effect of cold storage on carbohydrate metabolism

Mesophilic species of bacteria decrease utilization of carbohydrates as environmental temperature is reduced, while psychrotrophic species are able to continue to utilize carbohydrates as an energy source. Brown,[15] Ingraham and Bailey,[59] and Sultzer[60] reported that at reduced incubation temperatures, carbohydrate oxidation rates of psychrotrophic bacteria decrease to a lesser degree than oxidation rates of mesophilic bacteria. Temperature coefficient differences between mesophiles and psychrotrophs have been determined for the following catabolic processes: glucose oxidation, acetate oxidation, and formate oxidation by resting cells.[59] Maintenance of a high rate of carbohydrate metabolism when psychrotrophs are exposed to low temperatures may be one explanation for their ability to maintain their metabolic processes under adverse temperature conditions.

Bacterial "conditioning"

Culturing psychrotrophic bacteria under refrigeration has been demonstrated to increase their ability to grow at cold temperatures. Hess[61] found that culturing psychrotrophs (*P. fluorescens*) at 5°C produced strains that were more active at 0 and −3°C than other strains of *P. fluorescens* that had been incubated at 20°C. Chistyakov and Noskova[62] were able to successfully adapt a variety of bacterial strains to environmental temperatures as low as −2°C by growing them at 0 to −8°C for 2 years. Ingraham and Bailey[59] and Wells et al.[54] suggested that this "process of adaptation" may be the result of cellular reorganization. This "adaptation" is also important for understanding how bacteria react to very low temperatures, such as freezing.

Effect of freezing on species of psychrotrophic bacteria

MacFadyen and Rowland[63] summarized the unique ability of bacteria to survive freezing and thawing by stating the following:

> "It is difficult to form a conception of living matter under this new condition, which is neither life nor death, or to select a term which will accurately describe it. It is a new and hitherto unobtained state of living matter—a veritable condition of suspended animation."

The effects of freezing on the ability of bacteria to survive and reproduce have been studied as far back as the late 19th century. Burden-Sanderson[64] reported that all bacteria are not destroyed by freezing. Bacteria isolated from fish were found to be able to multiply at freezing temperatures, such as 0°C.[21] Fischer[65] isolated 14 different bacterial species that were able to proliferate at 0°C. Another researcher isolated 36 different bacterial species that multiplied at 0°C from sausage and fish intestines.[3] Microorganisms that are capable of multiplying at 0°C are widely distributed; however, their growth characteristics are described as being similar at 0°C as at higher temperatures; however, the rate of growth is decreased.[3] Bedford[66] determined that strains of *Achromobacter* were able to proliferate at temperatures as low as −7.5°C. Others revealed that −10°C is the lowest temperature at which bacteria are able to multiply.[67]

Survival of bacteria during storage

Berry and Magoon[67] observed that under specific conditions, moderately cold storage temperatures (−2 to −4°C) may negatively impact bacteria to a greater degree than storage at −20°C. This may be explained by the fact that when cells are frozen rapidly, both intra- and extracellular fluid freezes. However, when cells are frozen at a slow rate, an intra- and extracellular osmotic gradient occurs due to freeze concentration. This may result in cellular disruption.[68] After freezing various species of bacteria at −190°C for 6 months, MacFadyen and Rowland[36] found no difference in the vitality of the microorganisms. The normal functions of life cease at temperatures as low as −190°C. The authors hypothesized that intracellular metabolism must also cease as a result of withdrawal of heat and moisture.[36]

Although bacterial species are in "suspended animation" when frozen, a fraction of the microbial population is killed or sublethally injured during the freezing process.[69] During frozen storage, the individual cells that survive on meat surfaces can range from 1 to 100%, but averages 50%, depending on the type of food.[69] Straka and Stokes[70] observed that some nutrients that are required by bacteria for growth are rendered inaccessible by the freezing process, thereby preventing bacterial multiplication.

Studies have demonstrated that freezing and thawing may enhance the growth rate of surviving bacteria. Hartsell[71] reported that *E. coli* that are able to survive the freezing and thawing process were able to grow more rapidly than *E. coli* that had not been previously frozen. One reason why the growth of bacteria that have survived freezing may be accelerated is that tissue damage due to freezing may result in nutrient release and increased moisture, such that the tissue becomes a better growth medium.[72]

Because the rate of bacterial growth on meat surfaces is reduced by freezing, it would seem that freezing chicken would be an acceptable means of increasing its shelf-life. However, consumers presently have an aversion to buying frozen poultry. This may be attributed to the fact that in the early to mid-1900s, poultry was often held until it was about to spoil and was then frozen. People who purchased frozen poultry products found that when the product was thawed, it was of inferior quality and spoiled rapidly. Pennington[73] expressed concern about this when she remarked, "frozen poultry has too frequently been synonymous with carcasses held until they are just about spoiled and then frozen. Hence the consumer gets a low grade product and the reputation of the poultry business suffers." Thus, frozen poultry has not been widely accepted in the U. S.

Effects of freezing on shelf-life

The effect of freezing on the shelf-life of fresh poultry has been extensively studied. Spencer et al.[74] reported that carcasses that are frozen and held for two months and then thawed, had the same shelf-life as unfrozen controls. Similar observations were made by Spencer et al.[75] and Newell et al.[76] who observed no major increases or decreases in shelf-life of chicken carcasses as a result of freezing and thawing. Elliot and Straka[77] reported that chicken meat, which was frozen for 168 days at −18°C and subsequently thawed, spoiled at the same rate as unfrozen controls.

Eliminating psychrotrophic spoilage bacteria from poultry

Previous research has demonstrated that species of spoilage bacteria may be resistant to commonly used commercial sanitizing chemicals. Stone and Zottola[78] determined that *P. fragi* that were firmly attached to stainless steel were able to survive the combined effects of cleaning with a 2500 ppm alkali detergent for 7 min, cleaning with 500 ppm of an acidic detergent for 3 min, and then sanitizing with 100 ppm sodium hypochlorite for 3 min. Wirtanen and Mattila-Sandholm[79] demonstrated that sodium hypochlorite at a level of 0.1% was able to decrease *P. fluorescens* counts at 24, 48, 72, and 144 h in meat soup; however, 0.1% sodium hypochlorite increased *P. fluorescens* counts in milk after 48 h.

Hingst et al.[80] showed that *Pseudomonas putida* and *P. fluorescens* were highly resistant to quaternary ammonium compounds. *Pseudomonas aerugenosa* was not inhibited by 50 or 200 ppm of a basic quaternary ammonium compound at pHs ranging from 7.29 to 8.80, after 300 s of exposure.[81] Ouattara et al.[82] found that growth of *P. fluorescens* was inhibited in acetic acid at 0.1 and 0.2%, propionic acid at 0.1 and 0.2%, lactic acid at 0.3%, and citric acid at 0.2 and 0.3% for a period of 24 h, after which, the bacteria were able to multiply in the presence of the sanitizer. *P. fluorescens* was not inhibited by lactic acid at concentrations below 0.3% or by citric acid at 0.1%. Mountney and O'Malley[83] found that lactic acid at a concentration of 0.275% was least effective among the organic acids as a means of extending shelf-life of fresh poultry. In addition, longer exposure to 1,3-dichloro-2,2,5,5-tetramethylimidazolidin-4-one and 3-chloro-4,4-dimethyl-2-oxazolidinone sanitizer[84] was required to inactivate *P. fluorescens* than was required to inactivate *Salmonella enteritidis* or *Salmonella typhimurium*.

Most studies have focused on commonly used chemical sanitizers; however, there are some novel sanitizers that have gained popularity. Currently, there is little data regarding the effect of quaternary ammonium, trisodium phosphate (TSP), hydrogen peroxide, or Timsen on specific spoilage bacteria associated with poultry. A study by Hwang and Beuchat[85] demonstrated that psychrotrophs were significantly reduced by washing chicken skin with 1% TSP or 1% lactic acid. Recently, Russell[86] conducted a study to determine the effect of commercial and novel (not approved for use in poultry processing) sanitizers on bacteria associated with spoiled poultry. The author found that none of the *Pseudomonas* spp. were able to multiply after exposure to very low concentrations of Timsen (10 ppm), except for *P. putida*, which was significantly inhibited. *S. putrefaciens*, although significantly inhibited, was able to proliferate when exposed to as high as 100 ppm Timsen. Russell[86] concluded that, overall, Timsen appears to be much more effective at low concentrations than commercial quaternary ammonium compounds for eliminating the growth of spoilage bacteria.

Detecting populations of spoilage bacteria on poultry products

Traditional microbiological methods for enumeration of psychrotrophic spoilage bacteria

Elliott and Michener[41] stated that total bacterial populations are often used as a means of evaluating sanitation, adequacy of refrigeration practices, or to determine the rate at which poultry is distributed to consumers. To determine which of these factors are responsible for a excessive bacterial counts is impossible with only a total count on the product as a guide.[41] Other researchers emphasized the importance of incubating Petri plates at or near the temperature at which the product spoils when conducting counts, in order to obtain a true indication of bacteriological changes during spoilage and to enumerate the bacteria responsible for spoilage.[77] To enumerate psychrotrophic populations of bacteria, the incubation temperature must be low enough to preclude the multiplication of mesophilic species that may be present on the meat surface. Ayres[87] reported that, at temperatures of 0 and 4.4°C, many mesophilic species of bacteria are not able to proliferate. Senyk et al.[88] found that, in raw milk samples held at 1.7, 4.4, 7.2, and 10.0°C, mesophilic populations of bacteria increased after 48 h by 0.12, 0.13, 0.40, and 1.12 $\log_{10}$, respectively. Mesophilic bacteria are able to multiply more rapidly when milk is held above 4.4°C.[88] Barnes[9] reported that mesophiles, such as *E. coli*, are not able to grow at storage temperatures below 5°C. Hence, the literature indicates that the temperature at which mesophilic populations of bacteria are able to multiply and interfere with psychrotrophic bacterial counts seems to be between 4 and 7.2°C. Thus, if aerobic plate counts are performed at temperatures below 4°C, mesophilic bacteria should not contribute to the total plate count. Gilliland et al.[89] in the *Compendium of Methods for the Microbiological Examination of Foods* recommended conducting psychrotrophic plate counts at 7 ± 1°C for an incubation period of 10 days. At 7°C, mesophilic species are not able to multiply rapidly enough to interfere with psychrotrophic bacterial populations. Because this procedure requires 10 days to conduct, most fresh meat and poultry would have been purchased and consumed before psychrotrophic evaluations could be performed, making the procedure of little use to the industry.

Rapid microbiological methods

Very few rapid methods have been developed for enumerating psychrotrophic bacteria from meat and poultry. One particular method involving the measurement of electrical parameters has been widely researched for use as a means of rapidly enumerating psychrotrophs. Electrical methods have been used to measure the growth of bacteria since the late 19th century.[90] Parsons (as reported by Strauss et al.[91]) demonstrated that conductivity was a useful tool for measuring the ammonia produced by *Clostridium* spp. in various environments. In 1938, Allison et al.[92] used conductance to measure bacterial induced proteolysis. In the mid 1970s, Ur and Brown[93] proposed the use of monitoring impedance as a tool for enumerating microorganisms, and Cady et al.[94] investigated the ability of various microorganisms to produce detectable impedance changes when they were cultured in different microbiological growth media.

Impedance is defined as the opposition to flow of an alternating electrical current in a conducting material. As bacteria multiply, they convert large, complex molecules into smaller, more mobile metabolites that change the impedance of the medium. These metabolites increase the ability of the medium to conduct electricity and thereby decrease the impedance of the medium. When microbial populations reach a level of approximately 10^6 to 10^7 cells/ml, a change in the impedance of the medium is able to be detected using a sensitive impedance monitoring instrument. The time required for this electrical change to occur is termed the impedance detection time (DT).[95]

DT may be obtained in very short periods of time (<24 h) when compared to psychrotrophic plate counts (10 days). However, there are several fundamental differences between impedance microbiological techniques and psychrotrophic plate counts.[95] When performing psychrotrophic plate counts, all bacteria which are able to reach a visible biomass are counted,[96] whereas, the impedance technique relies on the measurement of metabolic byproducts.[95] Impedance measurements are based on the accumulation of metabolic byproducts produced by the fastest growing bacterium or group of bacteria in a sample. Factors such as media, time, and temperature become critical parameters in the assay because specific bacteria use different metabolic pathways depending on the media in which they multiply. Some end products of metabolism produce stronger impedance signals than others when bacteria are allowed to multiply and utilize different substrates in the media.[95] Therefore, the substrates on which bacteria are grown will determine the byproducts they produce and hence, are an important consideration when performing impedance assays.

Previous research using electrical methods

Bishop et al.[96] reported high correlation coefficients ($R^2 = 0.87$ and 0.88) between the shelf-life of milk and impedance readings taken at 18 and 21°C, respectively. Ogden[97] observed that conductance readings on fish samples diluted in brain heart infusion broth and incubated at 20°C correlated well ($R^2 = 0.92$ to 0.97) to H_2S producing bacterial counts. However, according to Firstenberg-Eden and Tricarico,[20] mesophilic bacterial populations, not *Pseudomonas,* would be enumerated using electrical measurements on samples of fresh chicken unless a selective medium is used at incubation temperatures of 18–21°C.

To rapidly predict the shelf-life of fresh fish, Jørgenson et al.[98] analyzed samples using conductance at 25°C in trimethylamine oxide nitrogen medium (TMAO). Using this method, H_2S-producing bacteria enumerated using conductance or the conventional method described by Gram et al.[99] were highly correlated ($R^2 = 0.96$) to the shelf-life of fresh fish. However, using an incubation temperature of 25°C, mesophilic and non-H_2S-producing bacterial populations would multiply and interfere with conductance analyses on fresh poultry samples.

Bishop et al.[96] described a rapid impedance based method for determining the potential shelf-life of milk. Milk samples were preincubated in plate count broth at 18 or 21°C for 18 h. After preincubation, the samples were placed in module wells containing modified plate count agar (MPCA) and incubated at 18 or 21°C. A previous study by Firstenberg-Eden and Tricarico[20] indicated that, at 18 and 21°C, the generation time of *E. coli* is less than the generation time for *Pseudomonas* spp. For samples containing mixed microflora where mesophilic bacteria represent 92% of the initial flora,[40] such as a broiler chicken carcass, mesophilic bacteria would be enumerated at 18 or 21°C instead of psychrotrophs. Bishop and White[100] reported high correlation coefficients (R^2 = 0.87 and 0.88) between the shelf-life of milk and impedance readings taken at 18 and 21°C, respectively. However, at 18 and 21°C, mesophilic species on chickens would be enumerated instead of psychrotrophs using impedance unless a selective medium is used.

Selective medium for psychrotrophic spoilage bacteria

Another consideration when selecting a medium for conducting impedance measurements is that some bacteria will multiply in a given medium, produce a detection time, exhaust the nutrients necessary for growth, and stop growing. Subsequently, another group of bacteria will use the remaining nutrients and begin to multiply, creating a bimodal impedance curve. The impedance curve represents the relationship of impedance change to incubation time[101] and the DT of these bimodal curves is difficult to assess.

If a selective medium and temperature are used, the bacteria or group of bacteria able to multiply most rapidly and reach the threshold level of 10^6 will be responsible for the DT. This feature of impedance microbiology makes it a useful tool in that for mixed samples, a particular bacteria or group of bacteria can be measured by selecting for its growth over the other competing microflora in the sample. For example, if a mixed sample contained 100,000 pseudomonads and 1 coryneform and was incubated at 30°C, the coryneform would be the bacteria responsible for the DT.[101] At 30°C, the generation time of the pseudomonads is 4 times that of the coryneform which allows the coryneform to multiply and reach 10^6 before the pseudomonads. Thus, selective media can be useful for enumerating one species of bacteria in mixed samples by selecting for its growth over other species present in the sample.

Russell[102] developed a method to selectively enumerate *P. fluorescens* from fresh chicken in less than 24 h using capacitance microbiology. Capacitance assays were conducted on whole carcass rinses at 25°C using brain heart infusion broth containing nitrofurantoin (4 μg), carbenicillin (120 μg), and Irgasan (25 μg) per milliliter. This medium was found to be optimal for selecting for the growth of *P. fluorescens* from among many other competing species and was termed *P. fluorescens* selective medium. The selective medium was found to be excellent for enumeration of *P. fluorescens* from broiler chicken carcass rinses using capacitance microbiology at 25°C. The time required to enumerate *P. fluorescens* for all samples, regardless of the initial concentration of *P. fluorescens,* was less than 22.4 h.

In a second study, Russell[103] conducted an experiment to determine if rapid enumeration of populations of *P. fluorescens* using standard methods and capacitance could be used to predict the potential shelf-life of fresh broiler chicken. For each carcass, psychrotrophic plate counts (PPC), *P. fluorescens* plate counts (PFPC) using the selective media with agar added, *P. fluorescens* detection times (PFDT) using capacitance, and subjective odor evaluations (ODOR) were determined. PPC, PFPC, and ODOR on groups of carcasses significantly increased after 3 days of storage at 3°C and every day thereafter throughout the storage period. Log_{10} *P. fluorescens* detection times (LPDT) significantly decreased, indicating a significant increase in bacterial populations, throughout storage from 0 to 12 days. Significant correlations were observed between PPC and day of storage (DAY), PFPC and DAY, LPDT and DAY, ODOR and DAY, PPC and ODOR, PFPC and ODOR, LPDT and ODOR, LPDT and PPC, and LPDT and PFPC. The author concluded that enumeration of *P. fluorescens* from fresh chicken at day of processing should be beneficial because: (1)

potential shelf-life of fresh chicken can be determined at day of processing, (2) processing sanitation and hygiene may be determined, and (3) carcasses exposed to temperature abuse may be identified.

Conclusion

In general, refrigeration of fresh poultry limits spoilage of fresh poultry to psychrotophs. However, instances of temperature abuse and other situations may allow mesophiles to complicate the situation. Of the many factors influencing the growth of spoilage on poultry, temperature is probably most important. Because of the slow growth rate of bacteria at refrigerated temperatures, classical culture methodology is not very useful in evaluating product before or during distribution. However, more rapid and useful methods have been developed that are based on changes of electrical parameters in the bacterial medium during growth.

References

1. Olson, J. C., Jr., Psychrophiles, mesophiles, thermophiles, and thermodurics—What are we talking about?, *Milk Plant Mon.*, 36, 32, 1947.
2. Greene, V. W. and Jezeski, J. J., Influence of temperature on the development of several psychrophilic bacteria of dairy origin, *Appl. Microbiol.*, 2, 110, 1954
3. Muller, M., Ueber das Wachstum und die Lebenstatigkeit von Bakterien sowie den Ablauf fermentativer Prozesse bei niederer Temperaturen unter spezieller Berucksichtigung des Fleisches als Nahrungsmittel, *Arch. Hyg.*, 47, 127, 1903.
4. Zobell, C. E. and Conn, J. E., Studies on the thermal sensitivity of marine bacteria, *J. Bacteriol.*, 40, 223, 1940.
5. Ingraham, J. L., Growth of psychrophilic bacteria, *J. Bacteriol.*, 76, 75, 1958.
6. Ayres, J. C., Mundt, J. O., and Sandine, W. E., *Microbiology of Foods*, W. H. Freeman, San Francisco, CA, 1980, 55.
7. Pooni, G. S. and Mead, G. C., Prospective use of temperature function integration for predicting the shelf-life of non-frozen poultry-meat products, *Food Microbiol.*, 1, 67, 1984.
8. Ayres, J. C., Ogilvy, W. S., and Stewart, G. F., Post mortem changes in stored meats. I. Microorganisms associated with development of slime on eviscerated cut-up poultry, *Food Technol.*, 4, 199, 1950.
9. Barnes, E. M., Microbiological problems of poultry at refrigerator temperatures—a review, *J. Sci. Food Agric.*, 27, 777, 1976.
10. Daud, H. B., McMeekin, T. A., and Olley, J., Temperature function integration and the development and metabolism of poultry spoilage bacteria, *Appl. Environ. Microbiol.*, 36, 650, 1978.
11. Baker, R. C., Naylor, H. B., Pfund, M. C., Einset, E., and Staempfli, W., Keeping quality of ready-to-cook and dressed poultry, *Poult. Sci.*, 35, 398, 1956.
12. Lockhead, A. G. and Landerkin, G. B., Bacterial studies of dressed poultry. I. Preliminary investigations of bacterial action at chill temperatures, *Sci. Agric.*, 15, 765, 1935.
13. Naden, K. D. and Jackson, Jr., G. A., Some economic aspects of retailing chicken meat, *Calif. Agric. Exp. Stn. Bull.*, 734, 107, 1953.
14. Spencer, J. V., Ziegler, F., and Stadelman, W. J., Recent studies of factors affecting the shelf-life of chicken meat, *Wash. Agric. Exp. Stnt. Inst. Agric. Sci. Stn. Cir.*, number 254, 1954.
15. Brown, A. D., Some general properties of a psychrophilic pseudomonad: the effects of temperature on some of these properties and the utilization of glucose by this organism and *Pseudomonas aeruginosa*, *J. Gen. Microbiol.*, 17, 640, 1957.
16. Allen, C. D., Russell, S. M., and Fletcher, D. L., The relationship of broiler breast meat color and pH to shelf-life and odor development, *Poult. Sci.*, 76, 1042, 1997.

17. Russell, S. M., Fletcher, D. L., and Cox, N. A., A model for determining differential growth at 18 and 42°C of bacteria removed from broiler chicken carcasses, *J. Food Prot.*, 55, 167, 1992.
18. Olsen, R. H. and Jezeski, J. J., Some effects of carbon source, aeration, and temperature on growth of a psychrophilic strain of *Pseudomonas fluorescens*, *J. Bacteriol.*, 86, 429, 1963.
19. Elliott, R. P. and Michener, H. D., Factors affecting the growth of psychrophilic microorganisms in foods, a review, *Technical Bulletin No. 1320, Agricultural Research Service*, USDA, Washington, DC, 1965.
20. Firstenberg-Eden, R. and Tricarico, M. K., Impedimetric determination of total, mesophilic and psychrotrophic counts in raw milk, *J. Food Sci.*, 48, 1750, 1983.
21. Forster, J., Ueber einige Eigenschaften leuchtender Bakterien, *Cent. Bakteriol.*, 2, 337, 1887.
22. Ayres, J. C., Temperature relationships and some other characteristics of the microbial flora developing on refrigerated beef, *Food Res.*, 25(6), 1, 1960.
23. Haines, R. B., The bacterial flora developing on stored lean meat, especially with regard to "slimy" meat, *J. Hyg.*, 33, 175, 1933.
24. Empey, W. A. and Vickery, J. R., The use of carbon dioxide in the storage of chilled beef, *Aust. Commonw. Counc. Sci. Ind. Res. J.*, 6, 233, 1933.
25. Haines, R. B., Microbiology in the preservation of animal tissues, Dept. Sci. Ind. Research Food Invest. Board (Gr. Brit.) Special Report No 45. London: Her Majesty's Stationery Office, 1937.
26. Empey, W. A. and Scott, W. J., Investigations on chilled beef. I. Microbial contamination acquired in the meatworks, *Aust. Commonw. Coun. Sci. Ind. Res. Bull.*, 126, 1939.
27. Kirsch, R. H., Berry, F. E., Baldwin, G. L., and Foster, E. M., The bacteriology of refrigerated ground beef, *Food Res.*, 17, 495, 1952.
28. Wolin, E. F., Evans, J. B., and Niven, C. F., The microbiology of fresh and irradiated beef, *Food Res.*, 22, 268, 1957.
29. Breed, R. S., Murray, E. G. D., and Smith, N. R., *Bergey's Manual of Determinative Bacteriology*, 6th ed., Williams & Wilkins, Baltimore, MD, 1948, 1094.
30. Bergey, D. H., *Bergey's Manual of Determinative Bacteriology*, 3rd ed., Williams & Wilkins, Baltimore, MD, 1930, 589.
31. Brown, A. D. and Weidemann, J. F., The taxonomy of the psychrophilic meat-spoilage bacteria: a reassessment, *J. Appl. Bacteriol.*, 21, 11, 1958.
32. Halleck, F. E., Ball, C. O., and Stier, E. F., Factors affecting quality of prepackaged meat. IV. Microbiological studies. A. Cultural studies on bacterial flora of fresh meat; classification by genera, *Food Technol.*, 12, 197, 1957.
33. Barnes, E. M. and Impey, C. S., Psychrophilic spoilage bacteria of poultry, *J. Appl. Bacteriol.*, 31, 97, 1968.
34. MacDonell, M. T. and Colwell, R. R., Phylogeny of the Vibrionaceae and recommendation for two new genera, *Listonella* and *Shewanella*, *Syst. Appl. Microbiol.*, 6, 171, 1985.
35. Breed, R. S., Murray, E. G. D., and Smith, N. R., *Bergey's Manual of Determinative Bacteriology*, 7th ed., Williams & Wilkins, Baltimore, MD, 1957, 1094.
36. Thornley, M. J., Computation of similarities between strains of *Pseudomonas* and *Achromobacter* isolated from chicken meat, *J. Appl. Bacteriol.*, 23, 395, 1960.
37. Russell, S. M., Fletcher, D. L., and Cox, N. A., Spoilage bacteria of fresh broiler chicken carcasses, *Poult. Sci.*, 74, 2041, 1995.
38. Barnes, E. M., Bacteriological problems in broiler preparations and storage, *R. Soc. Health J.*, 80, 145, 1960.
39. Schefferle, H. E., The microbiology of built up poultry litter, *J. Appl. Bacteriol.*, 28, 403, 1965.
40. Barnes, E. M. and Thornley, M. J., The spoilage flora of eviscerated chickens stored at different temperatures, *J. Food Technol.*, 1, 113, 1966.
41. Elliott, R. P. and Michener, H. D., Microbiological standards and handling codes for chilled and frozen foods. A review, *Appl. Microbiol.*, 9, 452, 1961.
42. Kraft, A. A. and Ayres, J. C., Post-mortem changes in stored meats. IV. Effect of packing materials on keeping quality of self-service meats, *Food Technol.*, 6, 8, 1952.
43. Dainty, R. H. and Mackey, B. M., The relationship between the phenotypic properties of bacteria from chill-stored meat and spoilage processes, *Ecosystems: Microbes: Food*, Board, R. G.,

Jones, D., Kroll, R. G., and Pettipher, G. L., Eds., S. A. B. Symposium Series Number 21, Blackwell Scientific Publications, Oxford, U.K., 1992, 103S.

44. Nychas, G.-J. E., Dillon, V. M., and Board, R. G., Glucose, the key substrate in the microbiological changes occurring in meat and certain meat products, *Biotechnol. Appl. Biochem.*, 10, 203, 1988.
45. Lampropoulou, K., Drosinos, E. H., Nychas, G.-J. E., The effect of glucose supplementation on the spoilage microflora and chemical composition of minced beef stored aerobically or under a modified atmosphere at 4°C, *Int. J. Food Microbiol.*, 30, 281, 1996.
46. Venugopal, V., Extracellular proteases of contaminant bacteria in fish spoilage: a review, *J. Food Prot.*, 53, 341, 1990.
47. Tarrant, P. J. V., Jenkins, N., Pearson, A. M., and Dutson, T. R., Proteolytic enzyme preparation from *Pseudomonas fragi.* Its action on pig muscle, *Appl. Microbiol.*, 25, 996, 1973.
48. Porzio, M. A. and Pearson, A. M., Degradation of myofibrils and formation of premeromyosin by a neutral protease produced by *Pseudomonas fragi*, *Food Chem.*, 5, 195, 1980.
49. Schmitt, R. E. and Schmidt-Lorenz, W., Degradation of amino acids and protein changes during microbial spoilage of chilled unpacked and packed chicken carcasses, *Lebensm. Wiss. Technol.*, 25, 11, 1992.
50. Nychas, G.-J. E. and Tassou, C. C., Spoilage processes and proteolysis in chicken as detected by HPLC, *J. Sci. Food Agric.*, 74, 199, 1997.
51. De Castro, B. P., Asensio, M. A., Sanz, B., and Ordonez, J. A., A method to assess the bacterial content of refrigerated meat, *Appl. Environ. Microbiol.*, 54, 1462, 1988.
52. Gill, C. O. and Newton, K. G., The ecology of bacterial spoilage of fresh meat at chill temperatures, *Meat Sci.*, 2, 207, 1978.
53. Schmitt, R. E. and Schmidt-Lorenz, W., Formation of ammonia and amines during microbial spoilage of refrigerated broilers, *Lebensm. Wiss. Technol.*, 25, 6, 1992.
54. Wells, F. E., Hartsell, S. E., and Stadelman, W. J., Growth of psychrophiles. I. Lipid changes in relation to growth-temperature reductions, *J. Food Sci.*, 28, 140, 1963.
55. Eklund, M. W., Biosynthetic responses of poultry meat organisms under stress. Ph.D. thesis, Purdue University, IN, 1962, 128.
56. Nashif, S. A. and Nelson, F. E., The lipase of *Pseudomonas fragi.* II. Factors affecting lipase production, *J. Dairy Sci.*, 36, 471, 1953.
57. Alford, J. A. and Elliott, L. E., Lipolytic activity of microorganisms at low and intermediate temperatures. I. Action of *Pseudomonas fluorescens* on lard, *Food Res.*, 25, 296, 1960.
58. Peterson, A. C. and Gunderson, M. F., Some characteristics of proteolytic enzymes from *Pseudomonas fluorescens*, *Appl. Microbiol.*, 8, 98, 1960.
59. Ingraham, J. L. and Bailey, G. F., Comparative study of effect of temperature on metabolism of psychrophilic and mesophilic bacteria, *J. Bacteriol.*, 77, 609, 1959.
60. Sultzer, B. M., Oxidative activity of psychrophilic and mesophilic bacteria on saturated fatty acids, *J. Bacteriol.*, 82, 492, 1961.
61. Hess, E., Cultural characteristics of marine bacteria in relation to low temperatures and freezing, *Contrib. Can. Biol. Fish.*, 8, 459, 1934.
62. Chistyakov, F. M. and Noskova, G., The adaptations of micro-organisms to low temperatures, *Ninth Int. Congr. Refrig. Proc.*, 2, 4.230, 1955.
63. MacFadyen, A. and Rowland, S., On the suspension of life at low temperatures, *Ann. Bot.*, 16, 589, 1902.
64. Burdon-Sanderson, The origin and distribution of microzymes (bacteria) in water, and the circumstances which determine their existence in the tissues and liquids of the living body, *Q. J. Microbiol. Sci.* n.s., 11, 323, 1871.
65. Fischer, B., Bakterienwachsthum bei 0 C. sowie uber das Photographiren von Kulturen leuchtender Bakterien in ihrem eigenen Lichte, *Cent. Bakteriol.*, 4, 89, 1888.
66. Bedford, R. H., Marine bacteria of the Northern Pacific Ocean. The temperature range of growth, *Contrib. Can. Biol. Fish.*, 8, 433, 1933.
67. Berry, J. A. and Magoon, C. A., Growth of microorganisms at and below 0°C, *Phyiopathology*, 24, 780, 1934.

68. Mazur, P., Freezing of living cells: mechanisms and implications, *Am. J. Physiol.*, 247, C125, 1984.
69. Elliott, R. P. and Michener, H. D., Review of the microbiology of frozen foods. Conference on Frozen Food Quality, USDA, ARS-74-21, Washington, DC, 1960, 40.
70. Straka, R. P. and Stokes, J. L., Metabolic injury to bacteria at low temperatures, *J. Bacteriol.*, 78, 181 1959.
71. Hartsell, S. E., The growth initiation of bacteria in defrosted eggs, *Food Res.*, 16, 97, 1951.
72. Sair, L. and Cook, W. H., Effect of precooling and rate of freezing on the quality of dressed poultry, *Can. J. Res.*, D16, 139, 1938.
73. Pennington, M. E., Studies of poultry from the farm to the consumer, USDA, *Bureau of Chemistry, Circular No. 64*, Washington: Government Printing Office, 1910.
74. Spencer, J. V., Sauter, E. A., and Stadelman, W. J., Shelf life of frozen poultry meat after thawing, *Poult. Sci.*, 34, 1222, 1955.
75. Spencer, J. V., Sauter, E. A., and Stadelman, W. J., Effect of freezing, thawing, and storing broilers on spoilage, flavor, and bone darkening, *Poult. Sci.*, 40, 918, 1961.
76. Newell, G. W., Gwin, J. M., and Jull, M. A., The effect of certain holding conditions on the quality of dressed poultry, *Poult. Sci.*, 27, 251, 1948.
77. Elliott, R. P. and Straka, R. P., Rate of microbial deterioration of chicken meat at 2°C after freezing and thawing, *Poult. Sci.*, 43, 81, 1964.
78. Stone, L. S. and Zottola, E. A., Effect of cleaning and sanitizing on the attachment of *Pseudomonas fragi* to stainless steel, *J. Food Sci.*, 50, 951, 1985.
79. Wirtanen, G. and Mattila-Sandholm, T., Effect of the growth phase of foodborne biofilms on their resistance to a chlorine sanitizer. Part II, *Lebensm. Wiss. Technol.*, 25, 50, 1992.
80. Hingst, V., Klippel, K. M., and Sonntag, H. G., Investigations concerning the epidemiology of microbial resistance to biocides, *Zentralbl. Hyg. Umweltmed.*, 197, 232, 1995.
81. Mosley, E. B., Elliker, P. R., and Hays, H., Destruction of food spoilage, indicator and pathogenic organisms by various germicides in solution and on a stainless steel surface, *J. Milk Food Technol.*, 39, 830, 1976.
82. Ouattara, B., Simard, R. E., Holley, R. A., Piette, G. J.-P., and Bégin, A., Inhibitory effect of organic acids upon meat spoilage bacteria, *J. Food Prot.*, 60, 246, 1997.
83. Mountney, G. J. and O'Malley, J., Acids as poultry meat preservatives, *Poult. Sci.*, 44, 582, 1965.
84. Lauten, S. D., Sarvis, H., Wheatley, W. B., Williams, D. E., Mora, E. C., and Worley, S. D., Efficacies of novel N-halamine disinfectants against *Salmonella* and *Pseudomonas* species, *Appl. Envirn. Microbiol.*, 58, 1240, 1992.
85. Hwang, C.-A. and Beuchat, L. R., Efficacy of selected chemicals for killing pathogenic and spoilage microorganisms on chicken skin, *J. Food Prot.*, 58, 19, 1995.
86. Russell, S. M., Chemical sanitizing agents and spoilage bacteria on fresh broiler carcasses, *J. Appl. Poult. Res.*, 7(3), 273, 1998.
87. Ayres, J. C., Some bacterial aspects of spoilage of self-service meats, *Iowa State J. Sci.*, 26, 31, 1951.
88. Senyk, G.F., Goodall, C., Kozlowski, S. M., and Bandler, D. K., Selection of tests for monitoring the bacteriological quality of refrigerated raw milk samples, *J. Dairy Sci.*, 71, 613, 1988.
89. Gilliland, S. E., Michener, H. D., and Kraft, A. A., Psychrotrophic microorganisms, *Compendium of Methods for the Microbiological Examination of Foods*, American Public Health Association, Washington, DC, 1984, 136.
90. Stewart, G. N., The changes produced by the growth of bacteria in the molecular concentration and electrical conductivity of culture media, *J. Exp. Med.*, 4, 235, 1899.
91. Strauss, W. M., Malaney, G. W., and Tanner, R. D., The impedance method for monitoring total coliforms in wastewaters, *Folia Microbiol.*, 29, 162, 1984.
92. Allison, J. B., Anderson, J. A., and Cole, W. H., The method of electrical conductivity in studies on bacterial metabolism, *J. Bacteriol.*, 36, 571, 1938.
93. Ur, A. and Brown, D. F. J., Impedance monitoring of bacterial activity, *J. Med. Microbiol.*, 8, 19, 1975.
94. Cady, P., Dufour, S. W., Shaw, J., and Kraeger, S. J., Electrical impedance measurements: rapid method for detecting and monitoring microorganisms, *J. Clin. Microbiol.*, 7, 265, 1978.

95. Firstenberg-Eden, R., Electrical impedance method for determining microbial quality of foods, *Rapid Methods and Automation in Microbiology and Immunology*, K. O. Habermehl, Ed., Springer-Verlag, Berlin, 1985, 679.
96. Bishop, J. R., White, C. H. and Firstenberg-Eden, R., Rapid impedimetric method for determining the potential shelf-life of pasteurized whole milk, *J. Food Prot.*, 47, 471, 1984.
97. Ogden, I. D., Use of conductance methods to predict bacterial counts in fish, *J. Appl. Bacteriol.*, 61, 263, 1986.
98. Jørgenson, B. R., Gibson, D. M., and Huss, H. H., Microbiological quality and shelf-life prediction of chilled fish, *Int. J. Food Microbiol.*, 6, 295, 1988.
99. Gram, L., Trolle, G., and Huss, H. H., Detection of specific spoilage bacteria from fish stored at low (0°C) and high (20°C) temperatures, *Int. J. Food Microbiol.*, 4, 65, 1987.
100. Bishop, J. R. and White, C. H., Estimation of the potential shelf-life of pasteurized fluid milk utilizing bacterial numbers and metabolites, *J. Food Prot.*, 48, 663, 1985.
101. Firstenberg-Eden, R. and Eden, G., *Impedance Microbiology*, Research Studies Press, Letchworth, Hertfordshire, England, 1984, 48.
102. Russell, S. M., A rapid method for enumeration of *Pseudomonas fluorescens* from broiler chicken carcasses, *J. Food Prot.*, 60(4),385, 1997.
103. Russell, S. M., Rapid prediction of the potential shelf-life of broiler chicken carcasses under commercial conditions, *J. Appl. Poult. Res.*, 6(2), 163, 1997.

chapter eleven

Functional properties of muscle proteins in processed poultry products

Denise M. Smith

Contents

Introduction

Proteins are required to do a variety of functions in poultry products. The typical characteristics of many poultry products are dependent on the successful manipulation of protein functional properties during processing. Yield, quality, and sensory attributes of processed poultry products are largely determined by the functional properties of the muscle proteins.

Functional properties are defined as physical or chemical properties of proteins that determine their behavior in foods during processing, storage, and consumption.[1] Functional properties of proteins contribute to many of the quality and organoleptic attributes of a food product as perceived by a consumer. The functional properties of poultry proteins must be understood for effective utilization of new ingredients, development of

0-8493-0120-3/01/$0.00+$.50

new products, modification of existing products, reduction of waste, and control of energy consumption during processing. Protein functional properties important in poultry meat products can be broadly classified into three categories: (1) protein-water interactions, (2) protein-fat interactions, and (3) protein-protein interactions.

The importance of a functional property varies with the type of product, source of meat, type and concentration of non-meat ingredients, type of processing equipment used, processing conditions, and stage of processing. The functional properties of muscle proteins are influenced by other ingredients in a formulation and by the processing conditions used. The functional properties of the muscle proteins are influenced by processing conditions and ingredients used in a formulation. This relationship is illustrated schematically in Figure 11.1. Functional properties are dictated by the molecular and biochemical properties of a protein. Thus, any changes in a product formulation or process requires an appreciation of the effect of that change on the muscle protein structure. Formulation changes can alter the pH, salt concentration, and protein concentration of a product, among other factors, all of which have an effect on the biochemical properties, and subsequently functional properties, of poultry muscle proteins. Changes in processing conditions, especially those that alter product temperature or the extent of comminution, can also affect the biochemical properties of muscle proteins. All of these changes, which affect protein structure ultimately affect the quality of the final product.

Often, the need for a particular functional property may change during processing. Solubility, water binding, and fat binding are the major functional properties of proteins

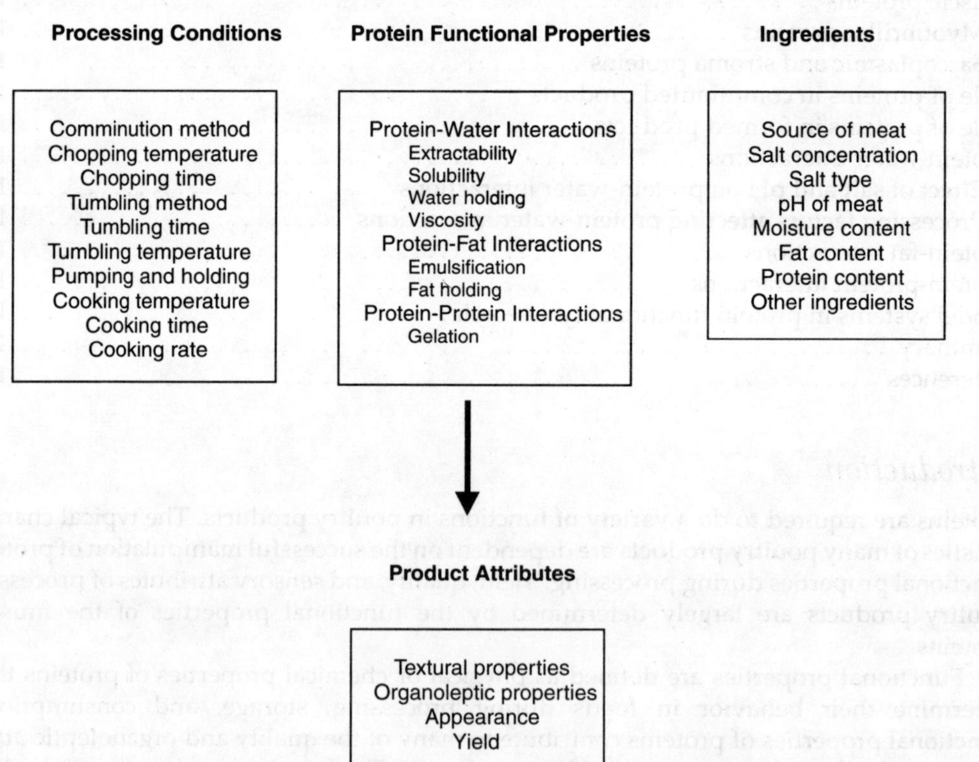

Figure 11.1 Diagram illustrating how processing conditions and ingredients affect the functional properties of muscle proteins and the resulting quality attributes of finished poultry products.

Table 11.1 **Protein Composition of Poultry Skeletal Muscle**

I. Myofibrillar proteins (55% of total protein)
  Contractile proteins
    Examples: myosin, actin
  Regulatory proteins
    Examples: tropomyosin, troponin
  Cytoskeletal proteins
    Examples: titan, nebulin
II. Sarcoplasmic proteins (35% of total protein)
  Glycolytic enzymes
  Mitochondrial/oxidative enzymes
  Lysosomal enzymes
  Myoglobin and other heme proteins
III. Stroma proteins (3–5% of total protein)
  Collagen
  Elastin
  Reticulin

required in raw poultry products. Water binding, fat binding, and gelation are some of the important functional properties in cooked meat products. Proteins are often required to be multifunctional. That is, each protein is expected to exhibit more than one functional property either simultaneously or sequentially during processing.

This chapter will begin with a brief description of muscle ultrastructure and an introduction to the major protein fractions of muscle. A short discussion of the role of proteins in comminuted and formed products is presented next, followed by a more in-depth look at the major functional properties of muscle proteins. This chapter concludes with a brief look at the role of model systems in protein functional property research.

Muscle proteins

Poultry meat is comprised of about 20 to 23% protein. Muscle proteins are divided into three categories based primarily on their solubility properties: myofibrillar, sarcoplasmic, and stroma (Table 11.1).

Myofibrillar proteins

The myofibrillar or salt-soluble proteins comprise about 50 to 56% of the total skeletal muscle protein and are insoluble in water, but most are soluble at salt concentrations above 1%. This group is comprised of about 20 distinct proteins organized within a myofibril of an intact muscle. Myofibrils extend the length of a muscle fiber or cell and are surrounded by the sarcoplasm (Figure 11.2). A single muscle fiber may contain 1000 to 2000 myofibrils. The repeating contractile unit of a myofibril is the sarcomere. Myofibrillar protein can be divided into three groups based on their function: (1) contractile proteins, which are responsible for muscle contraction, (2) regulatory proteins, involved in regulation and control of contraction, and (3) cytoskeletal proteins that support and maintain the structural integrity of the myofibril. For more information on skeletal muscle ultrastructure, the reader should refer to one of the numerous reviews on the subject.[1–5]

Myosin is the predominant protein in the thick filament of the sarcomere and comprises about 50 to 55% of the total myofibrillar protein. At physiological ionic strength and pH, myosin molecules aggregate spontaneously to form the thick filaments. Myosin is a

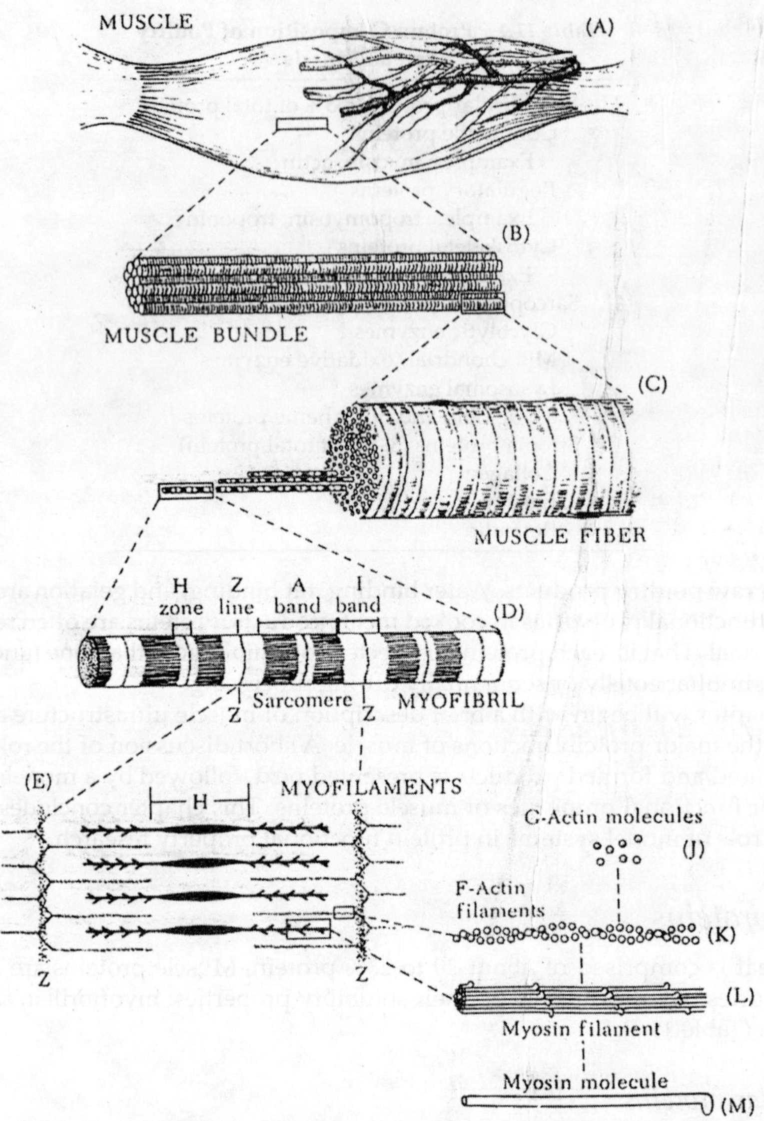

Figure 11.2 Organization of skeletal muscle structure. (From Hedrick, H. B., et al., *Principles of Meat Science*, Third Edition, Kendall/Hunt Publishing Company, Dubuque, IA, 1994. With permission.)

long thin molecule with dimensions of about 150 nm in length by 1.5 nm in width in the rod region and 8 nm in width in the globular head region. Poultry skeletal muscle myosin is a large molecule of about 520 kDa and is comprised of 6 polypeptide chains or subunits (Figure 11.3). The subunits include two heavy chains of about 222 kDa each and 2 pairs of light chains ranging from 17 to 23 kDa. Each heavy chain has a globular head region and a fibrous tail or rod region. The light chains are designated as alkali light chains or DTNB light chains and are associated with the globular head region. The globular head of myosin heavy chain also contains the actin binding site. The tail or rod region is comprised of a

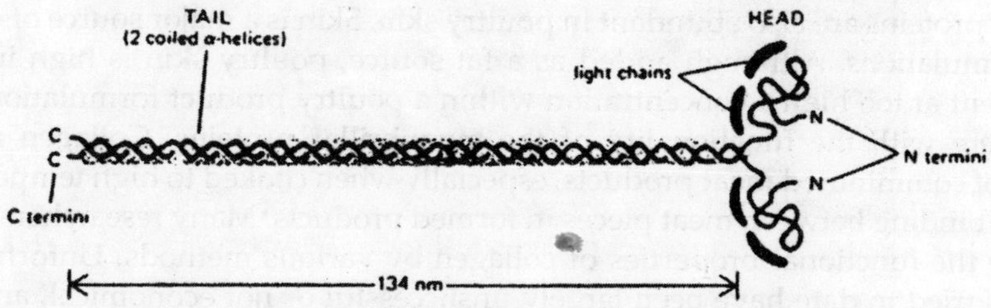

Figure 11.3 Schematic diagram of the myosin molecule.

coiled-coiled alpha helix. This is the region of myosin that is responsible for filament formation under physiological conditions. The head and fibrous tail regions of myosin exhibit distinct biochemical and functional properties. Chicken skeletal myosin contains 43 sulfhydryl groups and no disulfide bonds. The isoelectric point (pI) of myosin is about 5.3 and it is the pH at which the protein has no net charge in solution due to an equal number of positive and negative charges on the molecule.

Actin is the second most abundant myofibrillar protein and comprises about 20 to 25% of this fraction. G-actin is a globular protein with a molecular mass of about 42 kDa. The isoelectric point of actin is about 4.8. Actin, along with the regulatory proteins, troponin and tropomyosin, make up the thin filaments of the sarcomere. Myosin binds reversibly to actin in the thin filaments during muscle contraction. In post-rigor muscle, the globular head or subfragment-1 region of myosin binds irreversibly to actin to form a complex known as actomyosin. This cross-linking between actin and myosin in post-rigor muscle influences meat tenderness in intact muscle.

The contractile proteins, myosin and actin, have a large influence on muscle protein functionality. Myosin, in pre-rigor muscle, and actomyosin, in post-rigor muscle, are generally considered to contribute several functional properties to processed meat products and have been extensively studied.[3,6,7] Since actin is usually complexed with myosin in post-rigor muscle, actin modifies the functionality of myosin in both comminuted and formed poultry products. The ratio of actin to myosin, as well as the ratio of free myosin to actomyosin, influences the functional properties of a poultry product. Sarcoplasmic and stroma proteins modify the functional properties of the myofibrillar proteins.

Sarcoplasmic and stromal proteins

Sarcoplasmic proteins are located inside the muscle cell membrane in the sarcoplasm and comprise about 30–35% of the total muscle protein. These proteins are soluble in water or low ionic strength solutions (<0.6 μ). Proteins in this category include oxidative enzymes, myogloblin, and other heme pigments, the glycolytic enzymes responsible for glycolysis, and lysosomal enzymes. Myoglobin is the protein primarily responsible for meat color, but in general, these proteins play only a minor role in meat protein functionality.

The stroma proteins, often referred to as connective tissue proteins, hold together and support the muscle structure by surrounding the muscle fibers and entire muscle. Connective tissue surrounding the muscle is called the epimysium. Connective tissue surrounding bundles of muscle fibers is called perimysium, while that surrounding individual fibers is called endomysium. Stroma proteins usually comprise about 3 to 6% of the total protein of poultry skeletal muscle. The major stroma protein is collagen. Elastin and reticulin are minor constituents of the stroma fraction. All of these proteins are insoluble in water and salt solutions. Meat tenderness often decreases with age of the animal due to the increased cross-linking and other modifications that occur to collagen.[8]

Stroma proteins are also abundant in poultry skin. Skin is a major source of collagen in poultry formulations. Although added as a fat source, poultry skin is high in collagen. When present at too high a concentration within a poultry product formulation, collagen may interfere with the functionality of the myofibrillar proteins. Collagen may cause shrinkage of comminuted meat products, especially when cooked to high temperatures, or interfere in binding between meat pieces in formed products. Many researchers have tried to improve the functional properties of collagen by various methods. Unfortunately, all approaches tried to date have been largely unsuccessful or not economical, and thus the amount of skin that can be included in a processed poultry formulation must be kept below certain critical levels.

Role of proteins in comminuted products

To prepare a comminuted poultry product, meat, water, salt, phosphate and perhaps other ingredients are ground or chopped to form a paste-like batter. The meat batter is then stuffed into a casing of the desired shape and cooked. More details on the actual procedures used to prepare comminuted products are described in Chapter 12.

Meat batters are complex systems consisting of solubilized muscle proteins, muscle fibers, fragmented myofibrils, fat cells, fat droplets, water, salts, phosphates, and other ingredients. Comminuted products, such as frankfurters, bologna and sausages, typically contain about 17 to 20% protein, 0 to 20% fat, and 60 to 80% water. Thus, a relatively small amount of protein has to bind a relatively large amount of water and fat. In meat formulations about 1.5 to 2% salt is typically used to allow for the extraction and solubilization of the myofibrillar proteins.

Comminution, sometimes referred to as chopping, physically disrupts the muscle tissue by damaging the sarcolema (muscle cell membrane) and the supporting network of connective tissue. In the presence of salt, the muscle fibers swell, myofibrils are fragmented into shorter pieces, and myofibrillar proteins are extracted and solubilized. These events lead to the formation of a thick, paste-like batter which holds water and stabilizes fat. Upon cooking, the extracted and solubilized muscle proteins in the batter form a cross-linked gel matrix that binds the water and fat and forms the typical texture associated with cooked comminuted products.

Role of proteins in formed products

Formed poultry products are made from chunks or pieces of meat that are bonded or glued together. Turkey breast rolls and chicken cold cuts are common examples of these products (please refer to Chapter 12 for more details on processing). Similar events occur during the production of both comminuted products and formed products. The major exception is that during the production of formed products, most of the changes occur on the surface of the meat pieces. Tumbling, massaging, or mixing in the presence of salt are used to disrupt the muscle cells, disintegrate the muscle fibers, and extract the myofibrillar proteins from the surface of the meat pieces. A tacky myofibrillar protein exudate is formed on the surface of the meat pieces. This extracted protein exudate forms a gel on cooking that acts like a glue to hold the pieces of meat together. The myofibrillar protein, myosin, is thought to contribute most to the binding strength of the protein exudate. Collagen has been found to interrupt the binding of the meat pieces when present on the surface.

Protein-water interactions

In general all protein functional properties are influenced by the interaction of protein with water. However, three functional properties involving protein-water interactions are very important in raw poultry products. These are (1) protein extraction and solubilization, (2) water retention, and (3) viscosity.

Protein extractability is a term used to describe the amount of protein that is released or dissociated from the organized myofibrillar structure during processing. Under the proper environmental conditions, an extracted muscle protein is soluble. Solubility is primarily dependent on the distribution of hydrophobic and hydrophilic amino acids on the surface of a protein and on the thermodynamics of the protein-water interactions. Muscle protein extractability and solubility are affected by pH, salt concentration, type of salts, and temperature.

Water retention describes the ability of a protein matrix to retain water or absorb added water in response to an external force, such as during cooking, centrifugation, or pressing. The water may be chemically bound to the protein, held via capillary action, or physically entrapped within a protein structure. The proteins in the highly organized myofibrillar structure chemically bind water. Water is also physically held within the interfilamental spaces of the myofibril. The water-binding ability of a protein is also influenced by pH, salt concentration, the type of salts present, and temperature.

Viscosity, defined by rheologists as the resistance of a material to flow, has a large influence on the stability of the raw product prior to cooking. The viscosity of the meat batter increases during comminution when the muscle fibers swell and absorb water. Extracted proteins that are large, fibrous, and highly soluble, such as myosin, can increase solution viscosity, even at very low concentrations. Batter viscosity must be high enough to stabilize the raw product, but low enough to allow pumping and handling within the plant.

Effect of salt and pH on protein-water interactions

The effect of salt on the water-binding ability of a turkey muscle homogenate is illustrated in Figure 11.4. Water binding increases most rapidly as the salt concentration is increased from about 0.3 (1.8%) to 0.6 *M* (3.4%) NaCl in both breast and thigh meat.[9] The addition of salt reduces electrostatic interactions between protein molecules to increase protein extractability, solubility, and water binding. Chopping or tumbling of meat in the presence of salt disrupts the muscle tissue allowing the muscle fibers to absorb water and swell, which leads to an increase in viscosity of the batter. Also, the organized thick and thin filaments of the sarcomere are disrupted due to solubilization and extraction of the myofibrillar proteins. Individual myofibrils are released from the muscle fibers and are fragmented into shorter pieces. The extracted proteins, especially myosin, also bind water and increase the viscosity of a poultry meat batter which helps to stabilize dispersed fat. For these reasons, about 1.5 to 2.0% salt is added to most poultry product formulations. Although higher concentrations of salt may improve water binding, the salty flavor is undesirable.

The pH of the poultry meat batter also has a large influence on the extractability, solubility and water-binding ability of the muscle proteins.[9] The effect of pH on the water-binding ability of a turkey muscle batter is illustrated in Figure 11.5. Water binding is lowest at the isoelectric point of myosin and actin (near pH 5.0). The proteins have no net charge at the isoelectric point and tend to associate to form aggregates. The water-binding ability of the muscle homogenate is increased as the pH is adjusted away from this isoelectric point. As the pH is increased, the proteins become more negatively charged. A higher net negative charge leads to an increase in repulsive force between the proteins within the myofilament which subsequently allows the myofibril to swell and hold water.

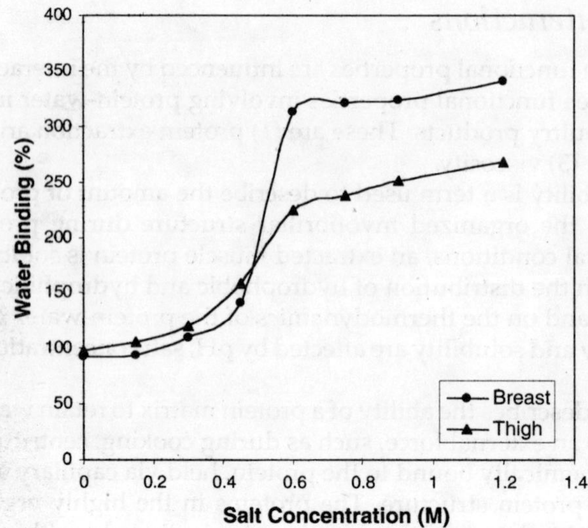

Figure 11.4 Salt concentration affects the water-binding ability of raw turkey meat batters at pH 6.0. (Adapted from Richardson, R. I. and Jones, J. M., *Int. J. Food Sci. Technol.*, 22, 683, 1987.)

The effect of pH on the concentration of extractable protein for a turkey meat homogenate is illustrated in Figure 11.6. Protein extractability and solubility are low near the isoelectric point of the myofibrillar proteins. As the pH is increased, the extractability and solubility of the myofibrillar proteins are increased as the proteins become more negatively charged. Alkaline phosphates are commonly used in poultry products. Alkaline phosphates increase the pH of the meat batter, usually by about 0.1 to 0.4 of a pH unit, to increase the water binding ability of the muscle proteins.

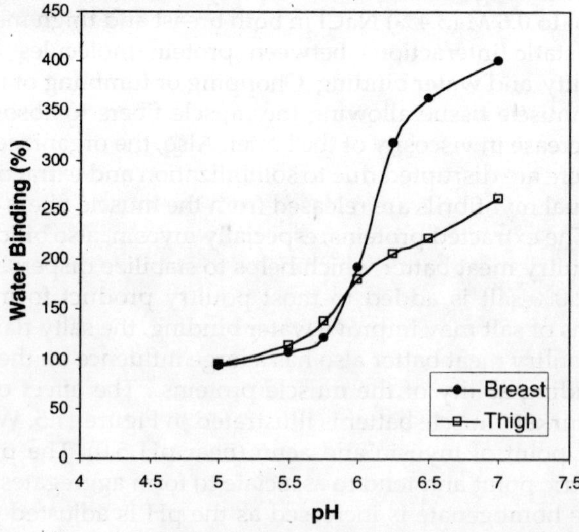

Figure 11.5 pH affects the water-binding ability of raw turkey meat batters containing 0.5 *M* NaCl. (Adapted from Richardson, R. I. and Jones, J. M., *Int. J. Food Sci. Technol.*, 22, 683, 1987.)

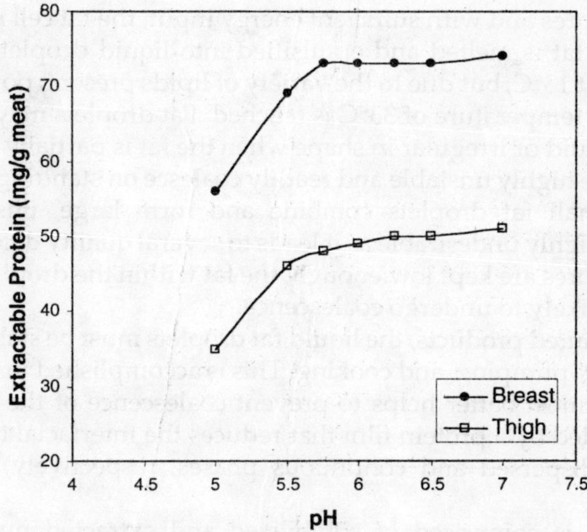

Figure 11.6 pH affects the extractable protein content of turkey meat batters containing 0.5 *M* NaCl. Water binding was defined in this study as the ability of raw meat batter to hold added water upon centrifugation. (Adapted from Richardson, R. I. and Jones, J. M., *Int. J. Food Sci. Technol.*, 22, 683, 1987.)

Processing factors affecting protein-water interactions

The time and temperature of chopping of comminuted products and tumbling of formed products must be carefully controlled during processing. Chopping and tumbling are required to disrupt the myofibril and to solubilize and extract the myofibrillar proteins as described above. However, excessive chopping or tumbling can lead to protein denaturation, usually due to increased temperature or excessive shearing. Thus, chopping and tumbling time must be optimized to maximize protein extraction while avoiding protein denaturation. Denaturation occurs when the native protein structure is destabilized and partially unfolded. Denatured muscle proteins usually form insoluble aggregates that have poor water-binding and film-forming abilities (see following section). Excessive chopping or tumbling may also lead to excessive disintegration of the muscle fibers and to a reduction in batter viscosity, which reduces the quality of the cooked gel network.

Protein-fat interactions

In coarsely chopped comminuted products, such as formed products and many sausages, fat is largely retained within intact fat cells. In these products, fat loss is not usually a problem during handling or cooking as fat is trapped within a cell membrane. The viscosity of the batter and the intact fat cell membrane prevent problems caused by fat instability.

In highly comminuted products, such as bologna and frankfurters, the fat cell is disrupted and fat droplets more typical of those found in emulsions may be formed. An emulsion is made of two immiscible phases, one of which is dispersed as fine droplets within the other continuous phase. In comminuted products, the fat droplets form the dispersed phase, while the continuous phase is comprised of water, protein, and salt. Energy is required to form an emulsion. This energy input occurs during comminution of the meat batter. In general, the greater the energy input, the smaller and more numerous are the fat droplets in the discontinuous phase of a meat batter.

At high temperatures and with sufficient energy input, the fat cell membranes are disrupted and the solid fat is melted and emulsified into liquid droplets. Most poultry fat begins to melt at about 13°C, but due to the variety of lipids present, poultry fat is not completely melted until a temperature of 33°C is reached. Fat droplets may be spherical when the fat is primarily liquid or irregular in shape when the fat is partially solid or crystalline. Liquid fat droplets are highly unstable and readily coalesce on standing. Coalescence is the process in which small fat droplets combine and form large, unstable fat droplets. Coalescence of fat is highly undesirable as it leads to several quality defects in comminuted products. If temperatures are kept low enough, the fat within the droplets may be partially crystallized and less likely to undergo coalescence.

In highly comminuted products, the liquid fat droplets must be stabilized to withstand the stresses of holding, pumping, and cooking. This is accomplished in two ways. First, the high viscosity of the meat batter helps to prevent coalescence of the fat. Second, the fat droplets are surrounded by a protein film that reduces the interfacial tension between the fat and water (the dispersed and continuous phases, respectively) and stabilizes the droplets.

The protein film is comprised of solubilized and extracted myofibrillar proteins. During emulsification, the solubilized and extracted proteins must diffuse to the surface of the oil droplet and then adsorb onto the surface of the droplet. Denatured proteins usually exist as large insoluble aggregates and do not diffuse as readily as smaller, soluble proteins. Once the protein is at the surface, its will unfold or rearrange such that polar regions of the molecule are oriented toward the water and non-polar or hydrophobic regions are oriented toward the oil droplet to minimize free energy. Also, the protein must be present in sufficient concentration so that the protein molecules can interact to form a continuous, stable film on the surface of the oil droplet. There must be a sufficient quantity of extracted protein so that all of the fat droplet surfaces are covered with a protein film. One reason highly comminuted batters are unstable is that very small droplets have a very large surface area and thus require more solubilized and extracted protein to form the stabilizing film. Myosin is the major component of the interfacial film surrounding the fat droplets and is thought to play a key role in stabilizing the fat droplets during holding and during the early stages of cooking.[10] An electron micrograph of the protein film at the surface of a fat droplet in a raw meat batter is shown in Figure 11.7.

Protein-protein interactions

Protein-protein interactions during cooking lead to the formation of a protein gel matrix. A protein gel is formed during heating when muscle proteins unfold and aggregate to form a continuous, defined solid cross-linked network or matrix. The formation of a continuous protein gel network has a large influence on the textural and sensory properties, as well as the cooking yields, of poultry products. The gelation of the myofibrillar proteins occurs during thermal processing of both comminuted and formed products and is probably the most important functional property in processed poultry products during cooking. However, connective tissue and sarcoplasmic proteins may interfere with the ability of the myofibrillar proteins to form a strong gel.

Myofibrillar proteins form thermally irreversible gels. This means that the cross-linkages or chemical bonds formed between proteins during heating are not appreciably altered by cooling or reheating. A schematic diagram illustrating the steps involved in the formation of a thermally irreversible myofibrillar protein gel is shown in Figure 11.8. When muscle proteins are heated they unfold or denature once a critical temperature is reached. In the second step, these unfolded molecules aggregate into small clumps to form an

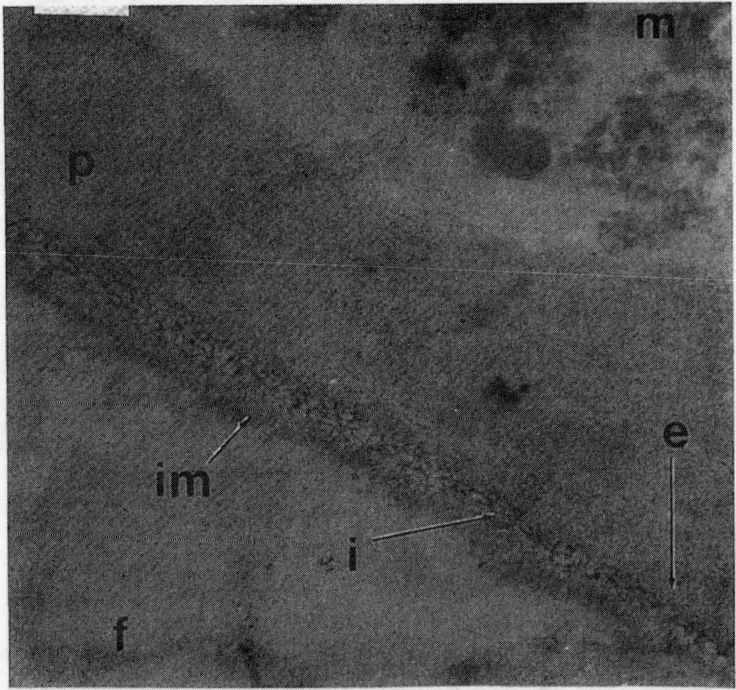

Figure 11.7 Electron micrograph showing the protein film formed on the surface of fat dropets in a highly comminuted poultry meat batter. f, fat droplet; p, proteinaceous material surrounding; i, interface between the fat droplet and the proteinaceous meat batter; m, matrix; e, outside of the protein film; im, inside of the protein film. (From Gordon, A. and Barbut, A., *Food Struct.*, 9, 77, 1990. With permission.)

increasingly viscous solution. The gel point is reached when the aggregates rapidly cross-link into a continuous gel matrix. Muscle protein gels are formed by a combination of hydrogen bonds, electrostatic interactions, hydrophobic interactions, and disulfide bonds. Upon cooling, slight changes occur in the relative importance of the chemical bonds forming the final gel matrix.

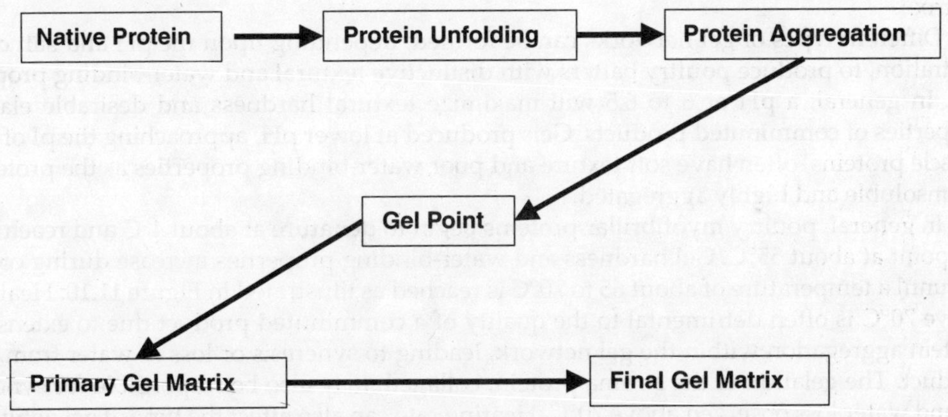

Figure 11.8 Diagram illustrating the steps necessary to form a heat-induced protein gel.

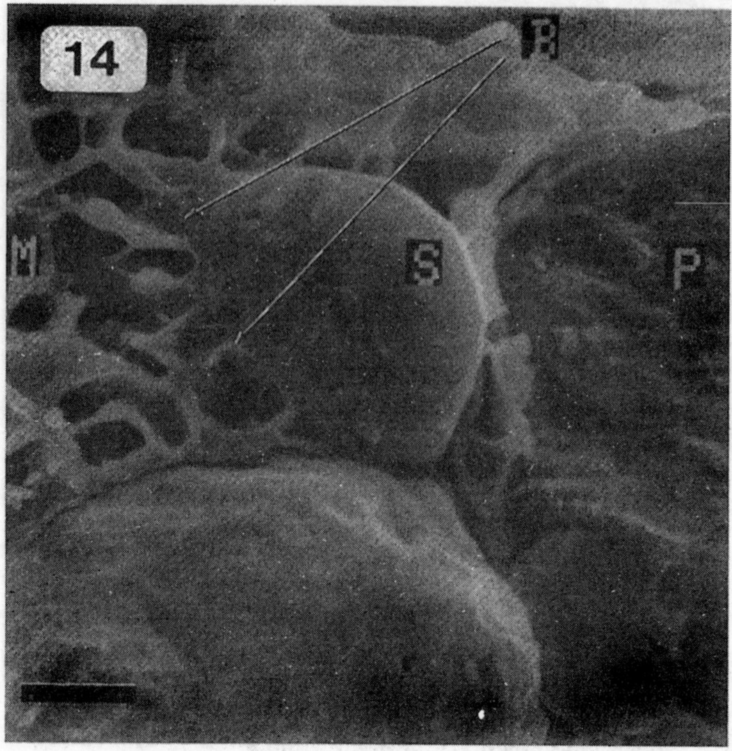

Figure 11.9 Electron micrograph of a cooked chicken meat batter made with 2.5% salt. M, protein gel matrix; S, fat droplet coated with protein; B, junction zone between protein film coating the fat droplet and the gel matrix. (From Gordon, A. and Barbut, A., *Food Struct.*, 9, 77, 1990. With permission.)

The microstructure of the gelled matrix of a poultry meat batter is shown in Figure 11.9. Protein gels hold large amounts of water within their network structure, bound by both chemical reactions and physical entrapment. The protein gel matrix physically restricts coalescence of fat within a cooked meat batter. Upon cooking, the interfacial protein film around the fat droplets also forms cross-links with the continuous protein gel matrix.

Different types of gel networks can be formed, depending upon the pH and salt concentration, to produce poultry batters with distinctive textural and water-binding properties. In general, a pH of 6 to 6.5 will maximize textural hardness and desirable elastic properties of comminuted products. Gels produced at lower pH, approaching the pI of the muscle proteins, often have soft texture and poor water-binding properties as the proteins are insoluble and highly aggregated.

In general, poultry myofibrillar proteins begin to denature at about 4°C and reach the gel point at about 55°C. Gel hardness and water-binding properties increase during cooking until a temperature of about 65 to 70°C is reached as illustrated in Figure 11.10. Heating above 70°C is often detrimental to the quality of a comminuted product due to extensive protein aggregation within the gel network, leading to syneresis or loss of water from the product. The gelation of the stroma protein, collagen, may also be responsible for syneresis and water loss observed above 70°C. Heating rate can also affect the type of gel network formed and subsequent quality of cooked poultry products. It is thought that a slower

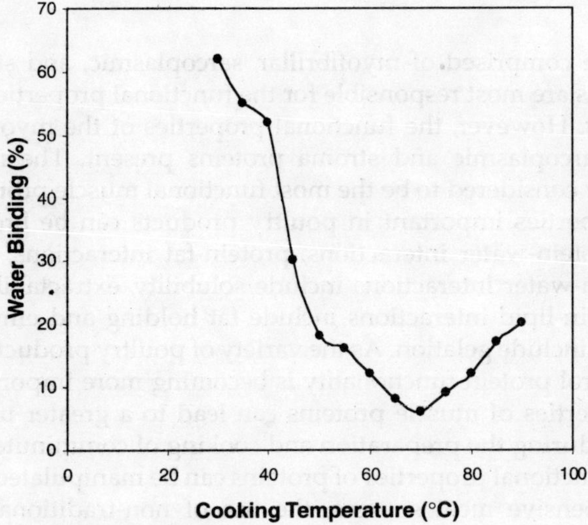

Figure 11.10 Effect of cooking temperature on the ability of ground meat to hold water added after cooking.

heating rate will result in the formation of more ordered gel structures with higher water-binding abilities. Thus, low fat frankfurters are cooked more slowly than their higher fat counterparts to form a protein gel network with higher water-binding ability.

Model systems in protein functionality research

Many different model systems have been used to study the functional properties of poultry meat proteins. Certainly, the easiest system to use is a whole muscle homogenate. However, it is often difficult to determine the true cause and effect in such a complex system, due to the myriad of ingredients and potential interactions. Simplified model systems have been used to limit the number of ingredients and decrease complexity. Researchers have used model systems comprised of fractionated proteins including: myofibrils, myofibrillar protein, salt-soluble protein, actomyosin, and even myosin, to try to understand how proteins function in a poultry product. Myosin has been studied extensively by biochemists, however, much of the work has been done under conditions of pH and salt concentration that are not typically found in poultry products.

In model systems, it is often difficult to compare work among researchers as the composition of the proteins in a particular fraction may change due to preparation procedures. For example, the salt-soluble protein fraction is comprised of a mixture of 15 or more proteins that all interact on heating. It is well known that the composition of the salt-soluble fraction can change depending on extraction conditions and starting material. Thus different amounts of total myosin or different ratios of actin to myosin in the salt-soluble fraction may affect the results obtained. Due to these limitations it is necessary to select a test system with care. For product development work, it may be best to work with the actual product or to select a system as similar to the product as possible. For more basic research, it may be best to start with a simplified system, such as pure myosin, and then move toward more complex systems to determine if the relationships discovered in a simple system are still true when other components are added.

Summary

Muscle proteins are comprised of myofibrillar, sarcoplasmic, and stroma fractions. The myofibrillar proteins are most responsible for the functional properties typically observed in poultry products. However, the functional properties of the myofibrillar proteins are modified by the sarcoplasmic and stroma proteins present. The myofibrillar protein, myosin, is generally considered to be the most functional muscle protein.

Functional properties important in poultry products can be broadly classified into those involving protein-water interactions, protein-fat interactions, and protein-protein interactions. Protein-water interactions include solubility, extractability, water retention, and viscosity. Protein-lipid interactions include fat holding and emulsification. Protein-protein interactions include gelation. As the variety of poultry products increases, the need to modify and control protein functionality is becoming more important. Understanding the functional properties of muscle proteins can lead to a greater understanding of the changes that occur during the preparation and cooking of comminuted and formed poultry products. The functional properties of proteins can be manipulated to allow for the utilization of less expensive meat sources, the use of non-traditional meat sources, the improvement of existing products, and the more efficient utilization of non-meat ingredients. The functionality of poultry proteins can also be manipulated to control processing and energy costs, as well as reduce production waste.

References

1. Kinsella, J. E., Functional properties of proteins in foods: a survey, *Crit. Rev. Food Sci. Nutr.* 7, 219, 1976
2. Forrest, J. C., Aberle, E. D., Hedrick, H. B., Judge, M. D., and Merkel, R. A., *Principles of Meat Science*, W. H. Freeman, San Francisco, CA, 1975.
3. Bechtel, P. J., *Muscle as Food*, Academic Press, New York, 1986.
4. McCormick, R., Structure and properties of tissues, in *Muscle Foods: Meat, Poultry, and Seafood Technology*, Kinsman, D. M., Kotula, A. W., and Breidenstein, B. C., Eds., Chapman and Hall Publishers, New York, 1994, 106.
5. Foededing, E. A., Lanier, T. C., and Haltin, H. O., *Characteristics of Edible Muscle Tissue*, Fennema, O. R., Ed., Marcel Dekker, New York, 1996, 880.
6. Damodaran, S., Amino acids, peptides and proteins, in *Food Chemistry*, Fennema, O. R., Ed., Marcel Dekker, New York, 1996, 322.
7. Smyth, A. B., O'Neill, E., and Smith, D. M., Functional properties of muscle proteins in processed poultry products, in *Poultry Meat Science*, Richardson, R. I. and Mead, G. C., Eds., CABI Publishing, Oxford , U.K., 1999, 337.
8. Bailey, A. J. and Light, N. D., *Connective Tissue in Meat and Meat Products*, Elsevier Applied Science, London, U.K., 1989.
9. Richardson, R. I. and Jones, J. M., The effects of salt concentration and pH upon water-binding, water-holding and protein extractability of turkey meat, *Int. J. Food Sci. Technol.*, 22, 683, 1987.
10. Gordon, A. and Barbut, A., The role of the interfacial protein film in meat batter stabilization, *Food Struct.*, 9, 77, 1990.

chapter twelve

Formed and emulsion products

Jimmy T. Keeton

Contents

0-8493-0120-3/01/$0.00+$.50

Introduction

Per capita consumption of chicken in the U.S. in the 1970s averaged approximately 26 lb (1975), but since then has risen to 54 lb in 1999.[1] This significant growth in consumption was the result of a host of new, convenient, brand-name products, items with added value and further processed products for food service. The poultry industry has benefited from the health-related concerns about animal fats, broadened its product array available to consumers, and offers a significant price differential between its products and other meat species. These products include portioned, seasoned cuts, batter-breaded patties and nuggets, sliced meats for delis, luncheon meats for sandwiches, and a large variety of low fat cured, cooked items like "turkey ham," "turkey bacon," frankfurters, and bologna. Further processed poultry will likely continue to grow due to the expansion of fast food outlets, home meal replacements from supermarkets and restaurants, and the continued growth of food service in the health care sector.

Product categories

Formed (sectioned and formed, restructured) products

Formed meat products may be produced by sectioning muscle pieces and combining with a ground or emulsified myofibrillar protein binder and a chilled brine. Restructured items have a smaller particle size which is reduced by grinding, flaking, dicing, chopping, slicing, or emulsifying. The particles are then mixed (Figure 12.1) with an appropriate binding material and formed into a specific portion size. Sectioned products are primarily intact muscles and have a more "whole-muscle" texture than restructured items. Because of the similarity of these two types of formed products, they will be discussed simultaneously with limited distinction between the sectioned and formed or restructuring processes. Examples of products in this category include poultry/turkey rolls, "fillets," poultry roasts, poultry patties, nuggets, loaf items, turkey bacon, and turkey ham. Some items may be coated with a batter-breading, precooked and packaged for reheating in the microwave, deep fat fryer, or conventional oven. Products such as poultry rolls and turkey ham have a Standard of Identity and must meet certain requirements as outlined in the Code of U.S. Federal Regulations[2] §381.159 and §381.171, respectively. For example, "turkey ham" specifies a labeling requirement of "chunked and formed" for thigh pieces ≥0.5" while pieces <0.5" are labeled "ground and formed" or "chopped and formed."

Sectioned and formed poultry products are prepared from well-chilled (−2.2 to 1.6°C or 28 to 35°F) whole muscle pieces or chunks that have been defatted and injected with or marinated in a salt brine containing alkaline phosphates. If cured, sodium nitrite and sodium erythorbate are added to the brine. A functional protein is required to coat the meat particle surfaces, form an interwoven network between meat pieces, and then coagulate when heated to form a solid tissue mass with a meat-like texture. Cold-set binders are available to bind meat pieces without the need for heat coagulation of the myofibrillar proteins. These include the hydrocolloid sodium alginate which is cross-linked with a calcium salt, transglutaminase, and a fibrinogen-thrombin combination. When using enzymatic binders, processing time may need to be reduced, product temperature kept near freezing to slow the enzymatic reaction, and meat surface moisture minimized to enhance the polymerization reaction occurring between the protein molecules.

Formed products offer the advantages of being: (1) boneless; (2) easily portioned into an appropriate size and shape; (3) lower in cook loss and higher in serving yield, having virtually no waste; (4) uniform in composition for better brine or cure distribution; (5) able to utilize whole muscle pieces with otherwise less utility; and (6) easier to heat, slice, and serve.[3] Obvious limitations are (1) low quality poultry pieces cannot always be improved; (2) formed products require more equipment, manufacturing technology, additional molds or casings, and handling considerations to avoid pathogen contamination; (3) shelf-life may not be as long as whole-muscle, non-marinated products; and (4) further processing requires a high input of labor and capital.

Emulsified (comminuted) products

Emulsified (comminuted) poultry products such as frankfurters, bologna, or loaf items are typically manufactured from chilled or frozen mechanically deboned poultry or turkey (MDP, MDT) as described in Chapter 14. These fully cooked products are more cost effective than their red meat counterparts, contain approximately half the maximum fat (30%) allowed by the USDA-FSIS, and are convenient to prepare. The USDA-FSIS permits up to 15% MDP or MDT in comminuted red meat products with the appropriate label declaration, but poultry products may contain 100% MDP or MDT as long as these conform to the CFR[2] specifications found in §381.173 and §381.174.

Red meat comminuted products, which are defined in the CFR[2] §319.180 have a Standard of Identity with a maximum fat content of 30%. Added water in the finished product is limited to no more than 4 × meat protein content % + 10% while the combination of added water + fat may not exceed 40% (this allows for varying amounts of fat and added water in the product, i.e., 5% fat + 35% added water = 40%). Additional binders

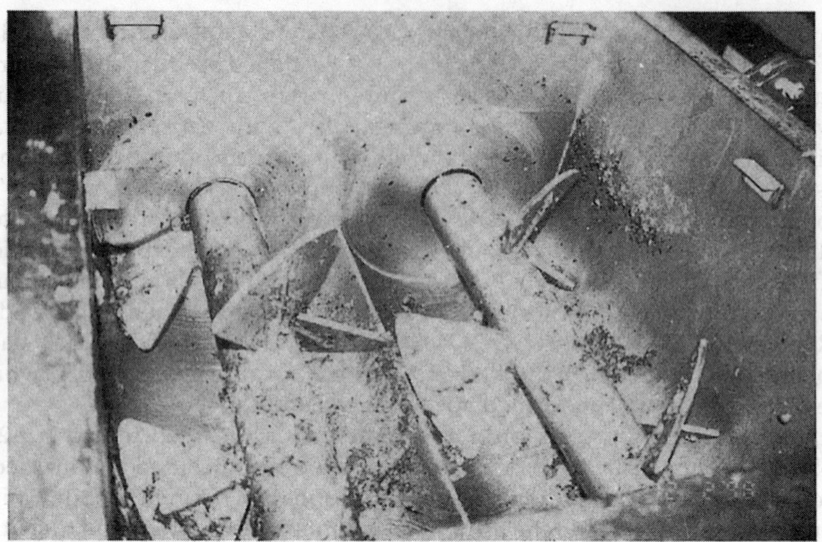

Figure 12.1 View inside a mixer/blender showing the mixing paddles.

such as soy protein concentrates and isolates are allowed at levels of 3.5 and 2.0% (dry weight basis, DWB), respectively, with appropriate label declarations. In comparison, poultry products do not have these limitations and have greater formulation flexibility with regard to meat type and content. If a product contains 50% or more poultry meat it may be labeled as a poultry item.

Emulsified poultry products are processed by homogenizing MDP or MDT in a bowl chopper with iced water, salt, cure, dextrose, alkaline phosphates, corn syrup solids, modified starch, spices, sodium erythorbate, and other additives to an end point temperature of approximately 10°C (50°F). Further processing may require passing the batter through an emulsion mill to further reduce particle size and obtain a smooth texture. Batter temperatures should not exceed 12.7°C (55°F) to avoid overheating the fat that would result in processing defects (fat caps, fatting-out) during thermal processing. Sausages are then vacuum encased in a cellulose casing (frankfurters) or moisture-proof fibrous casing (bologna) and fully cooked using a multiple-stage cooking cycle. Application of smoke to the product may be in the form of: (1) a liquid drench applied to the casing surface (sausages); (2) atomization of liquid smoke in the smokehouse; (3) incorporation of liquid smoke into the product formulation; or (4) natural smoke generated from hardwood sawdust. A minimum internal temperature of 68.3°C (155°F) is required if the product contains ≥100 ppm nitrite or 71.1°C (160°F) if <100 ppm of nitrite is present.

Raw materials

Formed (sectioned and formed, restructured) and emulsified products

Raw materials for formed and emulsified products include skeletal muscles such as boneless breasts, legs, thighs, desinewed drumsticks, and MDP/MDT (with or without skin). These materials may be fresh chilled or frozen, but should be of high quality (minimal purge, no off-color, no off-odor, no apparent microbial growth).

Specifications and sampling of raw materials

Specifications are typically developed for raw materials to minimize variations in quality, composition, and cost. These are product dependent and require some form of sampling and subsequent analysis to ensure that the products purchased meet the company's needs or a product's requirements. Net weights of raw materials and their composition — fat, moisture, protein content — should be monitored upon receipt from each supplier and variances noted for appropriate credits or debits.

Sampling procedures may involve randomly selecting boxes of fresh or frozen raw materials from a pallet and core drilling at several points throughout the meat block to obtain a composite sample. Other sampling procedures may include random sampling by quadrants. Samples may be recombined, ground to increase homogeneity, subsampled, and tested for fat content or some other specification parameter. As a general rule, a sample size of 10% of the batch or lot could be used, but in actual practice, only 1% of a batch or lot may be sampled. However, a statistically valid sample size may be determined by the following formula:[4]

$$n = (3s \div E)^2$$

Where: n = Number of samples to be taken

3s = Estimate of three standard deviations among all sampling units in the lot

E = Maximum allowable difference between the estimate of the sample parameter being measured and the true value of the parameter in the lot

For example, if a company purchases frozen MDP from different suppliers and the material ranges in fat content from 12 to 18% fat, how many 1 lb samples would be required to be taken from a 1000 lb pallet (lot) assuming the maximum allowable difference between the estimate of fat and the true fat content is ±1%?

Calculations:

$$s = \text{Range} \div 6 = [18 - 12] \div 6 = 6 \div 6 = 1$$

$$3s = (3)(1) = 3$$

$$E = \pm 1\%$$

$$n = (3 \div 1)^2 = (3)^2 = 9$$

Thus, nine, one-pound random samples would be taken from the 1000 lb pallet (lot). This assumes that the samples are randomly taken and that the material sampled has been homogenized to ensure a uniform sample for chemical analysis.

Raw material condition

Temperature. Fresh cuts or raw materials should not be above 4.4°C (40°F) when received and frozen materials should be below −17.8°C (0°F). Sample temperature checks should be made in the geometric center of boxes or combos (large containers of 1000 lb or more). If poultry muscles are received in combos, pull pieces from the center of the container and check the internal temperature. Pieces having temperatures of >7.2°C (45°F) may not have been chilled adequately prior to fabrication or have been temperature abused in transit. In either case, the higher receiving temperature means a potential decrease in shelf-life of the product, accelerated spoilage, or a potential pathogen risk. Real-time recording thermometers may be placed in refrigerated trailers and monitored for temperature deviations during transit.

Appearance/color/off-odor. Browning, graying, or two-toning of the muscle pigments may indicate prolonged storage after fabrication, temperature abuse, or early microbial spoilage of the raw material, while greening, slime formation, putrefaction, souring, musty aromas, or other off-color/aroma characteristics are signs of apparent spoilage. Accidental contamination by approved ingredients, lubricants, chemical compounds, cleaning agents, or holding conditions can also cause discoloration of muscle pieces and trimmings. These can include sanitizing chemicals (chlorine, iodine, or ammonium ions), sulfites (permanent "red" color), microbiological pigments (orange, brown, black, green), and a chalky, dry surface on frozen muscles that indicates freezer burn (freeze dehydration). Excessive purge present in boxes or combos can indicate poor freezing conditions or premature thawing of the product. Quality defects such as pale, soft, and exudative (PSE) tissue or dark, firm, and dry (DFD) muscles can result in reduced product yields and poor product quality (Lawrie, 1991). Likewise, lipid oxidation of the fat may produce rancid or stale off-odors due to inadequate packaging during storage or especially when precooked products are reheated.

Adulterated material. Adulterants can make raw materials illegal, inedible, or unwholesome and may include: denaturant dye on condemned tissues, cross contamination with other meat species, foreign matter (gloves, glass, wood, plastic, metal, bone, knives, meat hooks, paper, etc.), chemical residues, rodent droppings, insects, or other contaminants. Visual and physical inspection of raw materials should be conducted on a periodic basis to determine the extent of adulterants occurring in the product.

Refrigerated/frozen guidelines. Carcasses are chilled immediately after slaughter to reduce the temperature to <4.4°C (40°F) within 4 and 8 h for broilers and turkeys, respectively.[5] A weight loss of 0.5% will typically occur during fabrication of muscles used in further processing. Muscle pieces may be iced in combos or boxed and held in a cooler maintained at −2.2 to 2.8°C (28 to 37°F) for 24 to 36 h. Trimmings held for longer than 36 h should be boxed and frozen quickly to preserve quality. The high relative humidity (~85%) in most coolers assists in preventing carcass shrink and moisture loss.

Muscle pieces, MDP, or MDT destined for frozen storage are boxed in plastic-lined or waxed, cardboard containers for subsequent freezing immediately after fabrication. Generally, the lower the temperature and the more protection from atmospheric oxygen, the greater the reduction in oxidative rancidity and extension of storage life. At ≤−10°C (14°F) most microbial growth and enzymatic activity are reduced to almost zero because most of the cellular water molecules are fixed in a crystaline structure, but reactions may continue slowly down to −80°C (−112°F). Most commercial holding freezers range from −17.8 to −28.9°C (0 to −20°F) while air-blast or Instant Quick Freeze (IQF) freezers use high air velocity (2500 ft[762 m]/min at ≤−28.9°C [−20°F]) to rapidly remove the heat. Powdered carbon dioxide (CO_2) or CO_2 "snow" (−62.2 to −78.3°C [−80 to −109°F]) may be dusted among muscle pieces or MDP/MDT prior to boxing to accelerate the freezing process. However, caution should be used to avoid suffocation due to sublimation of the CO_2 if the boxed product is stored in a closed area such as a refrigerated trailer. If CO_2 snow is used in mixer/blenders, they should be properly vented to avoid the risk of displacing oxygen in the air.

In any freezing application, raw or finished products must be packaged to exclude air and protect the surface from excessive drying (freezer burn). Poultry muscle that is frozen

Table 12.1 **Maximum Recommended Length of Storage of Different Species of Meat at Various Temperatures for the Preservation of Optimum Quality**

Species	−12°C	−18°C	−24°C	−30°C
	Months			
Beef	4	6	12	12
Lamb	3	6	12	12
Pork (fresh)	2	4	6	8
Poultry	2	4	8	10

Source: Adapted from Hedrick, H. B., Aberle, E. D., Forrest, J. C., Judge, M. D., and Merkel, R. A., *Principles of Meat Science,* 3rd ed., Kendall/Hunt Publishing, Dubuque, IA, 1989.

and held at −17.8 to −28.9°C (0 to −20°F) should retain its quality for 6 to 10 months (Table 12.1). The least desirable temperature for holding frozen meat trimmings is −11.1 to −10°C (12 to 14°F), which is the point of phase transition between intercellular crystalline ice and a combination of ice and water. Frequent cycling of the refrigeration system through this temperature zone causes large ice crystal growth in muscle cells and excessive purge (water loss) when thawed.

Thawing and refreezing. Proper thawing prevents excess purge loss and the risk of microbial growth on muscle pieces and in MDP/MDT. Keeping the product in the packaging material during thawing prevents dehydration and drip loss. Raw materials are often thawed or tempered over a 2- to 3-day period in a cooler held at 0 to 2.8°C (32 to 37°F) until the product temperature reaches −3.3 to −2.2°C (26 to 28°F). Boxed product without metal staples or banding can be thawed more quickly by conveying through a microwave tunnel followed by holding approximately 8 h under refrigeration to allow for external-to-internal temperature equilibration prior to processing. Refreezing previously frozen product causes loss of proteins, flavor and juiciness, and excessive drip. It also poses some risk for subsequent microbial growth and increases product deterioration and is not recommended.

Raw materials should be dated and coded upon receipt, tracked through to final product form, and temperature/processing records should be kept with a designated lot number. Careful monitoring should be performed and raw products rotated through processing on a "first in, first out" basis.

Factors affecting functional properties. The most beneficial functional properties of raw poultry muscle tissues, MDP, and MDT are their ability to retain fluids (water-holding capacity; WHC) and bind (cohesiveness) meat pieces such that the finished product has a whole-muscle texture or that of a solid, homogenous emulsified product matrix. Critical to these properties is the amount of total myofibrillar protein available for binding and WHC, the ratio of moisture to total protein (M:P ratio), and the amount of myofibrillar protein in relation to sarcoplasmic and connective tissue proteins. Lean poultry muscle tissue contains approximately 19 to 23% protein while MDP or MDT without skin contains 14 to 16% protein and with skin 11 to 12% protein. Raw materials with approximately 16% protein or higher would be typically classified as good binders for water retention and meat particle binding. However, the type (myofibrillar, sarcoplasmic, connective tissue), physicochemical condition (PSE or DFD), and ratio of the protein types primarily determines their functional properties.

Sausage formulations use a bind index as an indirect measure of the amount of myofibrillar or salt-soluble protein available for binding and are often expressed in arbitrary units of 0.0 to 1.0, 0.0 to 30.0, 0 to 100, or 0 to 1000. Irregardless of the number, bind indices are based on pre-rigor bull meat being assigned the highest bind index value (i.e., 1.0, 30, 100, or 1000).[6,7] Whole-muscle poultry and turkey have bind indices of approximately 90 (pre-rigor bull meat = 100) while MDP may be only 50 to 60. The bind index along with color (the amount of myoglobin in lean tissue, 0.5 to 4 mg/g) and collagen (the amount of collagen in lean tissue, 2%) indices are combined with compositional characteristics (moisture, crude fat, total protein), processing constraints (fat content, moisture content, minimum bind, color or collagen value, etc.), and price constraints into least cost regression programs to formulate specific products such as frankfurters or bologna. Explanation of the use of Least Cost Analysis (LCA) may be found in Pearson and Gillett,[3] Romans et al.,[4] LaBudde,[7] and ROI.[8]

The M:P ratio can be used as a guide to WHC and binding ability. In general, raw materials with an M:P ratio of <3.6:1 are good binders while those with ratios of >4.0:1 are poor binders. Exceptions to this rule exist since a raw material can have a high protein content due to a high amount of collagen, but be a poor binder. Stable emulsions generally require >45% of the total protein in the formulation to be myofibrillar (high ionic strength or salt-soluble) protein with a maximum of 30% sarcoplasmic (low ionic strength or water-soluble) protein. Insoluble or connective tissue proteins should be limited to <25% of the total protein. Adjustments in raw materials may also be needed if the material has been frozen, has indications of being PSE, or if the pH is lower than expected (i.e., 5.3 vs. 6.0). Collagen has limited particle binding capacity but some WHC after converting to gelatin between 60 to 70°C (140 to 158°F).

The two factors that affect WHC and binding ability of muscle tissues most are final pH (net charge of the myofibrillar proteins) after resolution of rigor mortis and the degree of contraction of the muscle tissues (steric effect).[9] At ~pH 5.1 myofibrillar proteins have a net charge of 0 and retain the least amount of water. Ingredients and muscle conditions or treatments which tend to increase the muscle pH also increase the tissue's WHC. However, in muscle tissue exhibiting PSE characteristics, the denatured proteins do not respond well (no increase in WHC) to the increase in pH. Salt (NaCl) levels normally used in further processed or cured products (2 to 3%) increase protein solubilization and swelling to allow increased fluid retention. Alkaline phosphates in combination with salt and mechanical agitation in a mixer/blender, vacuum tumbler, or massager increase pH and myofibrillar protein extraction and solubilization.

The physical and chemical characteristics of poultry fat affect the processing characteristics of emulsified sausages and product stability. During the emulsification phase of processing, poultry batter temperature and chopping times should be monitored to avoid melting the fat globules. Listed in Table 12.2 are the fat melting ranges of species that might be incorporated into an emulsion product.

Table 12.2 **Fat Melting Ranges of Species that Might Be Incorporated into an Emulsion Product**

Specie/fat source	Melting point	Final chopping temp
Poultry/abdominal fat	80–110°F	52–55°F
Pork/back fat	86–104°F	58–62°F
Leaf fat	110–118°F	
Beef/subcutaneous fat	89–110°F	68–73°F
Kidney fat	104–122°F	
Lamb/subcutaneous fat	90–115°F	Same as beef
Kidney fat	110–124°F	

Non-meat ingredients

Preservation and curing

Non-meat ingredients for value-added poultry formulations serve to:

- Control cost of the product
- Extend product storage-life
- Increase meat particle binding to be more muscle-like
- Increase water holding capacity to improve yields
- Improve juiciness and succulence
- Modify textural characteristics
- Replace fat in the product (low fat, diet products)

Common non-meat ingredients used in further processed products include:

- Salt and alkaline phosphates (sodium tripolyphosphate)
- Sweetners like dextrose, sucrose corn syrup solids, and sorbitol
- Sodium or potassium nitrite (cured products) combined with sodium or potassium erythorbate or ascorbate (cure accelerators)
- Sodium or potassium lactate
- Sodium acetate and diacetate
- Liquid smoke
- Antioxidants like butylated hydroxy anisole (BHA), butlyated hydroxy toluene (BHT), propyl gallate (PG), alpha tocopherols, and spice extractives
- Seasonings, spices, and flavorants

Salt and alkaline phosphates (sodium tripolyphosphate)

Salt (NaCl or a NaCl/KCl blend) is the most basic curing ingredient and meat preservative that easily dissolves in water (26.4% = 100°S brine)[10] to form a brine or in the moisture provided by the meat tissues. It serves to: (1) flavor the product; (2) lower water activity and increase ionic strength which retards microbial growth; (3) assist with solubilization of muscle proteins that in turn serve as meat particle binders; (4) dehydrate muscle tissues at high concentrations (5 to 8%) for drying applications; and (5) act as a synergist in combination with sodium nitrite to prevent the outgrowth of *Clostridium botulinum*. Salt use is not regulated by USDA-FSIS because it is self-limiting. Formed and emulsified poultry products typically range from 1.5 to 3% salt. Pure salt should be used for processed products since contaminants (metals, halophilic bacteria, etc.) may cause off-flavors, interfere with the curing reactions, accelerate oxidative rancidity, and shorten product shelf-life. In curing brines, non-iodized salt is recommended.

Alkaline phosphates (sodium or potassium tripolyphosphate) are incorporated into poultry products or brines to: (1) increase WHC of muscle proteins, preserve juiciness, and increase product yield; (2) aid in the extraction of salt-soluble muscle proteins in synergy with salt for subsequent binding of meat pieces when cooked; (3) preserve the color of cured products; (4) enhance meat flavor; (5) retard oxidative rancidity by chelating metal ions; and (6) reduce expressed fluid (purge) in vacuum-packaged products. Combinations of sodium tripolyphosphate, sodium hexametaphosphate, or tetrasodium phosphate may be used, but total phosphate may not exceed 0.5% in the finished product.[2] Potassium salts, rather than sodium, may also be substituted to reduce total sodium content. Use of alkaline phosphates at levels near 0.5%, may cause a "slick" or "soapy" taste in products, decrease the rate of color development in small diameter products with fast cooking rates, and produce a rubbery texture in very lean products. Thus, most formulations incorporate 0.3 to 0.4% phosphate in the product. In some cases, phosphates will crystallize on product surfaces as diphosphate.

Alkaline phosphates are very corrosive in brines and should be kept in stainless steel or plastic containers. Pyrophosphates, diphosphates, tetrasodium pyrophosphate, and sodium acid pyrophosphate at pH 7 perform best in sausage emulsions while sodium tripolyphosphate and sodium hexametaphosphate are used in curing brines since these slowly hydrolyze to diphosphate in the presence of muscle phosphatases. Tetrasodium pyrophosphate is a good binder but highly caustic (pH 11). Because alkaline phosphates are not very soluble, they are the first ingredient added to a brine formulation followed by high shear mixing. Brine should then be refrigerated and held at 0°C (32°F) prior to pumping.

Curing salts

Curing salt (6.25% sodium or potassium nitrite) is typically bonded to salt crystals (93.75%) and colored pink to avoid confusion with other white crystalline ingredients such as salt or sugar. It is used to cure poultry products such as turkey ham, turkey bacon, frankfurters, and bologna. Nitrite serves to: (1) react with myoglobin to give cured meat its characteristic pink color and cured flavor; (2) prevent botulism by retarding the outgrowth of *C. botulinum* or other pathogenic microorganisms; and (3) retard lipid oxidation. In the U.S., nitrates are not allowed in poultry products, only nitrites. *Nitrites are controlled ingredients and must be used according to specific regulations.*

Different product categories require different levels of nitrite. For example, use of nitrite is not permitted to result in more than 200 ppm nitrite in finished products such as turkey ham and loaf items. Nitrite is allowed in sausages at levels not to exceed 156 ppm based on the meat weight of the product. Nitrite is limited to 120 ppm in "bacon" and must be accompanied by 550 ppm of sodium erythorbate or ascorbate (cure accelerator) to prevent the formation of carcinogenic compounds called nitrosamines. Products such as poultry rolls are not usually cured so that the meat will remain white, but turkey hams and turkey rolls may be cured to give the same color as a cured pork product.

Sweeteners — dextrose, sucrose, corn syrup solids, sorbitol

Sweeteners such as sucrose (sugar) and dextrose are added to poultry products, marinades, or brines to enhance flavor, increase moisture retention, reduce the harshness of salt, increase browning, and reduce costs. In the case of fermented meats, dextrose (0.5 to 1.0%) is added as a nutrient for starter culture bacteria and is converted to lactic acid to give semi-dry and dry sausages their characteristic "tangy flavor."

Sucrose has a high sweetness value relative to other sugars and will caramelize when heated. Because of the browning flavor during cooking, it is preferred in fresh sausages and grilled frankfurters, but is not desirable for products cooked on a rotisserie. The threshold level for taste of sucrose is 0.5 and 0.6% for dextrose, but most consumers prefer about 1% sugar in cured "ham-type" products. For pumping brines or marinades, processors may use 2.2 to 3.33 kg sucrose/100 kg of brine, but regulations allow up to 17.77 kg/100 kg of brine, depending upon the pump level.

Dextrose or glucose (corn sugar) is 70 to 80% as sweet as sucrose, a reducing sugar and the sweetener most often used. Dextrose is preferred as the sweetener of choice in the production of fermented sausages, because of its rapid utilization by bacteria.

Corn syrup, a byproduct of the corn sugar industry, is formed by the breakdown of starch resulting in dextrose, maltose, higher sugars, dextrins, and polysaccharides. Corn syrup is 40 to 50% as sweet as sucrose and will char or brown when exposed to high cooking temperatures. Corn syrup sweeteners are sold on the basis of their dextrose equivalent (DE) or percentage of dextrose. Most range from 40 to 50 DE and are used in sausages at levels of 2%.

Sorbitol is a non-browning, polyhydric alcohol that occurs in many berries as D-glucitol. It is 60% as sweet as sucrose and is used in skinless frankfurters to increase peeling ease and to retard caramelization and charring/browning of rotisserie-grilled products (frankfurters). Sorbitol is not to exceed 2% of the weight of the formula.

Liquid smoke

Liquid smoke is an aqueous smoke flavoring that contains acids, phenols, and carbonyls. Acids contribute to skin formation on sausages, accelerate the nitrite cure reaction, and provide tartness to flavor. Phenols are the primary flavor components and serve as antioxidants, antimicrobial agents, and minor coloring agents. Carbonyls are the primary color-forming compounds that contribute some flavor and serve to cross-link proteins. Use levels vary from 0.1 to 0.4% with an average value of 0.25% to contribute a desirable smoke flavor.

Antioxidants

Fat-soluble compounds such as butylated hydroxyanisole (BHA), butylated hydroxytoluene (BHT), tertiary butylhydroquinone (TBHQ), and propyl gallate (PG) retard fat oxidation in poultry products and cannot exceed 0.01% (singly) or 0.02% (combined) based on the fat content. Alpha-tocopherols (vitamin E) are limited to 0.03% (singly) or 0.002% (combined) based on the fat content of the product. Citric acid, ascorbic acid, and phosphoric acid serve to chelate heme iron and improve the effect of the antioxidants. Spice extractives such as deodorized rosemary, sage, and garlic also have natural antioxidant properties and are used to retard oxidative randicity.

Ingredients that enhance meat protein functionality

Salts — sodium and potassium chloride

Salt-soluble myofibrillar proteins (SSP) are the primary contributors to the fat- and water-binding abilities of meat tissues. The action of salt (1.5 to 3.0%) plus the moisture in the tissues and added water, combined with the mechanical energy of a mixer/blender or vacuum tumbler, serve to extract SSP to the surface of meat pieces. When the pieces are vacuum stuffed into a casing or mold and cooked, the protein undergoes coagulation, binding the product together into a cohesive mass with textural characteristics similar to whole-muscle products. Almost all further processed products have some salt added to the formulation.

Potassium chloride is most often used at levels up to 0.75% in a 60:40% NaCl:KCl combination to reduce the total sodium content of meat products. Higher levels of potassium chloride may produce a bitter or metallic flavor and should be adequately tested before complete product reformulation.

Alkaline phosphates

Alkaline phosphates are used primarily to improve moisture retention, retard oxidative rancidity, enhance color, and improve flavor. They act synergistically with salt to solubilize SSPs and increase the WHC of the proteins by increasing pH. Alkaline phosphates tend to preserve a pink color in meat and poultry even after cooking which is an advantage to cured products, but a disadvantage to uncured products, especially restructured poultry rolls with high pH. Phosphates overall are desirable for use in poultry products and are generally used at levels <0.4% to prevent a "soapy" or "slick" aftertaste.

Transglutaminase

Transglutaminase is an enzyme (protein) capable of cross-linking muscle proteins and binding pieces of poultry, meat, and seafood. Application of the enzyme directly to muscle surfaces enables the formation of a solid muscle mass through cross-linking of the muscle protein and contributes a whole-muscle like texture to the product. Normally, transglutaminase is dependent upon calcium to become active, but a non-calcium-dependent transglutaminase produced by *Streptoverticillium mobaraense*[11,12] has been found. This non-calcium-dependent enzyme allows cold-set (without heating) binding in a variety of products (restructured items, sausages, injected products, cheeses, and frozen desserts). Transglutaminase binding may be initiated by applying a powdered form directly to the surface of muscle pieces, by incorporation into liquid marinades (0.65 to 1.5%) and brines for injection, or by direct addition (0.1 to 0.3%) to emulsified sausages. Binding begins within 30 min after application and continues for a few hours at refrigeration temperatures. Thus, transglutaminase may be applied to a number of products to improve textural characteristics and product cohesiveness, and may also be used in combination with salt, alkaline phosphates, and cure ingredients.

Fibrimex® (bovine blood plasma proteins)

Fibrimex is a cold-set protein binding agent that is applied (7 to 10% of the meat weight) to the surface of meat pieces and allowed to cross-link or polymerize for 6 to 8 h to form a solid muscle mass. The binding system consists of two blood clotting components — fibrinogen (20 parts) and thrombin (1 part) — that are derived from bovine blood. The components are shipped frozen and when thawed and mixed together, the polymerization reaction is initiated. This reaction is temperature sensitive with the clotting components reacting slowly when cold (2 to 4°C) and becoming more active as the temperature increases (10 to 25°C). Fibrimex is applied to the surface of muscle pieces (loins) or other whole-muscle cuts and the pieces shaped by stuffing into a mold or casing (must be performed within 30 min). The pieces are then allowed to cross-link overnight at refrigeration temperatures (2 to 4°C). The resulting product is very similar to a whole-muscle cut that can be sliced raw and prepared as desired.

Alginate

Alginates are extracts from brown algae that are used as gelling agents, purge controllers, and texture modifiers. Means and Schmidt[13] used alginate to bind raw meat pieces together to form a restructured steak without the use of salt, alkaline phosphates, or the need for freezing prior to slicing. Alginate forms a heat stable gel that "cold-sets" at refrigeration or room temperatures. When used as a binder, sodium alginate (0.4%) may be slowly mixed with poultry breasts or muscle pieces. Encapsulated calcium lactate (0.4%) is added and the meat mass blended 3 to 5 min to distribute the ingredients. The product is then stuffed into a mold or casing and allowed to gel or cold-set 7 to 10 h. The product can then be sliced raw for grilling or cooked in a casing as desired. Alginate is unique in that binding is retained in products processed under retort (canning) conditions.

Ingredients to retain moisture and modify texture

Proteins

Soy protein flours, concentrates, and isolates. Soy proteins are the most utilized non-meat protein ingredients and are categorized as flours, concentrates, and isolates on the basis of their dry-weight protein content (50, 70, and 90%, respectively). Functional concentrates and isolates are nutritious proteins used mostly as binders to control purge, increase brine retention in injected products, and reduce cost, while maintaining a meat-like texture and appearance.[14] Concentrates may be incorporated into ground poultry products at levels up to 11% while isolate use levels range between 1 to 2% (DWB). These products have a low flavor profile and when heated, form a gel matrix comparable to the muscle tissues in appearance, texture, and color. Additional flavorants such as hydrolyzed proteins are recommended for highly extended products to avoid the effects of meat flavor dilution. Isolates are used mostly for marinated and brine-injected products requiring dispersibility, while concentrates are suited for fillings, sausages, and restructured products.

Milk proteins — whey and caseinate. Nonfat dry milk, sodium caseinate, and whey protein concentrates and isolates are nutritious milk proteins used as emulsifiers and water binders. Milk protein additives produce smooth textures and flavors as well as contributing to water and fat binding. Sodium caseinates have high viscosity in solution and do not gel as do soy proteins. Therefore, they do not bind meat pieces well, but contribute overall firmness to meat products such as hams due to their ability to hold water.[15] Whey protein concentrates have been used as meat replacers in sausages at levels between 0.5 and 2% (dry-weight basis).

Hydrolyzed plant and animal proteins. Hydrolyzed proteins result from the hydrolysis (breakdown) of soy bean, vegetable, gelatin, or milk proteins. This produces shorter chain proteins, peptides, and free amino acids that enhance meat and poultry flavors. Although particle binding is minimal, these proteins retain moisture and bind fat. Use levels range between 1 to 2% (dry-basis).

Gelatin. Gelatin is an inexpensive, commonly used water binder and gelling agent with minimal nutritional value. In canned meat products such as hams, loaves, frankfurters, Vienna sausages, and Spam® (cured, canned pork), gelatin is used to hold juices lost during cooking and to provide a good heat transfer medium during cooking. Gelatin is also used in emulsified meats and jellied products at levels ranging from 3 to 15%, but more typically from 0.5 to 3%.

Carbohydrates

Starch. Starches are the most widely used carbohydrate due to cost and availability. Starches bind two to four times their weight in moisture, provide freeze/thaw stability, can serve as fat replacements, and contribute to a firm texture. The most common starches originate from potato, corn, wheat, tapioca, and rice. Because native starches require high temperatures to gel and achieve their smooth texture and water-binding abilities, they are modified or pre-gelatinized to set at lower temperatures in the range of 60 to 75°C (140 to 167°F). Pre-gelatinized starches build viscosity rapidly in meat systems and are used in coarse and emulsified sausages or similar products rather than brine-injected products. Use levels range from 1 to 3.5% and up to 18%, depending upon the application and regulatory restrictions.

Hydrocolloids (gums). Carrageenan is a hydrocolloid (gum) derived from red seaweed that absorbs moisture to produce a firm gel texture. It can improve yield, control purge (improve water binding), improve finished product sliceability, enhance juiciness, and protect products from the effects of freezing and thawing. Carrageenan can be incorporated into a brine for injection into meat and poultry products or added directly to a mixer, blender, or tumbler. In most cases, carrageenan is used at levels <1.0% and needs to be heated to achieve complete solubility. Blends of carrageenans allow for modification of product texture. If used in a brine, alkaline phosphates should be dissolved first and then the salt followed by added sugar combined with the carrageenan. High quality carrageenans should be used to avoid premature gelation at needle injection sights in the product (an effect known as "tiger-striping").

Konjac is a flour material derived from the root of *Amorphophallus konjac* (elephant yam) that can swell and hydrate to form a highly viscous solution. It can be chemically modified to gel and remains stable at retort temperatures. Konjac is used at low levels in meat products to bind water and modify texture and may be combined with modified starches or soy proteins.

It is very rare that one non-meat ingredient will provide all the functional characteristics desired in a meat or poultry product. Usually combinations are required such as the inclusion of soy proteins, starches, and gums to give meat-like textures to poultry products.

Antimicrobial ingredients and antioxidants

Sodium/potassium lactate

Sodium or potassium lactate (sold as a 60% liquid solution) can be added to processed poultry products to: (1) extend shelf-life; (2) control pathogen growth; (3) enhance the salt flavor; and (4) improve texture by reducing moisture loss. Sodium lactate is incorporated into whole-muscle products, restructured poultry meats, ground patties, and coarse ground and emulsified sausages. The maximum level of use in the U.S. is 2.9% pure sodium lactate or 4.8% of a 60% lactate solution in fully cooked meat and poultry products. Levels of up to 4% (pure form) have been shown to suppress the growth of *Listeria monocytogenes* in an uncured chicken roll product and frankfurters. Salt (NaCl) should be reduced by approximately 20% to avoid making the product too salty.

Sodium acetate and diacetate

Sodium diacetate and sodium acetate are approved for use as flavoring agents in meat and poultry products at a level of up to 0.25% by weight of the total formulation. They serve as acidulants, flavoring agents, and as antimicrobial agents and may be included in marinades, brines, or as a dry ingredient. They are especially effective against *L. monocytogenes* at lower refrigeration temperatures and in low pH products. Potassium acetate or potassium diacetate are not yet approved for use.

Spices

Spices are aromatic substances taken from various plant parts or herbs. For example, the following spices come from various plant parts: cloves — flower bud; nutmeg and pepper — fruit; mace — aril (fleshy covering of the seed); cardamon, coriander, mustard —

aromatic seeds; cinnamon — bark; sage, thyme, marjoram — dried leaves; onion, garlic — vegetable bulbs; and ginger — rhizome. Natural spices come in whole and ground forms and the flavor is determined by the essential oil content. Grinding breaks down the cell structure releasing the essential oils, thus freshly ground spices are more flavorful. Particle size of the ground material also determines flavor release — the smaller the particle the faster the flavor released. Ground spices are sized between a No. 20 and 60 sieve. Spices not only contribute flavor, but color also; paprika contributes a red color while tumeric makes mustard yellow.

The aromatic properties of spices are found in the oleoresins and essential (volatile) oils (oleoresins = volatile oils + plant resins).[3] Piperine, for example, is the oleoresin from pepper (5 to 12%). Essential oils are the volatile aromatic fractions of spice derived by steam distillation. These compounds contribute to the odor and flavor value and include hydrocarbons, terpenes, and sequiterpenes (unstable) while alcohols, esters, aldehydes, and ketones are the main aroma carriers. Other compounds include non-volatile residues such as waxes and paraffins.

Oleoresins are thick resinous materials extracted with solvents (acetone, ethanol, isopropyl alcohol, ethylene dichloride, hexane, and petroleum ether) and contain both volatile and non-volatile fractions. These are more complete than essential oils, contain natural antioxidants, are free of enzymes and molds, and can be standardized for flavor and strength. Oleoresins can be made in liquid form when combined with emulsifiers (Polysorbate 80) and monoglycerides to make the oleoresin water soluble. Dry dispersed spices may be placed on soluble carriers like salt, dextrose, flour, or yeast or the oleoresin may be encapsulated with modified food starch, gum arabic, maltodextrins, gelatin, solid fat, or an oil.

Soluble spices (oleoresins and volatile oils) are frequently used in processed or canned meats to prevent darkening of the product during processing and reduce the risk of pathogen contamination from spices. Natural spices (whole, cracked, or ground) are most frequently used in sausages (cured, dry, and semi-dry) and should be sterilized to prevent bacterial contamination of the product (Table 12.3). For use of spices in meat products refer to specialized handbooks, the U.S. Dispensatory, or meat formulation textbooks.

Casings

Casings are flexible containers strong enough to give shape to sausages and portioned products and must be capable of withstanding thermal processing while encasing the meat mass. Casings may or may not be removed before slicing. Metal molds are also used to give shape to loaf items, but are removed before slicing. The process of placing products into containers is called "stuffing" and is usually performed with a vacuum stuffer that

Table 12.3 **Frequently Used Spices in the Meat Industry**

Black pepper	Allspice	Basil	Bayleaf
Cardamon	Cloves	Ginger	Fennel
Nutmeg	Mustard	Paprika	Pimento
Cayenne pepper	White pepper	Caraway	Coriander
Celery seed	Cumin	Marjoram	Thyme
Savory	Sage	Anise	Cinnamon
Capsicum	Onion	Garlic	Sesame

removes entrapped air resulting in a uniformly dense product without holes. Casings are classified as natural (intestinal or stomach lining) or manufactured (regenerated collagen, cellulose, fibrous, and plastic) and may be edible or inedible.

Natural casings

Natural casings are derived from the submucosa, a largely collagen layer of the gastrointestinal tract of cattle, swine, and sheep,[16] with the fat and inner mucosa lining removed. They are permeable to smoke and moisture and require humidity controlled heating to avoid "hardening" the casing. Casings are shipped in (1) a dry salt pack that requires flushing before use, (2) a slush or pre-flushed pack, or (3) a pre-tubed casing that does not require flushing.[17] The optimum storage temperature for casings is 4 to 10°C (40 to 50°F). "Whiskers" are small string-like capillaries that hold the intestine in the fat and provide blood flow to the intestine. If intestines are removed with a knife, these create a hair-like appearance on the surface of the casing, but generally disappear after cooking.

Hog casings are prepared from the stomach, small intestines (~20 yd), large intestine (2.5 yd), and terminal end of the large intestines (bungs, 2 yd). They are sold in "hanks" measuring ~91 meters or "shorts," 1 to 2 m in length, and are usually classified as 35 mm and down or >35 mm and up. Stuffing capacities for various sizes and types of natural casings may be obtained from the International Natural Sausage Casing Association, Washington, D.C. (http://www.insca.org). Hog casings are used for fresh and cooked sausages, pepperoni, Italian sausage, large frankfurters, Kielbasa, and bratwurst. Hog bungs are used for liver sausage, braunschweiger, genoa, thuringer, summer sausage, and cervelats.

Sheep casings are the highest quality small-diameter casings and range in size from 16 to >28 mm. They are prepared from the stomach, small intestines (~30 yd), and large intestines (bungs ~1 yd). Most often they are used for fine sausages such as bockwurst, frankfurters, Landjaeger, cabanosa, and port sausage.

Beef casings are derived from the entire length of the intestinal tract with beef bung caps (~2 yd), rounds (~35 yd), and middles (~9 yd) being the three most used casings. "Rounds" derive their name from their "ring" shape (35 to 46 mm) and are used for ring bologna, ring liver sausage, mettwurst, Polish sausage, blood sausage, Kishka, and Holsteiner. "Middles" (45 to 65 mm) are often used for bologna, dry and semi-dry cervelats, salami, and Leona-style sausage. "Bladders" hold 2.5 to 6.5 kg and are used most for minced sausages or Mortadella because of their flat to oval shape.

Regenerated collagen

Regenerated collagen casings are more delicate than natural casings, but offer consistent size, low microbial counts, and uniformity in the finished product. They are derived from the corium layer of beef hides beginning with an alkaline extraction, followed by swelling with an acid, forming into a tube through a die and then fixing to an appropriate size and shape with an alkaline bath to neutralize the collagen. The casings are then "shirred" into sticks for placing on a stuffing horn. These edible casings are most often used on fresh sausages and frankfurters ranging in size from 22 to 30 mm. Larger diameter collagen casings are treated with aldehydes to cross-link the collagen for strength, but these casings must be removed prior to eating.

Cellulosic

Cellulose casings are derived from cotton "linters" (fibers removed from the seeds), a high grade of alpha cellulose, or wood pulp. Chemical treatment produces cellulose xanthate that is treated with dilute caustic to form a viscose solution which is extruded through various sized nozzles into an acid solution to "fix" the cellulose polymer. Food grade glycerine, propylene glycol, mineral oil, surfactants, colorants, smoke, and water may be imbedded into the casing matrix to produce a pliable product for stuffing sausage products. Cellulose casings are inedible but offer similar advantages of the regenerated collagen casings. Casings are typically shirred into sticks from 40 to 160 ft in length and stuffed at a rate of 250 to 300 ft/min. After thermal processing, these casings are removed prior to packaging the product.

Fibrous

Fibrous casings are manufactured by extruding regenerated cellulose onto a paper base and forming into tubes. This produces a strong, inedible container for large diameter (2 to 6") products such as bologna, poultry rolls, fermented sausages, and turkey hams. Some casings are treated internally with a moisture/oxygen impermeable plastic barrier to prevent moisture release through the casing and subsequent oxygen penetration. This product is especially useful for water-cooked items such as bologna, liver sausage, and poultry rolls. These casings are typically removed prior to slicing, portioning, or packaging.

Plastic

Plastic tubes that are impermeable to smoke and moisture, but somewhat permeable to oxygen, are used to produce vacuum stuffed "chub" products such as fresh sausages or ground turkey packs. Metal clips are most often used to close the ends of the portioned tubes. These products are frequently sold at retail in 1, 2, 5, or 10 lb sizes and are portioned by the consumer. Because these packages have less oxygen present, their refrigerated shelf-life is extended beyond that of tray pack products prepared at the store level.

Processing procedures

Formed products

Formed products consist of whole poultry muscles and muscle trimmings that are encased or molded into a specified shape and fully cooked to yield a "whole" product suitable for portioning or slicing. Muscle pieces or meat homogenates are combined with a brine marinade (1.5 to 2.5% sodium chloride and <0.5% alkaline phosphates) or injected with a curing solution (brine + sodium nitrite and sodium erythorbate) to extract salt-soluble proteins that form a "tacky" meat surface and provide a natural protein-binding matrix when cooked. Up to 33% of the meat block may be finely communited or homogenized to provide particle-to-particle binding for a muscle-like texture and retention of juices. Non-meat binders such as soy protein isolate, hydrolyzed proteins, starch, and carrageenan may be incorporated into a brine/marinade for injection into a formed product. After brine marination/injection, vacuum tumbled products are held chilled for equilibration of the ingredients, stuffed into a mold or casing, and fully cooked to a safe temperature end point of 71.1 to 73.8°C (160 to 165°F). Outlined in Tables 12.4 and 12.5 are the processing sequences for two formed products, a boneless marinated poultry roll, and a cured turkey ham product.

Processing defects (formed products)

Some processing defects are given in Table 12.6.

Table 12.4 **Uncured Poultry Roll (Boneless Breast Fillets)**

Processing Procedures

Raw material

Fresh, chilled boneless breast fillets (32–35°F)

Maceration

Whole muscle macerated ~0.25″

Marination

Combine chilled (26–28°F) brine with macerated pieces and homogenized trimmings in paddle blender (15–30 min., 15 to 20 rpm, brine uptake ~15%)

or

Multineedle injection

Inject chilled (26–28°F) brine with seasonings, 15–20% pump level (sideport needles, 10 needles/in^2, 30–40 lb pressure)

Vacuum tumbling

Add seasonings/spices if required, tumble 45 min.–1.5 h, hold chilled 12 h in vats

Forming

Vacuum stuff into fibrous regular, pre-smoked (smokehouse) or moisture-proof fibrous (water cook) casing

Thermal processing

Smokehouse Product—Staged cook cycle with humidity control, cook to 160–165°F endpoint

Water cooked product—incremental temperature increases, keep water temperature ≤10°F above final endpoint, cook to 160–165°F

Chilling

Smokehouse cook—shower product to ≤100°F, place on cooler racks in blast chiller (−10°F, high airflow), chill product to 32°F, hold in tempering cooler (26–28°F) prior to slicing/portioning

Water cook—transfer to chilled water tanks, after removing molds place on cooler racks in blast chiller (−10°F, high airflow), chill product to 32°F, hold in tempering cooler (26–28°F) prior to slicing/portioning

Slicing/portioning/packaging

Remove casing, keep room and product chilled, high sanitation level, restricted access, frequent monitoring for microbial contamination

Pre-shipping storage

Box, code date, add to tracking inventory, verify HACCP pre-shipment review (hold at 32°F refrigerated or ≤0°F frozen)

Comments

Maceration severs connective tissue and increases muscle surface area for uptake of marinade.

Brine formulation—finished product
1.5–2% salt
0.4% alkaline phosphate
0.5% dextrose
3.0% potassium lactate
Seasonings and spices as specified

Emulsion products — sausages

Preblending

Poultry meat preblends allow for increased protein extraction and are used by some processors. Lean meats (fresh or previously frozen) may be ground through a 1/8 to 3/16″ plate while fat meats are ground separately through a 3/8 to 1/2″ plate and incorporated later in the processing sequence. Trimmings should not be held more than 24 to 48 h. After grinding, trimmings are transferred to a mixer/blender (Figure 12.1), sampled for fat and moisture analysis, and appropriate amounts of nitrite and salt added (if the preblend is for

Table 12.5 **Cured Turkey Ham (Boneless Thigh Muscles)**

Processing Procedures

Raw material

Fresh, chilled boneless thigh muscles and trimmings (32–35°F)

⇩

Maceration[1]/grinding

Whole muscle maceration ~0.25″ spacing, Comitrol® flake to 0.75″ or coarse grind muscles 0.5″, homogenize trimmings (use ≤33% in formulation)

⇩

Marination

Combine chilled (26–28°F) curing brine with macerated pieces and homogenized trimmings in paddle blender (15–30 min, 15 to 20 rpm, brine uptake ~15%)

or

Multineedle injection

Inject chilled (26–28°F) curing brine with seasonings, ≥20% pump level (sideport needles, 10 needles/in², 30–40 lb pressure)

Vacuum tumbling

Add seasonings/spices if required, tumble 45 min–1.5 h, hold chilled 12 h in vats

or

Massaging

Massage well-chilled injected product for 12–18 h, 4–5 rpm

Forming

Vacuum stuff into fibrous regular or pre-smoked (smokehouse) casing

⇩

Thermal processing

Smokehouse product—staged cook cycle with humidity control, liquid or natural smoke, cook to 155–165°F

Chilling

Smokehouse cook—shower product to ≤100°F, place on cooler racks in blast chiller (−10°F, high airflow), chill product to ≤32°F, hold in tempering cooler (26–28°F) prior to slicing/portioning

⇩

Slicing/portioning/packaging

Remove casing, keep room and product chilled, high sanitation level, restricted access, frequent monitoring for microbial contamination

⇩

Pre-shipping storage

Box, code date, add to tracking inventory, verify HACCP pre-shipment review (hold at 32°F refrigerated or ≤0°F frozen)

Comments

Maceration and grinding severs connective tissue and increases muscle surface area for uptake of marinade.

Curing brine formulation—finished product

- 2.2–2.5% salt
- 0.4% alkaline phosphate
- 1.0% dextrose
- 3.0% potassium lactate
- 200 ppm sodium nitrite
- 550 ppm sodium erythorbate
- Seasonings and spices as specified

a cured product), or salt alone an uncured product. A combination of 4 to 6 lb salt, 0.25 oz nitrite, 0.875 oz of sodium erythorbate, and 0.4% sodium tripolyphosphate per 100 lb of meat can be blended and stored in appropriate containers at 0 to 2.2°C (32 to 36°F) for no more than 72 h. If the preblend must be frozen for later use, it should be tempered [−3.3 to −2.2°C (26 to 28°F)], not thawed, for subsequent incorporation into the formulation. Each storage container should be identified with a lot number so it can be matched with the chemical analysis to make the final blend corrections.

Table 12.6 **Common Cured Meat Defects**

Observed defect	Possible causes	Comments
Odor, flavor problems Rancidity	Meat enzyme hydrolysis of fat, followed by oxidation; prooxidants include exposure to light (UV), oxygen, elevated temperature, peroxides, salt, ozone, cooked product	Salt may contain heavy metal impurities which are prooxidants: cooking can accelerate oxidative rancidity producing stale off-flavors.
	Air leak in vacuum package	
	Shelf-life extended too long; too long exposure to light during storage	
	Insufficient or incorrect antioxidant in product	
Spoilage	Sour odor from vacuum product indicates spoilage by lactic acid bacteria	
	Putrefactive odors from over wrapped product (non-vacuum) indicates spoilage by psychrotrophic bacteria	
Color problems—uncured products Pinking	Undercooking of product	May indicate potential concern for pathogen survival and food-borne illness
	Residual nitrite contamination of equipment, formulation, processing environment, cooking oven	Thoroughly clean equipment, room, etc. prior to processing
	Water-soluble nitric oxides (NO_x, NO_2) produced by incomplete combustion of natural gas in oven or smokehouse	Use high temperature drying step (180–200°F) for the product surface at the beginning of the heat cycle
Color problems—cured products	Insufficient nitrite in cure	Processing too rapidly after brine curing can often produced undercured product
Interior is undercured or faded in appearance	PSE muscles	Denatured myoglobin protein causes light pink or gray color
	Cured pigment has been oxidized by light, oxygen and accelerated by increased temperature	Oxidants include vacuum package leakers, strong lighting (UV), elevated storage temperature, bacteria
	High pH	
	Anaerobic storage combined with high cooking temperature	

Table 12.6 (continued)

Observed defect	Possible causes	Comments
Green patches in product interior	Nitrite burn due to excessive use of nitrite or improper distribution of nitrite Undercured due to too short curing time or cold curing room termperature	
Brown turkey hams	Oxidation of meat pigment	Dehydration due to low relative humidity and high storage temperature
Pale color, faded interior	Muscle not thoroughly cured Curing rate and efficiency retarded due to abnormally low curing temperature Elevated curing and storage temperature have permitted sufficient bacterial growth to fade meat pigments	
Iridescence—unnatural shiny appearance; mother-of-pearl-appearance	Diffraction of incident white light due to striated or fibrous nature of muscle Bacterial growth	
Pink or green discoloration	Metabolic products of halophilic bacteria Bacterial growth due to salt concentration too low, moisture content too high	
Dark frying bacon	Inversion of sugar by bacteria Poor sugar quality Excessive heat during frying	
Pumping problems Pickle pockets	Use of high pump pressures to overcome poor injection distribution due to wide muscle spacing Overpumped	
Hemorrhage problems Pinpoint hemorrhage in muscles—small blood spots	Capillary rupture due to excessive time between electrical stunning and bleeding or excessive voltage	
Large muscle hemorrhage	Due to moldy feedstuffs, e.g., toxins or blood thinning agents	
Joint hemorrhages	Blood vessels ruptured during shackling	
Post-processing problems Broken pieces during slicing of processed meats, e.g., boiled hams	Off-conditioned muscles used in curing Failure to extract salt-soluble, binding protein Meat or pickle was too acid Product overcooked Failure to press pieces together tight enough	

Meat emulsion theory

Meat batters (frankfurters and bologna) are complex emulsions in which microscopic fat droplets are the *discontinuous phase* and the myofibrillar (salt-soluble) proteins are the *continuous phase* and coat the fat droplets. They are suspended in a complex matrix of water, proteins (myofibrillar, sarcoplasmic, and connective tissue), lipid droplets, and non-meat ingredients (spices) to form a stable gel matrix when heated. The process of emulsion formation has three phases or stages, although the actual processing steps are continuous.

Processing phases

Protein extraction and swelling. Formation of a meat batter or emulsion initially consists of chopping lean meats (myofibrillar proteins predominantly) such as chicken breasts or mechanically deboned chicken without skin in a bowl chopper (Figure 12.2); (vacuum preferred) with 4 to 6% salt and cure ingredients (sodium nitrite and sodium erythorbate). Approximately half of the water is added as an ice slush keeping the temperature near 0°C (32°F) and chopping continued until the muscles are homogenized and the batter is ≤4.4°C (40°F). During this stage, sarcoplasmic and myofibrillar proteins solubilize and swell due to ionic forces (partial unfolding to allow more interstitial space), thus enhancing water absorption. Inclusion of alkaline phosphates increases pH of the meat batter and further enhances swelling. Acid phosphates (sodium acid pyrophosphate) do the opposite and are used to "slacken" the emulsion for greater ease of pumping. Allowing the chopped lean to "rest" for a short period of time (5 min) will enhance the extraction of myofibrillar proteins and aid in binding capacity. Collagen proteins are not readily solubilized due to their triple helical structure and are not good emulsifying agents, thus, low binding raw materials should be limited to 15% of the total meat block or 25% if high connective tissue trimmings are used.

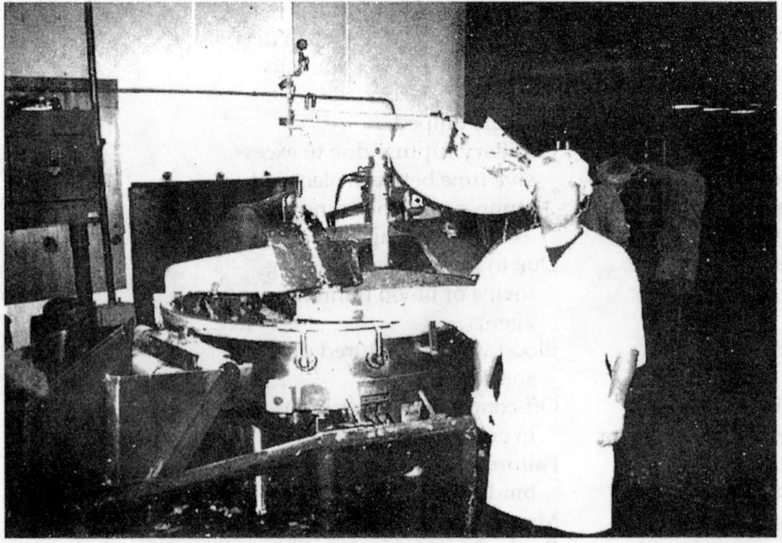

Figure 12.2 Bowl chopper with rotating blades contained under the hooded cover at the rear of the bowl.

Fat encapsulation (emulsion formation) and fat entrapment. After extraction of the salt-soluble proteins, the remaining ice or water, fat tissues, non-meat ingredients, and other additives are combined in the chopper (under vacuum) and homogenization continued to a final temperature of 10 to 12.8°C (50 to 55°F) depending on the fat source. During chopping, the emulsifying agent (myofibrillar proteins) undergoes a conformational change orienting hydrophobic portions of the protein toward the lipid droplets while hydrophilic portions are pointed toward the water phase. This results in the protein surrounding the lipid droplet and enabling its dispersion within the water/protein/ingredient phase. Myofibrillar proteins are preferentially absorbed onto the surface of the microscopic fat particles and lose their water-binding ability. Once the fat particles are coated with the protein, the "emulsion" is formed and is later stabilized by cooking. Water is both entrapped in the emulsion matrix and bound to negatively charged protein groups. Once the proteins interact to form the fat/water interface, "new" protein must be available for further emulsification or water binding. Subcellular fat particles must remain in a somewhat plastic state to form a stable emulsion. If the temperature increases above the melting point of a specific fat, then the fat particle liquefies and cannot be encapsulated. Thus, temperature control is critical to stable emulsion formation.

Final batter temperatures of 10 to 11.7°C (50 to 53°F) for poultry, 15.6 to 17.8°C (60 to 64°F) for pork, and 21.1 to 22.2°C (70 to 72°F) for beef may be achieved without harming the batter. If mixtures of fat are used, the composite final temperature will be adjusted to the proportion of the predominant fat present. If an emulsion mill is used after the chopper, slightly higher temperatures can be tolerated (i.e., for an all-beef product, 23.9 to 24.4°C [75 to 76°F]), but excessive chopping or high temperatures can "break" (cause separation of the fat and liquid phases) the emulsion. Very hard crystalline fats such as beef kidney fat do not generally produce good emulsions and result in a grainy texture.

Formation of a heat-set gel. Excessive physical handling or long holding times can reduce emulsion stability, thus the product should be encased and heat processed as soon as possible. The product should be heated to an internal temperature of 68.3 to 73.9°C (155 to 165°F) to denature or "set" the myofibrillar protein causing the formation of a "meat gel" entrapping the fat and water in a solidified matrix. Coagulation of the proteins begins at about 57.2 to 60°C (135 to 140°F) and continues up to temperatures of 90°C (194°F). The "skin" formation on franks is the result of protein denaturation. Collagen fibers shrink to one-third their length upon heating to 64.4°C (148°F) and with the continued application of moist heat, will form gelatin. In a stable emulsion, gelatin is entrapped and holds some water. Heat processing typically follows a stepwise schedule until the target end point temperature is reached. Products are then showered to decrease the temperature to ≤37.8°C (100°F) and chilled overnight to ≤4.4°C (40°F) prior to peeling and packaging. Heating to 75°C (167°F) or higher causes more fiber shrinkage, excessive moisture loss, and fat melting. A processing outline for frankfurters and bologna is presented in Table 12.7.

Processing defects (emulsion products)

Processing defects of emulsion products are presented in Table 12.8.

Table 12.7 **Poultry Frankfurter/Bologna (Boneless Breast, Thighs, Drumstick Meat, Mechanically Separated Poultry, Trimmings)**

Processing Procedures

Raw material

Select fresh/tempered chilled boneless breasts, thighs, drumstick meat, mechanically separated poultry, trimmings or pre-blend (26–30°F)

⇩

Analysis/formulation

Analyze poultry meats for moisture, fat, protein (AOAC Tests)

Formulate to compositional endpoint constraints with least cost programming (15% fat or 0.5% fat)

⇩

Grinding

Grind lean through 0.125″ plate not required for MSP

⇩

Chopping/homogenization

1. Combine pre-blend or lean meats with salt, alkaline phosphates, cure (nitrite/erythorbate) with 1/2 water as iced slush in vacuum bowl chopper (80% vac.)
2. Emulsify to paste, maintain ≤40°F, use powdered CO_2 if needed, rest for ~5 min to extract protein
3. Combine fat trimmings, additives, seasons, spices with 1/2 iced water slush, homogenize (60% vac.) to paste, maintain ≤50°F

⇩

Emulsification

Pass through emulsion mill to ensure smooth emulsion, endpoint temperature ~55°F

Forming

Vacuum stuff sausage into peelable collagen casings (24–30 mm) or bologna into moisture-proof fibrous casing for heat processing

⇩

Thermal processing

Smokehouse product—staged cook cycle with humidity control, higher humidity for low fat products, liquid or natural smoke, cook to 155–165°F end point

Chilling

Smokehouse cook—shower product to ≤100°F, place in brine chiller or on cooler racks in blast chiller (−10°F), chill product to ≤40°F, hold in tempering cooler (26–28°F) prior to peeling or slicing

⇩

Peeling/slicing/packaging

Remove casing, keep room and product chilled, high sanitation level, restricted access, frequent monitoring for microbial contamination

Pre-shipping storage

Box, code date, add to tracking inventory, verify HACCP pre-shipment review (hold at 32°F refrigerated or ≤0°F frozen)

Comments

Formulation—finished product

2.2–2.5% salt

0.4% alkaline phosphate

1.0% dextrose

2.0–3.0% modified starch

3.0% potassium lactate

200 ppm sodium nitrite

550 ppm sodium erythorbate

Seasonings and spices as specified

Table 12.8 **Defects in Sausage and Processed Meats**

Observed defect	Possible causes	Comments
Fat caps and jelly pockets	Due to borderline or unstable emulsion	
	Air incorporated during comminuting or stuffing; these air pockets will fill with gelatin if emulsion has borderline stability	Sausage products cooked in water are more likely to exhibit gelatin pockets than sausages cooked in dry heat.
	Short meat, i.e., too much collagen protein and insufficient salt-soluble myosin protein	Collagen protein should contribute less than 33% of the total protein, and preferably less than 25% in fine cut, small diameter cured/cooked sausages; final chopping temperatures should be no higher than 50 to 55°F if the meat trimmings are high in collagen
	High fat, high collagen ratios in the product	
	Heating too fast and cooking to excessive final product temperatures	During heating, fat globules expand while the proteins coagulate and shrink slightly; thus, fat ruptures the protein matrix and rises to the surface or top of the product as a "fat cap"
Fat rendering, fat pockets, greasing out	Emulsion breakdown	
	Too much collagen protein	
	Too much frozen meat	If frozen meat is held at −4 to −2°C (25–28°F), the resulting formation of large ice crystals will rupture cell structure and denature protein, thus reducing bind capacity and emulsion stability
	Too much frozen fat	
	Too much edible byproduct	
	Too much product rework	Limit product rework to 5–10% of formulation
	Emulsion chopped too long and too fine	As a result of overchopping, not enough salt-soluble protein is available to coat and stabilize the fat globules; finely chopped emulsions require more salt soluble protein than coarse chopped or ground emulsions
	Emulsion held at elevated temperature	
	Emulsion held too long under pressure before stuffing	

Table 12.8 **Defects in Sausage and Processed Meats** *(continued)*

Observed defect	Possible causes	Comments
	Emulsion kept too long in vats before stuffing	
	Pressure or distance too great during transfer of emulsion from emulsifier to chopper or stuffer	
	Emulsion overworked during transfer through pumps, pipes and transfer augers	
	Product understuffed	
	Understuffed, loose ends of stuffed strands	
	Air pockets due to improper stuffing	
Fat rendering near smokehouse rack contact	Too much heat during cooking Too rapid heating	During heating, fat globules expand while the protein firms and shrinks; thus the fat ruptures the protein matrix.
Surface may have a slight greasy touch	Heated too high Too high relative humidity due to use of a steam cook for 2–5 min at the end of a cookhouse cycle	
Thin greasy film on surface of emulsion sausage	Insufficient salt soluble protein to stabilize the emulsion	This problem may be reduced or eliminated by reducing relative humidity in the smokehouse.
	Quick release coated casings held too long after stuffing and prior to cooking	
Fat separation and jelly pockets in liver sausage	Too high of cooking water temperature	
	Prior to cooking raw emulsion temperature was in excess of 21°C (70°F)	
	Chopping period was too long, and emulsion temperature was not kept low enough	
Dark rings in liver sausage	Water cooked liver sausage was too cool before being placed into smokehouse	
Poor peelability of wieners	Improper initial surface protein coagulation during first stage of smoking	Dehydration hinders peelability. On the other hand, sweating (i.e., cold production, warmer environment) improves peelability

Table 12.8 Defects in Sausage and Processed Meats *(continued)*

Observed defect	Possible causes	Comments
	Overdrying of wieners during chilling, i.e., too much air circulation	
Sour or slimy casings	Natural casings held in stagnant water	
Casings burst on cooked liver sausage	Overcooking	Liver tends to expand during cooking
	Raw livers used	
	Emulsion overstuffed	
Casings break during cooking of large diameter emulsion products	Product heated too quickly	In the case of too quick heating, the surface sets and shrinks while the inside remains wet
	Batter held overnight in cooler so the interior temperature was lower than normal	For dry casings, breakage generally occurs at the ends; for wet casings, breakage generally occurs in the center
	Clip closure puncture of second tie on casing	
Case hardening of fermented sausage	Sausage dried too quickly	Case hardening may be accompanied by surface ridges and excess shriveling
Floating emulsion products such as frankfurters	Air trapped in emulsion	This air which has been trapped in the emulsion may seep into the vacuum package, causing the appearance of a leaker
Excessive shrinkage on cooking	Excessive fat	
	Soft fats such as poultry or pork overchopped	Soft oily fats such as internal fat will experience greater shrinkage than carcass fats
	Excessive moisture as a result of inaccurate estimate of flaked ice to be added	
	Low water holding capacity of emulsion	
	Excess moisture due to binder or bread crumbs stored in room having high relative humidity	
	PSE poultry or pork	
	Ingredients not mixed long enough	

Table 12.8 **Defects in Sausage and Processed Meats** *(continued)*

Observed defect	Possible causes	Comments
Gassy processed meats wieners	Carbon dioxide producing facultative anaerobic bacteria	Gassiness may appear as vacuum failure since packages generally bloat; the product may have an acid flavor; generally gassinéss presents no health concern
	Underprocessing	
	Leaker	
Gassy, dry sausage	Growth of heterofermentative lactic bacteria; in a few cases there may be growth of yeast which produces CO_2	To avoid gas production, lactic acid starter cultures should be used
Musty, weedy, parsnips, cheesey, off-odor	Bacterial growth	
	Insufficient salt	
	Poor sanitation	
	Abusive storage temperature	
Lack of cured flavor; chicken feather flavor	Incomplete cure	
	Poor distribution of cure	
Sour flavor and odor		Similar bacteria as those which cause development of green care except there is no pigment discoloration
Flat aroma and acid taste	PSE poultry and pork	
Rancidity	Meat enzymes hydrolyze fat; followed by oxidation	Rancidity may be accompanied by acidity; or a fish odor
	Prooxidants include exposure to light, elevated temperature, peroxides, salt, ozone	
	Salt may contain heavy metal impurities which are prooxidants	
	Air leak in package	
	Storage life extended too long	The use of phosphate, where permitted, will tie up metal impurities
	Too long exposure to light during storage	
	Bacterial enzymes	
Weak color, color fading	Insufficient nitrite in cure	The surface color fades rapidly while the interior may range from faded pink to gray or light green
	Interior is undercured	Reducing agents such as ascorbic acid should be added
	PSE poultry or pork	The amount of PSE pork should be minimized
	Cured pigment has been oxidized by light and oxygen and accelerated by temperature	Curing procedures should be improved to guarantee at least 70% of the meat pigment being cured

Table 12.8 **Defects in Sausage and Processed Meats** *(continued)*

Observed defect	Possible causes	Comments
	Vacuum package leakers	The quality of vacuum should be improved
	Strong lighting	The use of opaque packages and reduced exposure to light will reduce the potential for pigment to fade
	Elevated storage temperature	
	Bacteria oxidizing pigments	
Discoloration in emulsion products	Inclusion of air in the casing	Gray spots in large emulsion products are common with this defect
	Faulty stuffer	
	Damaged stuffing horn	
	Careless handling of casings	
	Air trapped in meat emulsion as it is placed in stuffer	
Discoloration on large emulsion products	Cold showered too long	
Color faded or smeared at time of stuffing	Meat too warm during grinding	
Light marks on small emulsion products	Smoke sticks wrong size and interfere with smoking	May appear at tips or on sides
	Contact between wieners during smokehouse operations.	
Light color and dry flavor	PSE pork	In addition to dryness, product may have undergone an additional 3–5% shrinkage during processing
Central discoloration in large emulsion products	Insufficient thermal process	Cooking to a minimum internal temperature of 68°C (155°F) will reduce discoloration
White salty crystal appearance	Lactose crystals due to excessive whey or milk powder use	
Green patches	Problem intensifies in an acid medium	
	Nitrite burn due to excessive use of nitrite	
	Nitrite burn due to improper distribution of nitrite	

Table 12.8 **Defects in Sausage and Processed Meats** *(continued)*

Observed defect	Possible causes	Comments
	Undercured due to too short of curing time Undercured due to too cold curing room temperature	Better distribution of ingredients, longer curing time, temperature of curing room about 2–3°C (35–38°F), and use of cure accelerators such as ascorbate will reduce greening due to undercuring
Surface greening in large emulsion products	Relatively salt-resistant bacteria, capable of growing at refrigerator temperatures Improper hygiene of racks and working surfaces Improper cooking cycle Surface contamination after processing Inadequate refrigeration of finished product Freshly prepared product exposed to product returns	A greenish gray discoloration may be accompanied by slime; surface greening may appear soon after processing or late in retail; it generally does not show up until at least 5 days after processing and up to a few weeks; surface greening will increase and spread if product held at elevated temperatures Common in summer Increased salt concentration, increased smokehouse temperature, avoiding excess moisture in packages will reduce the incidence of surface greening
Green rings in large emulsion products	Same cause as green cores, due to facultative anaerobic bacteria which may be relatively heat resistant Heavily contaminated emulsion Insufficient thermal process Remixing of off-conditioned meat Dubious quality of raw materials Abusive storage of finished product to permit growth of surviving bacteria	Green rings generally appear as continuous rings at 2–4 mm depths beneath surface at time of cutting; for development they require exposure to oxygen; generally after several hours the entire core may fade Processing to a minimum of 71°C (160°F) in addition to checking and controlling quality of raw materials will reduce the incidence of green rings

Table 12.8 (continued)

Observed defect	Possible causes	Comments
Green cores in large emulsion products	Same bacteria as surface greening, contamination of emulsion prior to process	Green cores appear a few hours after slicing product after surface is exposed to air; they are not apparent at time of slicing
	Poor quality of raw materials Emulsion held too long before cooking, or heavily contaminated Improper storage temperature of finished product	*Note:* because the bacteria are relatively heat resistant, they are alive in green cores; as a result, they are capable of cross-contaminating process lines and products
	Insufficient thermal process; smokehouse may have cold spots	Process to a minimum of 72°C (162°F) will destroy these bacteria
	Smokehouse was overcrowded	
Slime	Due to high bacterial counts of lactic acid bacteria, micrococci, and yeast	White or yellow slime is the visible presence of the microorganisms themselves rather than the metabolites
	Accentuated by moisture condensation, e.g., product permitted to sweat	These bacteria may appear as whitish, milky liquids in vacuum packages
	Post-process surface contamination	
	Abusive storage temperatures	
	Leakers or loose vacuum	
Mold or dried sausage	Surface mold, yeast Too moist surface Too slow drying	In addition to the mold the outside surface may be soft
Mold on wieners	Require oxygen for growth High moisture content Leaky packages	

Source: From Research Bulletin Meat Packers Council of Canada, Islington, Ontario, Canada and Terrell, R.N., *Sausage and Cured Meat Operations: An Instruction Manual*, Texas Food Research, Bryan, TX, 1981. Used with permission.

Summary

Formed and emulsified poultry products continue to offer consumers variety and value. Modified versions of traditional products that do not violate the product integrity or nutritional value will likely be the "new" products in the future. Convenience, taste, and safety will continue to be important and drive the development of products for consumers.

References

1. United States Department of Agriculture — ERS, Major Trends in U.S. Food Supply, 1909–99, Judy Putnam, Food Review, January–April 2000, 8.

2. United States Department of Agriculture, Code of Federal Regulations, Animal and Animal Products, Title 9, Part 200 to End, January 1, 1999, Office of Federal Register, National Archives and Records Administration.
3. Pearson, A. M. and Gillett, T. A., *Processed Meat,* 3rd ed., Chapman and Hall, New York, 1996.
4. Romans, J. R., Costello, W. J., Carslon, C. W., Greaser, M. L., and Jones, K. W., *The Meat We Eat,* 13th Ed., Interstate Publishers, Danville, IL, 1994.
5. Addis, P. B., Poultry muscle as food, in *Muscle as Food,* Bechtel, P. J., Ed., Academic Press, New York, 1986, chap. 9.
6. Saffle, R. L., Meat emulsions, *Adv. Food Res.,* 16, 105, 1968.
7. LaBudde, R. A., *Least Cost Formulator,* Least Cost Formulations, Ltd., Virginia Beach, VA, 1993.
8. ROI, ROI Formulation System, Resource Optimization, Inc., Knoxville, TN, 1999.
9. Hedrick, H. B., Aberle, E. D., Forrest, J. C., Judge, M. D., and Merkel, R. A., *Principles of Meat Science,* 3rd ed., Kendall/Hunt Publishing Company, Dubuque, IA, 1994.
10. Claus, J. R., Colby, J.-W., and Flick, G. J., Processed meats/poultry/seafood, in *Muscle Food: Meat, Poultry and Seafood Technology,* Kinsman, D. M., Kotula, A. W., and Breidenstein, B. C., Eds., Chapman and Hall, New York, 1994, chap. 5.
11. Ando, H., Adachi, M., Umeda, K., Matsuura, A., Nonaka, M., Uchio, R., Tanaka, H., and Motoki, M., Purification characteristics of a novel transglutaminase derived from microorganisms, *Agric. Biol. Chem.,* 53, 2613, 1989.
12. Washizu, K., Ando, K., Koikeda, S., Hirose, S., Matsuura, A., Takagi, H., Motoki, M., and Takeuchi, K., Molecular cloning of the gene for microbial transglutaminase from *Streptoverticillium* and its expression in *Streptomyces lividans, Biosci. Biotechnol. Biochem.,* 58, 82, 1994.
13. Means, W. J. and Schmidt, G. R., Algin/calcium gel as a raw and cooked binder in structured beef steaks, *J. Food Sci.,* 51, 60, 1986.
14. Hand, L. W., Purge Controllers, Protein Technologies International, St. Louis, MO, 1999.
15. Van den Hoven, M., Functionality of dairy ingredients in meat products, *Food Technol.,* October, 72, 1987.
16. Rust, R. E., Advances in meat research, in *Edible Meat By-Products,* Pearson, A. M. and Dutson, T. R., Eds., Elsevier Applied Science, London, 1988, 5, 261.
17. International Natural Sausage Casing Association, *Natural Sausage Casing,* International Natural Sausage Casing Association, Washington, D.C., 1997.

chapter thirteen

Coated poultry products

Casey M. Owens

Contents

Introduction

The consumption of chicken and turkey meat has increased dramatically over the last several decades. This increase in consumption can be attributed to the marketing and innovation of the poultry industry. Value-added processing is a term that means adding convenience and variety for the consumer while increasing profits for the processor. The poultry industry has been very accommodating to consumers' needs with the development of more ready-to-cook and ready-to-eat products. Consumers want foods that are easy to prepare. The introduction of chicken nuggets in the early 1980s opened a whole new market for the food industry. Today, the chicken nugget and patty are some of the most popular convenience poultry items available, being sold at virtually every fast food restaurant and grocery store in the nation.

0-8493-0120-3/01/$0.00+$.50

Forming the product

Meat source

Nuggets and patties can be made from a variety of meat sources. They are usually made from whole-muscle trimmings and usually reflect the preference of locality.[1] For example, in the U.S., white meat is preferred by consumers and has higher value. However, in other regions such as Pacific Asia, dark meat is preferred and is therefore considered the higher value meat.[1]

In the U.S., the most common formulation of chicken nuggets is breast meat and skin.[1] Breast meat is often chosen because of its uniformly soft texture and its light color. However, other meat sources such as thigh, drums, and rib meat can also be incorporated. Mechanically deboned poultry meat (MDPM) is also another source of meat that can be used in the production of nuggets and patties. The dark meat and MDPM are used to help reduce production costs and also to improve flavor due to its higher fat content. When dark meat is used in combination with white meat, the ratio of white to dark meat is typically 70:30.[1] The problem with the use of dark meat and MDPM in formulations is their susceptibility to oxidative rancidity due to the high fat and iron contents. Dark meat and skin can also compromise texture. Dark meat can have a softer texture, which can be improved by the addition of an added protein gel such as isolated soy protein.[1] Use of dark meat also tends to darken the color of the product. Consumers expect to see a light color in nuggets and patties (Figure 13.1). To overcome the color problem of dark meat, processors can use ingredients such as isolated soy protein and sodium caseinate to help lighten the color.[1] Processors must label their products according to USDA standards. Table 13.1 provides the meat content standards for poultry meat products such as nuggets and patties.[2]

Ingredients

There are many ingredients that can be added to nuggets and patties during production for various reasons. One of the most important ingredients added is salt. Salt has two main

Figure 13.1 Cut nuggets showing color and texture differences.

Table 13.1. **Poultry Meat Content Standards for Certain Poultry Products**

Label terminology	Light meat (%)	Dark meat (%)
Natural proportions	50–65	50–35
Light or white meat	100	0
Dark meat	0	100
Light and dark meat	51–65	49–35
Dark and light meat	35–49	65–51
Mostly white meat	≥66	≤34
Mostly dark meat	≤34	≥66

Source: Adapted from United States Department of Agriculture, Code of Federal Regulations, Title 9, Sec. 381.156.

functions in the production of nuggets. Salt adds flavor and aids in myofibrillar protein extraction (see Chapter 11), a necessary step for meat particle binding in forming a nugget. It is added at a concentration of less than 2% of the formula; however, formulations usually contain less than 1% in industry. Sodium tripolyphosphate (STP) is another ingredient used to aid in protein extraction. In addition, STP helps to increase water-holding capacity by increasing the pH and unfolding muscle proteins to allow for more water-binding sites.[3] Furthermore, STP helps to retard oxidative rancidity. Sodium tripolyphosphate can only account for 0.5% in the finished product.[3] However, because meat contains approximately 0.1% phosphate, processors usually add 0.3 to 0.4% STP in formulations. Water is also often added for moisture and to aid in mixing of the product. Other ingredients such as starches and soy proteins are also used as binders, extenders, and fillers. Isolated soy proteins can also retard oxidative rancidity, improve water-holding capacity, and lighten dark meat.[1] Furthermore, a variety of spices and seasonings can also be added depending on product specifications.

Prepare formulation

The first step in the production of nuggets and patties is to prepare or develop a formulation for the product (Figure 13.2). Consumer demands, marketing, technology, and creativity are all important considerations. Proper amounts of meat and other ingredients should be measured and ready to use. It is important that the ingredients are precisely measured so that the product will remain consistent.

Reduce particle size

The next step is to reduce the particle size of the meat in order to increase the surface area for protein extraction.[1,3] Muscle is covered by an epimysium connective tissue layer. When this layer is present and intact, little or no protein extraction can occur. Therefore, by chopping or grinding the meat using a bowl chopper or grinder to reduce particle size, the epimysium layer is disturbed and more surface area becomes available for protein extraction. This is an essential step because if there is no protein extraction, the meat pieces will not bind together upon cooking, resulting in a product with inconsistent texture (Figure 13.1). During particle reduction, the ingredients such as salt and STP are added. These ingredients will aid in myofibrillar protein extraction.[3] It is important that these ingredients are added after some particle reduction has occurred. The goal is that the salt and STP will contact the meat surface and aid in protein extraction. Water is added to solubilize the salt

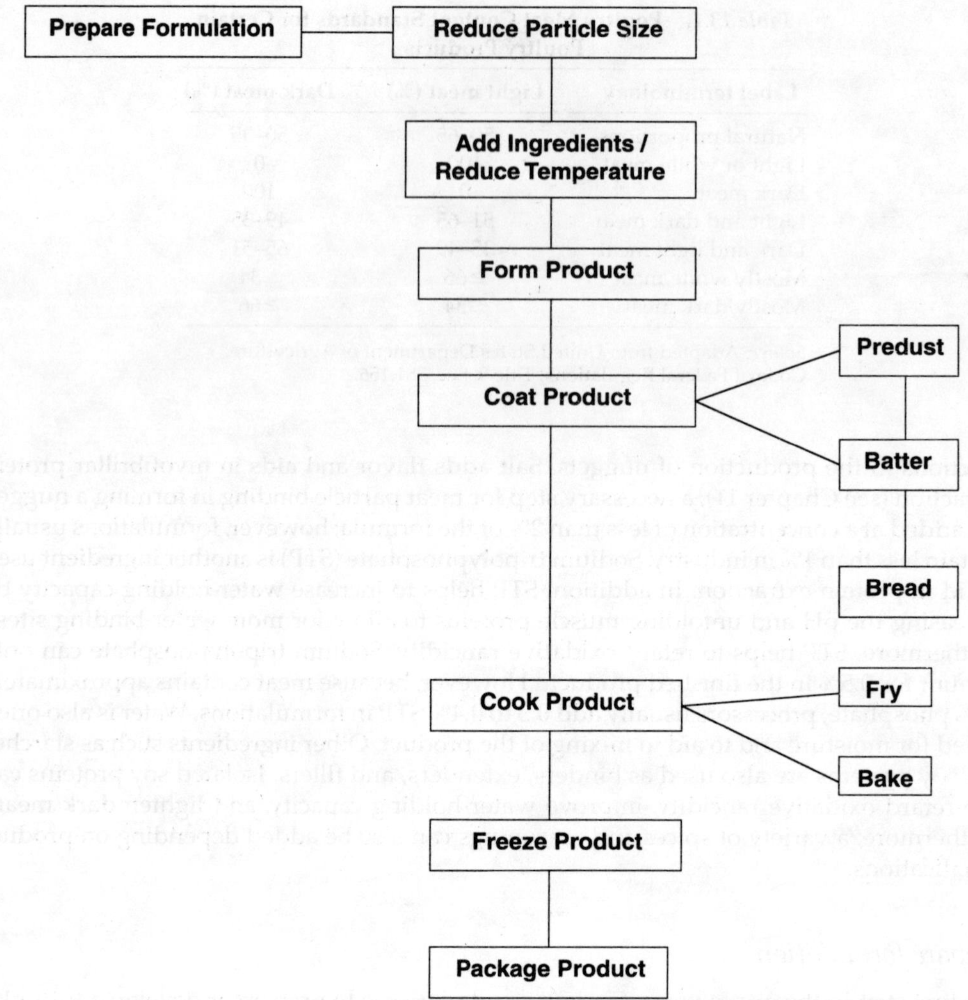

Figure 13.2 Flow diagram of coated poultry nugget and/or patty production.

and STP so that maximum protein extraction can occur. Water is added in the form of ice in order to maintain the low temperatures of the meat. If the meat temperature increases too much, protein denaturation can occur and result in poor product binding. It is also important not to overchop or overblend the meat because protein denaturation can occur from this process as well.

Reduce temperature

During particle reduction and before forming, the temperature of the meat formulation must be reduced to aid in product forming. If the meat temperature is not cold enough, the meat batter will be too soft and will not retain the desired shape when formed. The formed nuggets will not be "knocked out" properly, which can result in an oddly shaped product and the product can fall apart. In addition, problems with batter adhesion can occur with formed meat that is above −2.2°C because the meat surface is wet. If the meat temperature is too cold, the formed product can break producing a defective nugget or patty. The temperature should be reduced to −3.3 to −2.2°C. The temperature can be reduced during

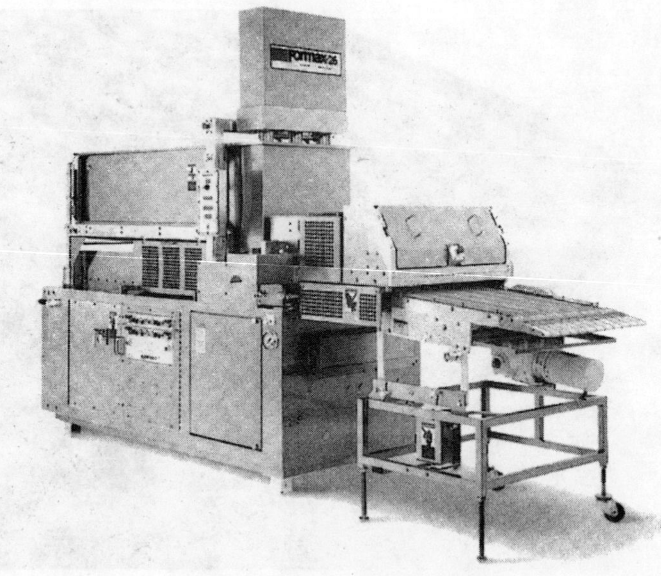

Figure 13.3 Formax® F • 26® forming machine. (Courtesy of Formax, Inc., Mokena, IL.)

the chopping process using ice and by using a frozen meat source. Processors often use a blend of frozen and chilled meat sources. Carbon dioxide snow can also be used in the forming machine hopper to reduce product temperature, but this process can become expensive.

Form the product

After the meat is chopped, blended, and cooled, the meat is ready to be formed. There is forming equipment available to processors to form the comminuted meat product (Figure 13.3). The meat mixture is placed in a hopper where the meat is then augured to the forming apparatus. The meat is pressed into mold plates that resemble the desired shape of the product. Once the meat fills the mold, the plate then slides out where a knock-out apparatus (Figure 13.4) pushes the formed meat onto a conveyor belt (Figures 13.5 and 13.6). The

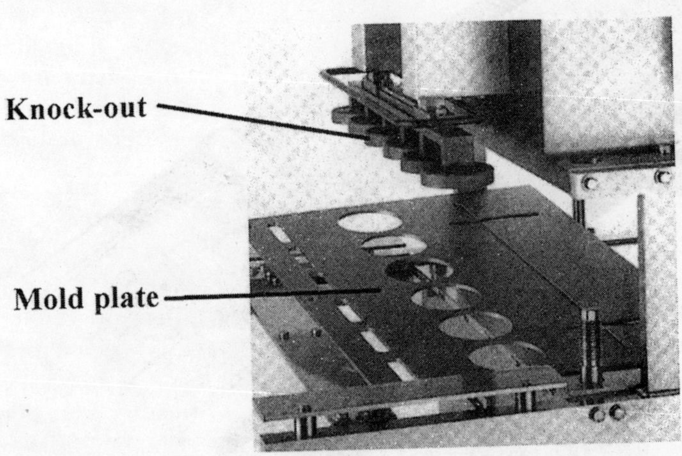

Figure 13.4 Formax® forming machine mold plate. (Courtesy of Formax, Inc., Mokena, IL.)

Figure 13.5 Formax® F • 26® conveyor with Port-Fill®. (Courtesy of Formax, Inc., Mokena, IL.)

plate then slides back and is refilled, completing the repeating cycle. Once the product is formed, it is moved on the conveyor belt to the next step, coating.

There are many shapes that can be made using forming equipment. The most common and probably among the first shapes formed were the round and oval nugget/patty (Figure 13.7). Recently, nuggets have been formed in various shapes including dinosaurs, stars, cartoon characters, rings, and athletic balls. Furthermore, instead of the common round-shaped patty, a breast-shaped mold has been used to produce "breast-shaped patty" to mimic a whole-muscle breast (Figure 13.8). These newer shapes have been very successful on the market. New technology has led to three-dimensional shapes such as surface indentations like ball logos or animal facial features to add even further interest for the consumer.

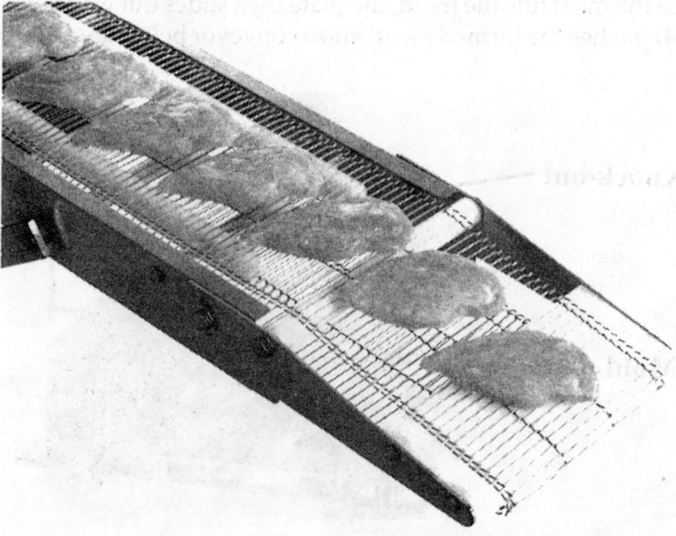

Figure 13.6 Formax® F • 6® Conveyor with Poultry-Plus®. (Courtesy of Formax, Inc., Mokena, IL.)

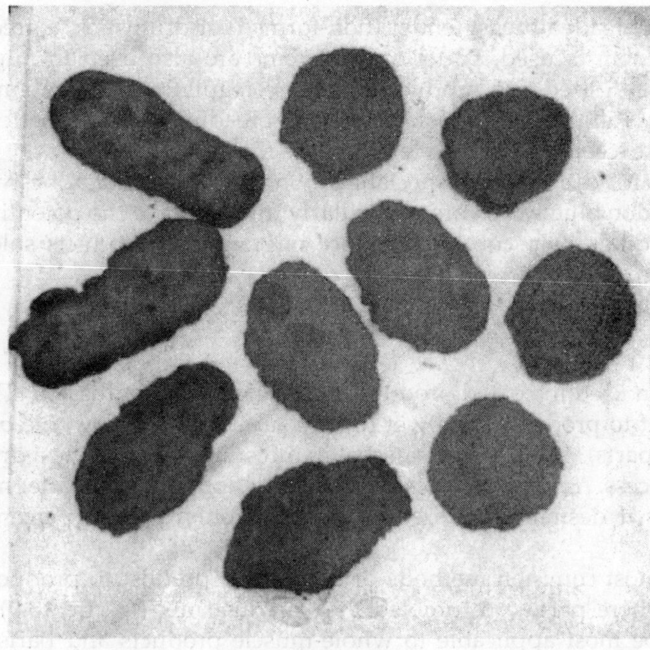

Figure 13.7 Nuggets in various shapes and colors.

Coating the product

The next step in the production of nuggets and patties is to coat the product. The three parts in a coating system are the predust, the batter, and the breading steps.[4] However, various combinations of these steps can be incorporated into nugget and patty production. For example, one product can have all three steps while another product may include just one step. Typically, in a formed and breaded product, a batter and bread system is used. In a

Figure 13.8 Formed chicken breast patties in two different shapes.

product such as a breaded breast tender (non-formed) or drumstick, a predust, batter, and bread-coating system is used. Batter-only systems are also used in nugget production depending on the desired end product. There are regulations on the amount of coating uptake a product is allowed. USDA states that the breading (predust, batter, and breading) amount shall not exceed 30% of the weight of the finished product.[4, 5]

Coating systems can be quite specialized. Some have strong adhesive properties for products like cordon blue which are particularly thick or have the potential to leak. Other batters and breadings help control moisture migration in products sold fresh (raw or cooked, but not frozen).[6]

Predust

Predusts are often used in coated products to improve batter adhesion. This predust step is very important to products with wet or oily surfaces such as whole-breast tenders or drumsticks (i.e., parts). The predust can seal in moisture and provide a dry, rough surface for the batter process. A predust typically consists of flour or a dry batter mix and possibly some seasonings if desired. The predust provides only a small percentage of coating pickup.

One of the most common methods of applying a predust to products is by using a drum breader where parts are tumbled in a predust mix (Figure 13.9). However, this method would be most applicable to whole-muscle products and parts because of the impact caused from the tumbling mechanism of the machine. Another method of application is a sprinkle applicator (Figure 13.10). This type of method would be more applicable to formed products because the product would be less disturbed. The final step in predusting is to remove excess predust, because excessive predust on products can cause problems in the batter process.[7] This can be done by blowing off excess predust or shaking the product using vibrators.

Batter

Batters play a very important role in the coating process. There are two classifications of batters that can be used in products: leavened or unleavened.[7, 8] The use of one type or the other depends upon product specification. These batter types can be used for either coating

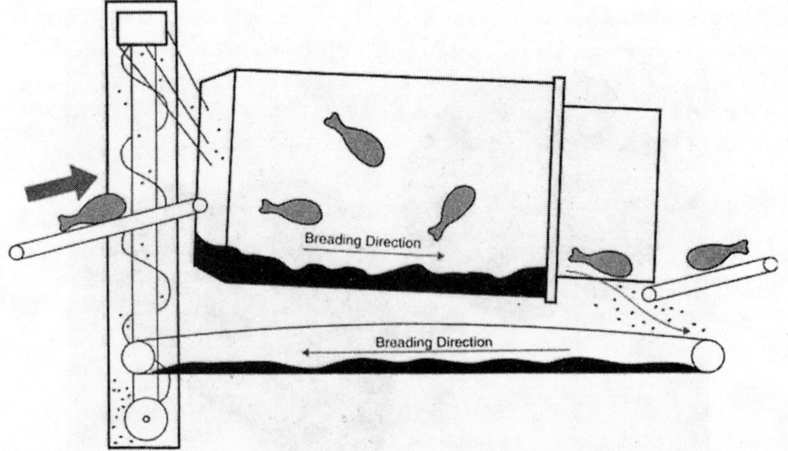

Figure 13.9 Drum breader. (Courtesy of STEIN • DSI, a business of FMC FoodTech, Sandusky, OH.)

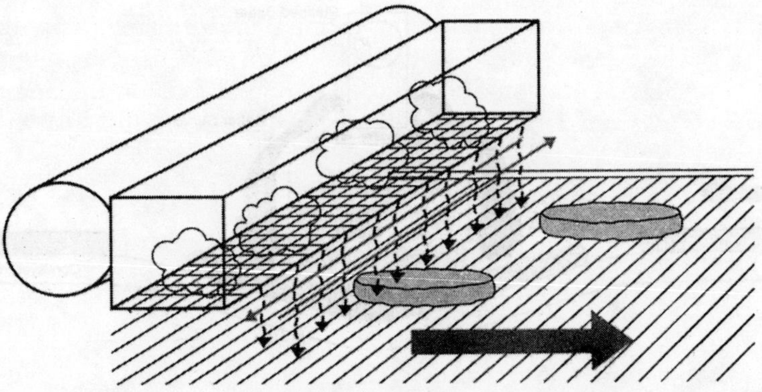

Figure 13.10 Sprinkle applicator for predust. (Courtesy of STEIN • DSI, a business of FMC Food-Tech, Sandusky, OH.)

or adhesion purposes. Batters consist of a mixture of various ingredients which can include flours, starches, eggs, milk, spices and seasoning, leavening agents, and stabilizers.

Typically, the leavened, or tempura, batters are used for coating. If a coating batter is used, it is the final step in the coating system (i.e., breadings are not used). A coating batter provides a protective outer covering to the product.[8] Tempura batters are leavened, which means that upon cooking, the batter will rise creating a fluffy appearance and cake-like texture.[8] Low leavened coating batters can also be used to result in a product with a ridge-like appearance and a crunchy texture.[8] Tempura batters are used at high viscosity levels so that the product is well coated. Because of this, special processes must be used when applying the batter. The batter is applied using a "still" system so that pumping of the batter is minimized.[7] If the batter is excessively stirred or pumped, it will lose trapped gas and cause the batter not to rise upon cooking.[7] In the batter application process, the nuggets or other formed products are transferred to the batter machine where the product is moved along the conveyor through a pool of batter (Figure 13.11). The product is submerged in the batter so that sufficient coating is achieved.

Adhesion batters are used in combination with breadings. They serve to bind the breading to the meat product as well as to add flavor and texture to the product.[8] These batters are unleavened and are used at various viscosity levels. Adhesion batters can be applied using a top submersion system where the product moves along the conveyor through a pool of batter, similar to the tempura style (still system) applicator; however, the batter in this system can be recirculated. Another system known as the overflow batter application method is more appropriate for low to medium viscosity batters.[7] In the overflow process, the batter flows over the product and is then recirculated (Figures 13.12 and

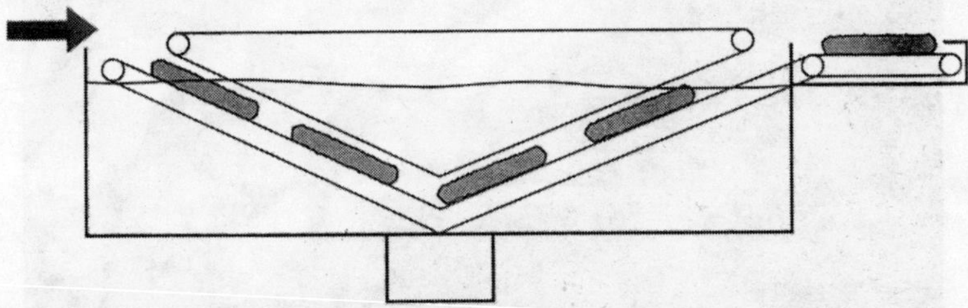

Figure 13.11 Batter application: top submersion system. (Courtesy of STEIN • DSI, a business of FMC FoodTech, Sandusky, OH.)

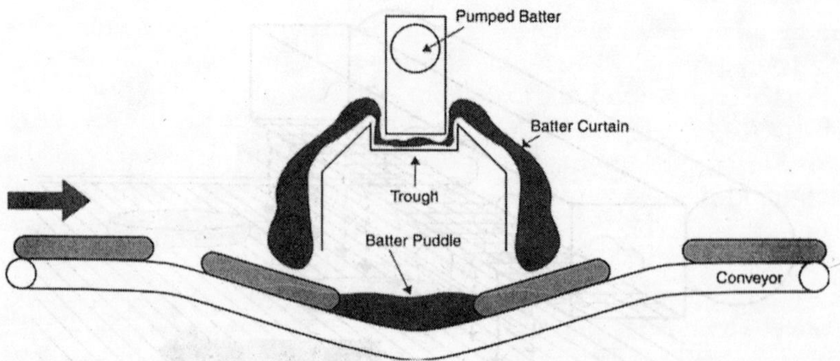

Figure 13.12 Batter application: overflow system. (Courtesy of STEIN • DSI, a business of FMC FoodTech, Sandusky, OH.)

13.13). The product also travels through a small batter puddle so that the product is completely coated.

Breading types

There are many different types of breadings that can be used in a coating system.[4] The five major types of breadings are American bread crumbs, Japanese bread crumbs, crackermeal, flour breaders, and extruded crumbs. They can vary in size, shape, texture, color, and flavor.[7,8] These breadings can be used alone or in combination with other ingredients such as spices and seasonings.

Figure 13.13 Batter application on chicken wings using overflow system. (Courtesy of STEIN • DSI, a business of FMC FoodTech, Sandusky, OH.)

American bread crumbs are somewhat round in shape and are similar to homemade bread crumbs. There is usually a two-tone appearance to this crumb representing the inner crumbs and surface crumbs from the loaf of bread. It is a durable breading and is generally medium in cost.

Japanese bread crumbs are made from crustless bread and have a sliver shape. The crumbs are long and flake-like in appearance and, because they are made from crustless bread, more uniform in color. These crumbs are available in a range of fine to coarse granules. Due to its characteristic shape, special breading equipment must be used so that damage to the crumbs is minimized.[7]

Cracker meal is a fine, flat, dense crumb that is similar to cracker crumbs. It is made of flour and water, which is baked, dried, and ground. It often gives a chunky appearance to the breaded product. It is very popular in the U.S. and is a low-cost material. It is available in various sizes from fine to coarse and can be used as a predust or breading.

Flour breadings are very popular and traditional because they provide a flaky, homemade appearance to the product. These breadings are often used in combination with spices, seasonings, and other ingredients. Flour breadings can be used to achieve many different surface textures. For example, when combined with batter, breading balls are formed which result in the unique homestyle appearance.

Extruded crumbs are generally used as a low cost alternative to the other types of breadings. Various shapes can be achieved by extruding crumbs. Extruded crumbs, made from starchy materials, can be formed in the shapes of American bread crumbs, Japanese bread crumbs, and crackermeals. These crumbs may be used as a predust or breader, but must be carefully formulated so that texture and flavor are not negatively affected.

Breading characteristics

There are several breading characteristics that can affect the final outcome of the product's appearance and texture. "Pickup" is a term used to describe the amount of breading that a product picks up upon application. The thickness of the batter and the size of the crumb particles can affect the pickup of breading onto the product.[8] Thicker, more viscous batters will tend to pick up more breading than a thin, less viscous batter. Coarse crumbs will provide good pickup compared to a finer crumb; however, there are trade-offs that must be considered when selecting the appropriate crumb size. Coverage (of breading on the product) can be affected by crumb size. Fine crumbs will provide a uniform coverage of breading to the product; however, more coarse crumbs that would offer good pickup do not always offer good coverage.[8] Appearance of the breading is also affected by the crumb. Fine crumbs provide a smooth texture whereas coarse crumbs provide an irregular, nonuniform texture.[8] A medium crumb could provide a uniform texture with some highlights in the breading. Finally, texture is also affected by crumb size. Finer crumbs offer a tender texture whereas the coarse crumbs offer a crisp, crunchy texture.[8] Medium crumbs offer a mixture of the two properties.

Breading application

Breading is applied using a recirculating system. With most breading applicators, the product travels along a moving crumb bed so that the underside of the product is covered (Figure 13.14 and 13.15). The battered product passes under a curtain of flowing crumbs so that the top surface of the product is coated. Pressure rollers are further down the conveyor to apply pressure to the coated product so that the crumbs will "embed" into the batter. After this process, excess breading is blown off and the fully coated product is then transferred to the cooking step. The application of Japanese bread crumbs is a similar process;

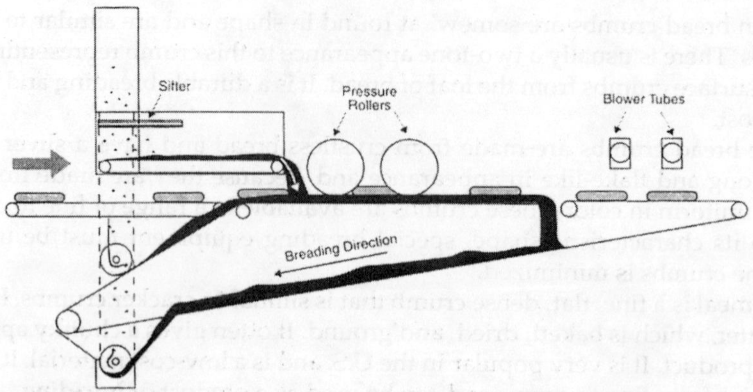

Figure 13.14 Breader application. (Courtesy of STEIN • DSI, a business of FMC FoodTech, Sandusky, OH.)

however, the crumbs are subjected to a more complex sifting process so that the large and small Japanese bread crumb particles are uniformly distributed onto the product (Figure 13.16). Furthermore, the crumbs are sifted onto the product rather than the product traveling under a curtain of crumbs.

Cooking

After the product is coated, it is then cooked. The formed product is typically fully cooked by either frying or baking, depending on the product specifications. Frying the formed products is probably the most popular cooking method; however, with more consumers concerned about their eating habits, baked products have also become popular. The cooking process causes the product to change to a golden color that can also have color highlights depending on the breading used. During frying, the product remains on the conveyor belt which is lowered into a pool of hot oil to cook the product (Figure 13.17). There are two methods that the processor can use to cook the products: full cook and prefrying. Processors can either fully cook the product in one step, partially cook the product and distribute the product as raw, or use a combination of methods (pre-fry and fully cook the product).

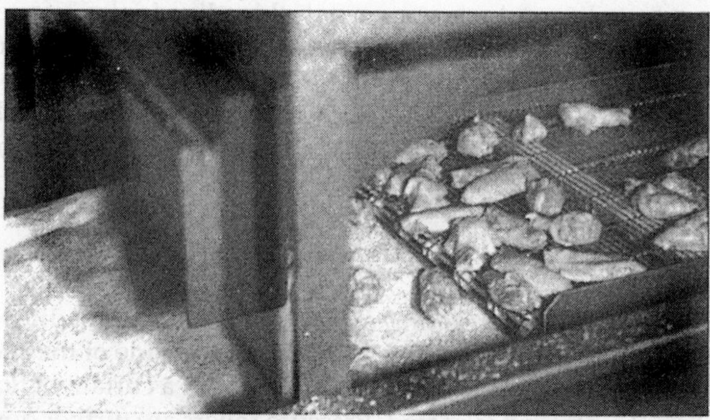

Figure 13.15 Breader application on battered chicken wings. (Courtesy of STEIN • DSI, a business of FMC FoodTech, Sandusky, OH.)

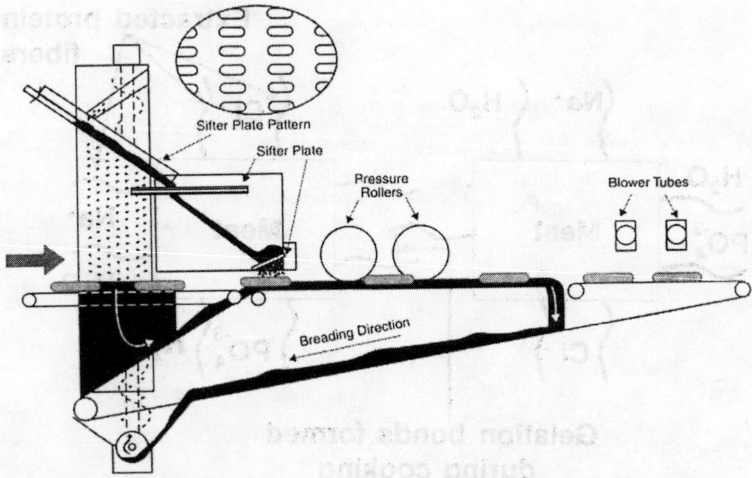

Figure 13.16 Breader application: Japanese bread crumbs. (Courtesy of STEIN • DSI, a business of FMC FoodTech, Sandusky, OH.)

One method is to cook the product in one step, or full cook. However, by cooking in one step, the number of product defects can increase. Nuggets can stick together during the cooking process. This can result in the product's not being cooked to the proper temperature, especially at the adjoining surfaces. In addition, coating adhesion can be compromised and voids in breading on the product can occur. Therefore, most processors use a two-step cooking process. The nuggets are first cooked in oil at 179.4 to 198.8°C for 30 to 45 s and then removed for a short period of time.[7] This first cook "sets the coating," or cooks the surface of the meat and batter/breading. The extracted myofibrillar proteins at the surface will bond with proteins in the batter. This first cook, known as "par frying" or "flash frying," also helps to reduce the number of nuggets that touch (i.e., to avoid products cooking together). After the first cook, the nuggets are again submerged into a second fryer and cooked at 165.5 to 179.4°C for a varying amount of time, depending on the product. The purpose of the second fryer is to fully cook the product. If an operation were only par frying, the product would be transferred to the freezing and packaging step after the pre-fry. When fully cooking, the extracted myofibrillar proteins at the surface of the meat particles will form tight bonds, resulting in a consistent, cohesive product (Figure 13.18). One drawback to using a full fry system by the processor is that frying drives off moisture, which decreases product yield.

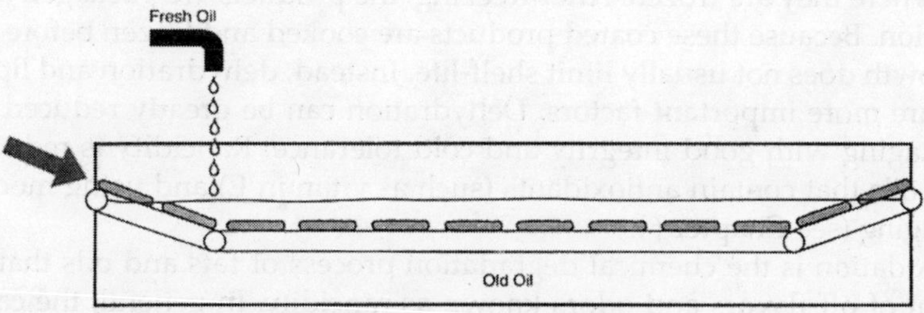

Figure 13.17 Diagram of an inline fryer. (Courtesy of STEIN • DSI, a business of FMC FoodTech, Sandusky, OH.)

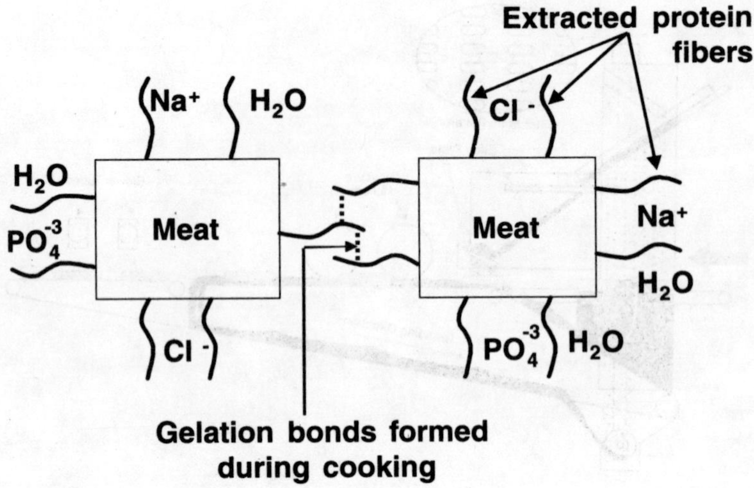

Figure 13.18 Diagram of extracted protein fibers and gelation bonds formed during cooking.

An alternative to frying is baking the product. This helps to reduce the amount of fat in the product. Instead of the product being cooked in oil, the coated product is baked. The current challenge with baking coated products is to produce a baked product with the same crunchy coating texture and golden color as is produced with frying. Innovative combinations of ingredients and cooking conditions/equipment (time, temperature, air flow, and humidity) are making progress toward this goal.

The Fry Shield™ system developed by Kerry Ingredients uses a pectin starch surface layer to reduce oil uptake during frying.[6] The product is coated as normal, then dipped in a 1% pectin solution for a few seconds, and then partially or fully fried. This system reduces oil uptake by 20 to 50%, depending on the product and coating.

In the No Fry™ system from Morton Foods, the product is coated with a special product, batter, and breading system.[6] Then, the coated product is sprayed with an emulsion of oil, water, protein, and flavorings. The products are then heated in an infrared oven at 900°C for 40 to 60 s. The amount of emulsion sprayed on depends on the desired product and therefore allows more precise control of the product characteristics.

Freezing and packaging

As soon as the cooking process is completed, the nuggets/patties travel on the conveyor to the freezer where they are frozen. After freezing, the products are packaged and prepared for distribution. Because these coated products are cooked and frozen before distribution, bacterial growth does not usually limit shelf-life. Instead, dehydration and lipid oxidation (rancidity) are more important factors. Dehydration can be greatly reduced by moisture barrier packaging with good integrity and cold tolerance. Rancidity is reduced by using fresh frying oils that contain antioxidants (such as vitamin E) and using modified atmosphere packaging (see Chapter 6).

Lipid oxidation is the chemical degradation process of fats and oils that leads to the development of off-flavors and odors known as rancidity. In general, the carbon double bond of an unsaturated fatty acid is attacked by an activated form of oxygen, known as a peroxidide (Figure 13.19). The fat then breaks at the site of the double bond, resulting in a variety of degradation products with a range of off-flavors and odors. The formation of

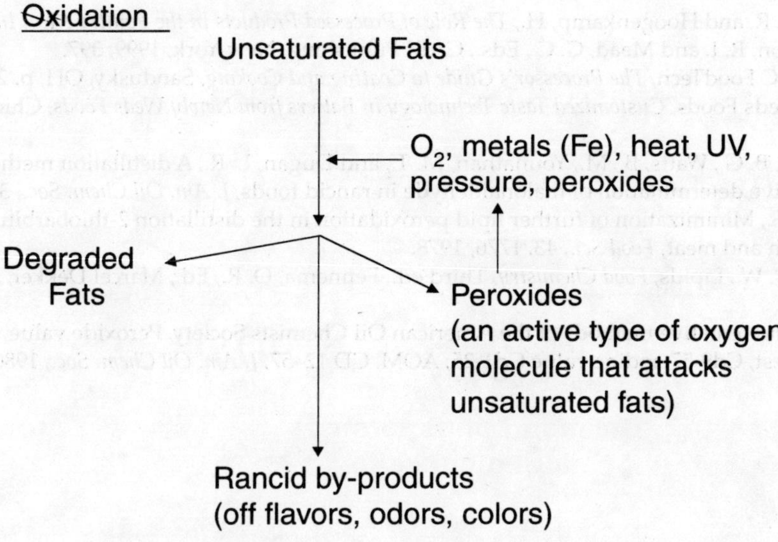

Figure 13.19 Diagram of the oxidation of unsaturated fats.

peroxides is catalyzed by ultraviolet light, heat, pressure, and metals so the exclusion of these factors (as well as oxygen and the use of unsaturared oils) would reduce rancidity development. Lipid oxidation is determined by sensory evaluation, by measuring the degradation products with the TBA or TBARS method (thiobarbituric acid reactive substances[9,10]), or by measuring the level of peroxides with the peroxide value.[11,12]

Summary

Coated poultry products have been a large part of the growth of the further processed poultry industry in recent decades. These products provide virtually unlimited versatility in shape, texture, and appearance to appeal to the ever-changing consumer demand. Production of these patties, nuggets, and sticks is a complex procedure involving particle size reduction, blending, forming, coating, and cooking. The possibilities existing in each of these steps adds to the potential variety of these products but also increases the potential for problems if done incorrectly. A major concern in the production and distribution of these products is lipid oxidation, which starts with cooking and continues through frozen storage to produce off-flavors and -odors. However, the tremendous presence of these products in today's marketplace is evidence of their success and appeal to consumers.

References

1. Bowers, P., Golden nuggets pan out globally, *Poultry,* 1994.
2. United States Department of Agriculture, Code of Federal Regulations, Title 9, Sec. 381.156.
3. Pearson, A. M. and Tauber, F. W., *Processed Meats,* Second ed., Van Nostrand Reinhold, New York, 1984, 46.
4. Cunningham, F. E., Developments in enrobed products, *Processing of Poultry,* Mead, G. C., Elsevier Science Publishers, U.K., 1989, 325.
5. United States Department of Agriculture, Code of Federal Regulations, Title 9, Sec. 381.166.

6. Mandara, R. and Hoogenkamp, H., *The Role of Processed Products in the Poultry Meat Industry*, Richardson, R. I. and Mead, G. C., Eds., CABI Publishing, New York, 1999, 397.
7. Stein, FMC FoodTech, *The Processor's Guide to Coating and Cooking*, Sandusky, OH, p. 2.
8. Newly Weds Foods, *Customized Taste Technology in Batters from Newly Weds Foods*, Chicago, IL, 1998.
9. Tarladgis, B. G., Watts, B., M., Younathan, M. T., and Dugan, L. R., A distillation method for the quantitative determination of malonaldehyde in rancid foods, *J. Am. Oil Chem. Soc.*, 37, 1, 1960.
10. Rhee, K. S., Minimization of further lipid peroxidation in the distillation 2-thiobarbituric acid test of fish and meat, *Food Sci.*, 43, 1776, 1978.
11. Nawar, W. W., Lipids, *Food Chemistry*, Third ed., Fennema, O. R., Ed., Marcel Dekker, New York, 1996, 276.
12. Official and tentative methods of the American Oil Chemists Society, Peroxide value, Cd8-53; oxirane test, Cd9-57; iodine value Cd1-25; AOM, CD 12–57, *J. Am. Oil Chem. Soc.*, 1980.

chapter fourteen

Mechanical separation of poultry meat and its use in products

Glenn W. Froning and Shelly R. McKee

Contents

Introduction

In the late 1950s and early 1960s marked changes in the poultry processing industry began. At that time, the poultry industry began marketing more cut-up and further processed poultry meat products. As the popularity of these consumer choices grew along with the increased consumption of poultry meat, more parts such as frames, backs, necks, drumsticks, wings, etc. became available for mechanical separation. In the process of mechanical separation, meat is removed from the skeletal bone tissues by grinding the starting material (frames, necks, etc.) and passing it through a sieve under high pressure. Most of the bones and cartilagenous materials are removed based on a differing resistance to shear.

0-8493-0120-3/01/$0.00+$.50

Mechanical separation provides a means of harvesting functional proteins which can be used in the preparation of a variety of further processed meat products. Mechanically separated poultry meat (also reported as mechanically deboned poultry meat prior to 1995) has been widely utilized in further processed poultry meat products such as bologna, salami, frankfurters, turkey rolls, restructured meat products, and soup mixes. This low cost meat source has led to poultry meat products being more cost effective in the market place.

Yields of mechanically separated poultry range from 55 to 70%.[1] The meat-to-bone ratio largely influences the yield from specific parts. In 1994, the USDA indicated that approximately 1 billion pounds of raw poultry material produces about 700 million pounds of mechanically separated poultry.[2] This mechanically separated poultry has been formulated into about 400 million pounds of sausages (bologna, salami, and franks) and 300 million pounds of nuggets and patties. Some mechanically separated poultry is combined with other species (e.g., beef and pork) in various sausage products.

Regulations

The use of mechanically separated poultry meat is regulated by the U.S. Department of Agriculture Food Safety Inspection Service,[3] while the Food and Drug Administration (FDA) regulates fish and fishery products.

Changes in labeling regulations

Regulations regarding mechanically separated poultry meat were first established by the Food Safety Inspection Service (FSIS) in 1969. There have been significant changes in the regulations relating to the labeling of mechanically separated poultry meats. Prior to 1996, mechanically separated poultry meat was generically referred to as "mechanically deboned poultry" or "comminuted poultry," but the label needed only to state "chicken" or "turkey." As far as labeling was concerned, there was no discernable difference in poultry meat that was processed using a deboning machine or poultry that was hand-deboned.

Disparaging differences existed in the labeling requirements for poultry in comparison to other livestock (primarily beef and pork) meats. Specifically, mechanically separated livestock products had to be labeled as mechanically separated species (MS [Species]) with the species being defined as beef, pork, or lamb. Red meat sausage manufacturers alleged that the poultry industry possessed an unfair marketing advantage, because of the differences in labeling requirements. This allegation resulted in a lawsuit, Bob Evans Farm, Inc., et al., v. Mike Espy, Secretary of Agriculture.[4] In response to the lawsuit regarding these labeling inconsistencies, poultry regulations were re-evaluated by FSIS. FSIS suggest that mechanically separated poultry is not different (in consistency and form) from the product resulting from the mechanical separation of other livestock products. Moreover, the final texture and form of mechanically separated poultry is different from hand-deboned poultry, even if the hand-deboned poultry is further processed through a grinder. As a result of these conclusions, FSIS revised the regulations regarding mechanically separated poultry meat which became effective in November of 1996. Specifically, the poultry products inspection regulations were amended so that mechanically deboned poultry would be required to be labeled as "mechanically separated (kind of poultry)" as opposed to labeling it "ground chicken or turkey" without indication of the mechanical deboning process used. However, this labeling requirement is dependent on the starting materials. If the starting materials are frames, trim, or parts where most of the meat has been removed, then the labeling should indicate the meat is mechanically separated poultry. However, if the starting materials are parts with the majority of meat still attached or whole birds such as spent fowl, rooster, and mature breeder hen, then the standard of identity on the label can still indicate "ground chicken."

Use of mature fowl in baby foods

Another change in regulations included the use of mature fowl in baby foods. Many of the original regulations were based on an USDA review of a report on the health and safety of the use of mechanically separated poultry. At that time, there was a concern about the high fluorine (from bone) content in mechanically separated meats from mature fowl. As a result, the past regulations prohibited the use of mechanically separated mature fowl in baby, junior, and toddler foods and limited the amount in other poultry products to 15%. USDA recently re-evaluated the health report of 1979[5] and concluded that the fluorine content in mechanically separated poultry was not a health concern. This change of attitude was based on discussions with dentists, medical doctors, and baby food companies, who agreed the chances of developing fluorosis (fluorine toxicity) from overconsumption of poultry products containing mechanically separated mature fowl were negligible. Therefore, the 1996 regulations removed the prohibition of the use of mechanically separated mature fowl in baby foods.

Kidney and reproductive organ removal

Currently, the only restrictions associated with mature fowl are that the kidneys and reproductive organs must be removed from the carcass prior to the mechanical separation process. Specifically, the regulation calls for removal of the kidneys and mature reproductive organs from mature fowl during evisceration in the slaughter facility. Immature reproductive organs in young birds (broilers) processed at their usual market age are not prohibited in any poultry products. However, kidneys from mature fowl contain heavy metals, such as cadmium, which are accumulated in the tissue over time and may pose a health or safety concern. In contrast, kidneys from young birds that are processed (broilers 6–8 weeks, turkey 19–21 weeks) do not contain high levels of heavy metals compared to the mature hens. When mechanically separated poultry represents a significant portion of a meat product such as hot dogs, then the kidneys must be removed regardless of bird age. Otherwise, kidneys from young birds do not necessarily have to be removed. In the industry, it is common practice to remove kidneys from all birds that will be used for mechanical separation.

Bone and calcium content

Because of the mechanical separation process, bone material is a component that is measured and limited in mechanically separated poultry. Bone solids content is restricted to no greater than 1%. In meat products, this equates to a calcium content no greater than 0.235% in products made from turkey or mature fowl or 0.175% in products made from broilers processed at the traditional ages (6–8 weeks). Because mature fowl have more brittle bones and turkeys have larger bones, slightly higher calcium content from residual bone is expected in their mechanically separated meats. Residual bone particle size is also restricted so that 98% of the bone particles may be no larger than 1.5 mm in their greatest dimension and no bone particles may be larger than 2.0 mm in their greatest dimension.

Limitations on the bone particle size were thought to reduce the likelihood of any physical hazard from bone particles and to limit the amount of bone material that may be incorporated into the separated poultry meat as a result of the mechanical process.

Limitations on the use of mechanically separated poultry in products

Limitations on the amount of mechanically separated poultry used in various poultry products are defined by the standard of identity for that particular product. For example, in the standard of identity for frankfurters, mechanically separated poultry is limited to 15% of the final product composition. Other poultry products may have different restrictions depending on the product requirements as defined in their standard of identity.

Cholesterol, protein, and fat

FSIS decided that cholesterol was not an issue in mechanically separated poultry because this meat is primarily used as an ingredient in further processed products where cholesterol levels must be declared on the label. It was suggested that people needing to limit cholesterol in their diets would be able to make educated decisions based on product labels.

Protein and fat were addressed based on the standard of identity of mechanically separated poultry. Mechanically separated poultry cannot contain greater than 25% fat and not less than 14% protein for it to be deemed as mechanically separated poultry. While mechanically separated meat may contain slightly higher amounts of collagen than hand-deboned poultry, protein quality is not greatly affected. Moreover, Froning[6] reported that protein efficiency ratios of mechanically separated poultry were comparable to that of casein, a high quality protein. Mechanically separated poultry that is labeled as such must have a minimum protein efficiency ratio of 2.5. Product derived from the newer "advanced recovery meat/bone separating systems" which can be labeled as "meat" (chicken, turkey, beef, etc.) also has particular protein quality standards as defined by a minimum protein efficiency ratio.

Equipment

Currently, the most readily used process for mechanical separation of poultry consists of chopping the starting materials and then separating the bone, sinew, and tendons from meat by passing it through a sieve under high pressure (Figure 14.1). There are two basic categories of mechanical separators. One type forces meat from an outer chamber through slots of perforated drum, leaving the bone material in the outside of the drum. In a similar design, meat is forced outside through a perforated cylinder while the bone residue is maintained in the interior.[6] Deboning machines can process anywhere from 500–20,000 lb of product per hour depending on the size and capacity of the machinery, and all automated mechanical deboning equipment for poultry, fish, and red meats must be USDA approved.

Many factors related to the equipment can affect end product quality. For instance, yield is affected by the amount of pressure that is applied when pushing product through the sieve. However, when pressure is increased, the separation process can become slightly less efficient by allowing more bone, sinew, and other non-meat residues in the final product. Processors determine the optimum machine settings to achieve high yield and product quality. Maintenance of the equipment is another factor that affects product quality. Maintaining sharp edges on cutting surfaces greatly influences the end product texture

Figure 14.1 Example of machine for mechanical separation of poultry meat. (Courtesy of Beehive.)

and consistency. Poor equipment maintenance can cause product to smear and become pasty in texture. Texture can also be altered by changing screen or sieve sizes. Large pore sizes in sieves result in a course textured product.

Product temperature is another factor that can alter end product quality. Most equipment can process meat that is chilled, but not frozen. One Midwest processor has modified the separation equipment so that they are able to process meat that is frozen. This is a tremendous advantage because the product has a superior texture (Figure 14.2), longer shelf-life, and lower bacterial counts. Because of the uniqueness of this process, this company can label the product as "ground chicken" or "turkey" as opposed to "mechanically separated poultry."

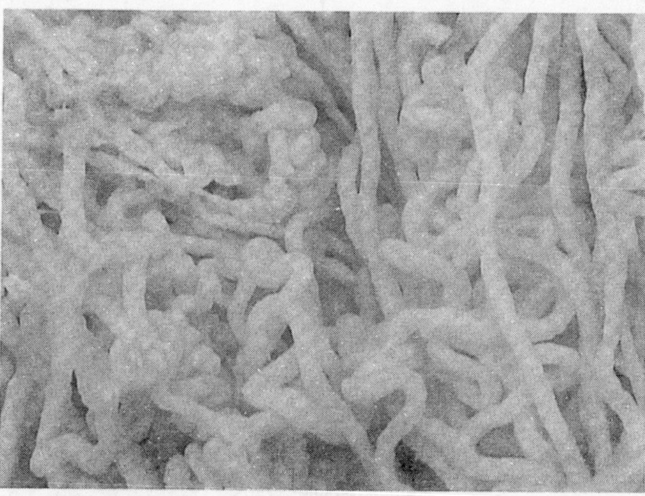

Figure 14.2 Example of product from a modified separation process that can be labeled as ground chicken.

Newer separating equipment referred to as advanced recovery separating systems can mechanically separate meat that also bears "chicken" or "turkey" on the label as opposed to "mechanically separated (poultry)." Advanced recovery meat/bone separating systems are thought not to pulverize or grind bones thereby lessening the amount of calcium in the final product. Mechanically, the system uses a piston instead of an agar to push the product through the separating sieve. Currently, these types of systems are more common in beef and pork processing than in the poultry industry.

Types of mechanically separated poultry

As the growth of further processed turkey and chicken products has grown, more parts have become available for mechanical separation. Chicken broilers are now routinely cut-up, or hand deboned. After cut-up or hand-deboning the frame, back, neck, drumsticks, and wings are often mechanically separated to be used in various further-processed meat products (Figure 14.3). Turkeys are also now sold as cut-up parts. More commonly, turkeys are hand-deboned and further processed. After hand-deboning of turkey carcasses, the frame, drumsticks, backs, and necks are also mechanically separated.

Mechanical separation of meat from the whole carcass has not been a common practice. However, Froning and Johnson[7] mechanically separated spent fowl meat from the whole carcass after pregrinding of the carcass. Also, the industry has sometimes precooked leghorn spent fowl and hand-deboned the cooked meat for utilization in soup or other further-processed poultry products. After hand-deboning the cooked fowl meat, the remaining stripped carcass is mechanically separated.

After mechanical separation, the bone residue is often utilized in animal feed. Scientists have found that the bone residue has some excellent potential as a feed ingredient or could be used to produce a protein isolate.[8–10] With increased environmental concerns, utilization of the bone residue has become an important priority.

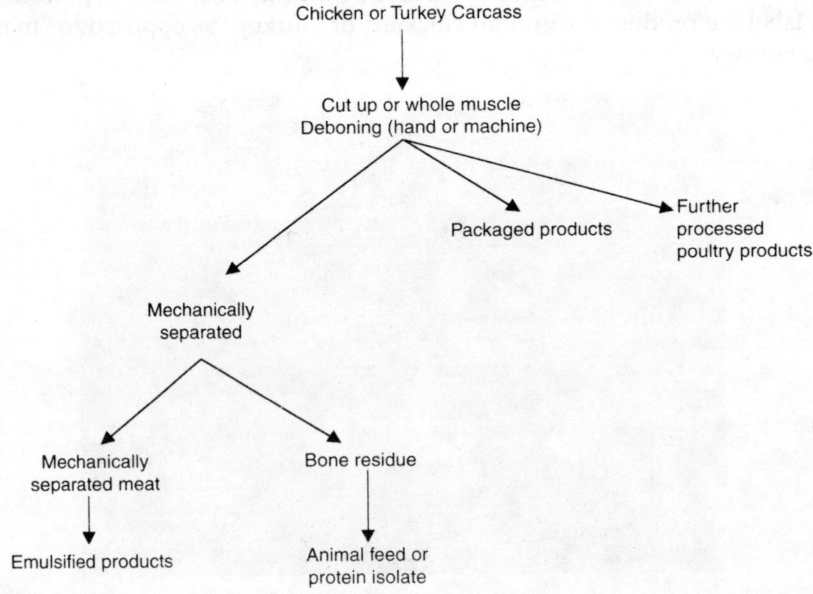

Figure 14.3 Mechanical separation of poultry meat.

Composition

When poultry meat is mechanically separated, considerable shearing action causes marked cellular disruption. The extent of the cellular damage is largely affected by the screen size utilized. Schnell et al.[10] observed that small screen sizes used in the separator will reduce the size of the myofibrils. Breaks were noted at the Z or M bands. Also, the bone marrow is released from broken bones during the separation process, thereby contributing increased lipids and heme components into the separated meat. Lipid and heme fractions further dilute the amount of protein in separated meat.

Proximate composition of various sources of mechanically separated poultry meat is shown in Table 14.1.[7,11–16] As indicated, there is considerable variation in the composition. Factors influencing the composition include bone-to-meat ratio, age of the bird, skin content, cutting methods, deboner settings, and species. Younger birds generally will have more heme and lipid components from the bone marrow influencing the proximate composition. Skin content may greatly increase the fat content of the resulting separated meat while the collagen from the skin is largely found in the bone residue.[17] However, if cooked carcasses or parts are mechanically separated, the collagen is likely gelatinized thereby increasing the collagen content of the separated meat. Deboner settings can affect the yields and the proximate composition substantially. If the settings are set for high yields, the fat and ash content in the resultant mechanically separated meat may be largely increased. High settings may also increase the temperature resulting in protein denaturation, which may ultimately affect functionality.

Protein quality of mechanically separated poultry meat has received considerable emphasis. Several scientists have observed that the protein quality of mechanically separated poultry is comparable to that found from hand-deboned sources.[14,16,18,19]

One concern has been the fatty acid and cholesterol content of mechanically separated poultry meat. Moerck and Ball[20] observed that the bone marrow from chicken broilers contained a higher percentage of phospholipids and cholesterol than that found in other broiler meat. However, the fatty acid composition of chicken bone marrow and mechanically separated poultry meat was quite similar to that from hand-deboned meat sources.

With the advent of mechanical separation of poultry meat, the issue of possible bone content came under close scrutiny. Bone particles from hand-deboned and mechanically separated poultry meat have been characterized.[21] Bone particles isolated from hand-deboned sources were actually somewhat larger than that obtained from mechanically

Table 14.1 **Range of Composition of Mechanically Separated Poultry from Different Sources**

Source	Protein (%)	Moisture (%)	Fat (%)	Ref.
Chicken backs and necks	9.3–14.5	63.4–66.6	14.4–27.2	Froning[7] Grunden et al.[8] Essary[9]
Chicken backs	13.2	62.4	21.1	Froning[7]
Skinless necks	15.3	76.7	7.9	MacNiel et al.[10]
Turkey frame	12.8–15.5	70.6–73.7	12.7–14.4	Froning[7] Grunden et al.[8] Essary[9]
Spent layers	13.9–14.2	60.1–65.1	18.3–26.2	Grunden et al.[8] Froning and Johnson[3]
Cooked spent layer	18.3	63.2	16.5	Babji et al.[12]

separated meat. Any bone particles found in mechanically separated poultry meat were indicated to be of a "powdery" form presenting no hazard to the consumer. Calcium content in terms of bone equivalents is closely monitored today.

Several minerals in mechanically separated poultry have been investigated as they may affect health and safety. Murphy et al.[22] analyzed for several minerals including arsenic, fluoride, cadium, strontium 90, selenium, iron, nickel, copper, lead, and zinc. None of these were indicated to be a health hazard in mechanically separated poultry meat. This may, in part, be due to the exclusion of body parts (e.g., kidneys) from the production flow as has been previously discussed.

Functional properties

As more mechanically separated poultry has been used in further processed meat products, functional attributes have become an important consideration. Much of the mechanically separated poultry has been used in emulsified products. Thus, the effects of the mechanical separation process on salt-soluble proteins and fat content have been found to influence functional properties. Mechanically separated turkey meat has been found to have less salt-soluble proteins.[23] These authors further observed superior emulsion capacity in hand-deboned turkey as compared to mechanically separated turkey. However, water-holding capacity appeared to be higher in mechanically separated turkey. Mayfield et al.[24] indicated that mechanically separated poultry with 12% protein produced more viscous emulsion batters and emulsion stability was superior to that noted from sources containing 11% protein. Others have reported that the mechanically deboned poultry from different sources may exhibit variability in emulsion characteristics and water-holding capacity.[25]

Skin content has influenced the functional properties of mechanically separated poultry.[26,27] Higher skin levels decreases emulsion stability and emulsion capacity, which are largely related to the higher fat content contributed by skin. However, Schnell et al.[27] reported that higher skin content increased organoleptic tenderness of frankfurters.

There have been attempts by some researchers to texturize mechanically separated poultry. Acton[28] forced mechanically separated chicken through a grinder path (cutting blade omitted) with a 4-mm orifice. The meat strands were heat set at 100°C for 1, 3, 5, 7.5, and 10 min. Longer heating times and higher skin content increased the shear resistance. Although heating caused a loss of extractable protein, emulsion stability of extruded strands was improved. Lampila et al.[29] also texturized mechanically separated turkey meat by extrusion and heat setting. They proposed its utilization in restructured meat products.

Modifying mechanically separated poultry by centrifugation has been investigated.[7,30] Centrifugation reduced the fat content and improved the water-holding capacity and emulsifying capacity. Commercial-scale centrifugal separators are available.

Several additives have been observed to affect functional attributes of mechanically separated poultry meat. Froning and Janky[31] reported that salt preblending improved the emulsifying stability of mechanically separated poultry meat. Schnell et al.[27] found that the addition of 3% sodium caseinate or 0.5% Kena (polyphosphate) increased the viscosity of frankfurter emulsions made from mechanically separated chicken meat. Froning[32] chilled spent fowl in 6% polyphosphate prior to deboning and observed improved emulsification stability and emulsion capacity of the resultant mechanically separated fowl meat. McMahon and Dawson[23] reported that a combination of 0.5% polyphosphate and 3% sodium chloride improved extractable protein, water-holding capacity, and emulsifying capacity of mechanically separated turkey meat.

The use of structured soy protein in combination with mechanically separated poultry improved textural attributes.[33–35] However, others have observed decreased emulsion stability of mechanically separated poultry with added soy protein.[36]

Color and heme pigments

Mechanical separation of poultry meat influences the color of the resultant meat. The process releases heme pigments from the bone marrow into the mechanically separated meat. Froning and Johnson[7] found that the mechanical separation of poultry meat will increase the heme protein content approximately three times that found in hand-deboned poultry. This increase is primarily due to hemoglobin from the bone marrow. Hemoglobin is more subject to abnormal color problems since it is more easily oxidized and more susceptible to heat denaturation during processing and storage. Abnormal brown, green, and gray color defects have been reported in further-processed poultry meat products containing mechanically separated poultry. During the separation process the meat is exposed to considerable air, which may accelerate the oxidation of heme pigments.

Composition and processing variables have been shown to affect the color characteristics of mechanically separated poultry meat. Froning et al.[26] investigated the effect of skin content prior to deboning on the color of mechanically separated poultry meat. Higher skin levels generally increased the lightness and decreased redness of the resultant mechanically separated poultry meat. These color changes were attributed to the dilution of the heme pigments by the additional fat from the skin.

Researchers have attempted to modify color characteristics of mechanically separated poultry meat by centrifugation.[7,30] Froning and Johnson[7] observed that centrifugation increased redness of the mechanically separated poultry meat while Dhillon and Maurer[30] reported less redness due to centrifugation. The discrepancy may be partially explained by the use of mechanically separated fowl meat in Froning and Johnson's study while Dhillon and Maurer utilized mechanically separated chicken and turkey meat in their study.

Cryogenics have been investigated by some researchers as a faster method to cool mechanically separated poultry meat.[37–40] Carbon dioxide snow produced a darker and redder meat, which became more dark and gray during subsequent storage. Cooling with CO_2 snow apparently increased the oxidation rate of the heme pigments during storage of the mechanically separated poultry meat.

Certain processes and formulations may affect the color of products containing mechanically separated poultry meat. Dhillon and Maurer[41] found that 50/50 mixtures of mechanically separated poultry (chicken or turkey) and beef produced summer sausages with excellent color scores. Froning et al.[15] found that the addition of 15% mechanically separated turkey meat to red meat franks decreased redness as compared to that observed from 100% beef franks. Also, franks containing 15% mechanically separated turkey meat had a slightly higher rate of color fading during storage. However, it was felt that this fading would not be noticed by the consumer. Today, mechanically separated poultry meat is utilized routinely in emulsified meat products in combination with other species.

Flavor stability

The mechanical separation process produces considerable cellular disruption and releases hemoglobin and lipids from the bone marrow. Also, heat produced from the separation process may accelerate lipid oxidation if not controlled. Therefore, quality assurance programs must be especially rigorous to reduce flavor oxidation problems during processing and storage. Mechanically separated poultry meat produced today is much better than that marketed 20 to 25 years ago. This is largely due to improved equipment and a better understanding of factors related to improved handling of mechanically separated poultry meat.

Several studies have emphasized the storage stability of mechanically separated poultry meat and approaches to improve its flavor stability. Dimick et al.[42] reported that minimal lipid oxidation of mechanically separated meat occurred during 6 days of storage at 3°C. Mechanically separated turkey meat was the least stable during storage at 3°C. Froning et al.[15] indicated that mechanically separated turkey meat stored at −24°C for 90 days exhibited high 2-thiobarbituric acid (TBA) values and unexceptable flavor scores. On the other hand, Dhillon and Maurer[30,41] reported that summer sausages made from mechanically separated poultry stored for 6 months were highly acceptable. Also, Johnson et al.[43] reported that mechanically separated turkey meat had minimal lipid oxidation up to 10 weeks of storage. Janky and Froning[44] observed the interaction of lipid and heme components in mechanically separated turkey meat. Heme oxidation decreased as storage temperatures were reduced from 30 to −10°C. There was a strong interaction of heme and lipid oxidation, particularly between 10 to 15°C. This strong interaction likely contributes to the increased lipid oxidation being experienced during frozen storage of mechanically separated turkey meat. Heme pigments are known to be strong catalysts for lipid oxidation in meat products. During mechanical separation, oxygen incorporation, temperature increases, high pressure, and metal contact with the deboner may further contribute to the lipid oxidation problem.

Antioxidants have been investigated by several scientists as a potential means to control lipid oxidation in mechanically separated poultry meat. Froning[32] chilled spent fowl in 6% polyphosphate prior to mechanical separation and observed lower TBA values than the controls after storage at −29°C for two months. Polyphosphates chelate metal ions which likely explains their effectiveness during the separation and storage process. MacNeil et al.[45] used rosemary extractives, polyphosphates, and BHA + citric acid and reported that they were effective antioxidants in simulated mechanically separated poultry meat. Moerck and Ball[46] used Tenox II at 0.01% by weight of fat present and extended the induction period for lipid oxidation. TBA values were below 1.0 after storage at 4°C.

Washing or surimi-like processing

The fish industry has been utilizing a washing process for mechanically separated fish to produce a protein ingredient known as surimi. This process removes odorous substances and soluble sarcoplasmic proteins (mostly hemoglobin and myoglobin) while concentrating myofibrillar proteins.[47,48] Because of its excellent binding and gelation properties and light color, surimi is widely used in various fish analogs. With the success of surimi, considerable interest was generated in the possibility of applying this process to mechanically separated poultry meat. Aqueous washing of mechanically separated poultry has been found to remove heme pigments and fat while concentrating myofibrillar proteins.[49–54]

Various washing media have been utilized for washing mechanically separated poultry meat. Generally, one part of mechanically separated poultry meat is washed in three parts of washing media. Yang and Froning[52] investigated tap water, 0.1 *M* sodium chloride, sodium phosphate (ionic strength = 0.1), or 0.5% sodium bicarbonate as possible washing solutions. Washing pH as well as mixing time were studied. More heme pigment and soluble protein were removed as mixing time and pH were increased. Optimal concentration of myofibrillar proteins occurred at pH 7 to 8 and at a mixing time of 20 min.

Table 14.2 **Moisture, Fat, Protein, and Collagen Content of Raw and Cooked Mechanically Deboned Chicken Meat that was Washed with Selected Washing Solutions**

Treatment	Moisture (%)	Protein (%)	Fat (%)	Collagen (mg/g of dry meat)
Uncooked				
Unwashed MDCM	68.1[d]	46.6[c]	14.5[a]	67.8
Tap water	84.4[c]	74.2[a]	1.2[b]	109.3
0.1 *M* NaCl	85.8[b]	74.8[a]	1.1[b]	116.6
Sodium phosphate buffer	87.8[a]	70.1[b]	1.3[b]	156.6
0.5% $NaHCO_3$	88.7[a]	71.2[b]	0.8[b]	142.6
Cooked				
Unwashed MDCM	75.5[c]	45.6[c]	9.9[c]	68.6
Tap water	83.7[b]	59.4[b]	1.6[b]	109.3
0.1 *M* NaCl	84.3[c,b]	68.6[c]	0.5[b]	102.9
Sodium phosphate buffer	86.1[a,b]	69.8[a]	0.6[b]	111.4
0.5% $NaHCO_3$	86.7[a]	67.9[a]	0.7[b]	131.2

[a-c] Means within columns having different superscripts are significantly different ($p < 0.05$).

Final composition of the resultant washed meat indicates that fat content is drastically reduced while total protein content and collagen is increased by the washing process (Table 14.2) Later Yang and Froning[54] developed a screening process to reduce the collagen content in the resultant washed meat. The washing process also substantially lightens and decreases the redness of the washed meat. In fact, the washed meat is quite similar in appearance to white poultry meat.

Perhaps one of the biggest advantages of washed mechanically separated poultry meat is its exceptional textural attributes. Yang and Froning[53] observed an improvement of the textural profile of washing mechanically separated chicken meat as compared to the unwashed control (Table 14.3). Hardness, gumminess, springiness, and chewiness were all significantly increased by washing. Scanning election microscopy of cooked washed mechanically separated meat indicated that meat gels had a dense fibrous protein network.

Although the fish industry has been highly successful in marketing fish analogs from surimi, the poultry industry has not utilized this technology. One concern is the water usage and the fat removed which may become environmental issues. However, the water could likely be recycled using such technology as reverse osmosis ultrafiltration. Fat utilization would need to be addressed. Another concern with the washing procedure is that it has been reported to increase lipid oxidation.[55]

Table 14.3 **Textural Profile Analysis Values from Mechanically Deboned Chicken Meat that was Washed in Tap Water, Sodium Phosphate Buffer Solution, $NaHCO_3$ Solution, or NaCl Solution**

Treatment	Hardness (kg)	Cohesiveness	Gumminess (kg)	Springiness (mm)	Chewiness (kg-mm)
Unwashed MDCM	1.5[c]	0.74	1.1[c]	8.9[b]	9.6[c]
Tap water	2.5[a]	0.70	1.8[a]	9.0[b]	15.8[a]
0.1 *M* NaCl	2.5[a]	0.70	1.8[a]	9.3[a]	16.4[a]
Sodium phosphate buffer	2.0[b]	0.71	1.4[b]	9.4[a]	12.9[b]
0.5 % $NaHCO_3$	1.7[c]	0.75	1.3[b]	9.3[a]	12.3[b]

[a-c] Means within colums having different superscripts are significantly different ($p < 0.05$).

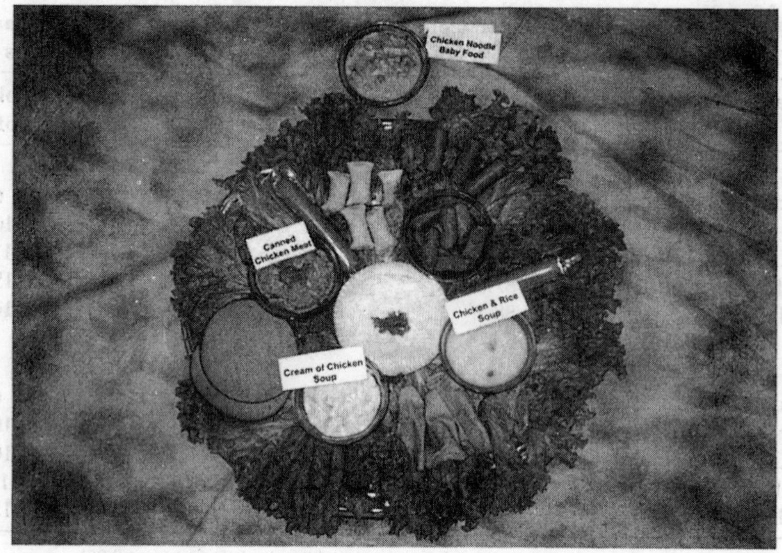

Figure 14.4 Examples of food products made with mechanically separated poultry meat.

Utilization in poultry products

Mechanically separated poultry is utilized in a wide range of emulsified and restructured meat products including frankfurters, bologna, breakfast sausage, nuggets, roasts, etc. (Figure 14.4). Several of these products are mixtures of meat from various species. Its use requires good quality assurance guidelines to avoid rancid meat. Parts to be deboned should be fresh and held at near freezing temperatures (−1 to 2°C). After separation, mechanically separated poultry must be incorporated into a formulated product within a one day period unless frozen and should not be held longer than 90 days in frozen storage.

Summary

Mechanically separated poultry meat continues to have wide usage and provides an economical meat source. With the improvements in mechanical deboners and better quality assurance, it has provided a major contribution to the popularity of poultry meat in our diets. There is a continuing need for advancements in mechanical deboners, which limit the amount of bone marrow in the final mechanically separated poultry meat. Heme pigment and other deleterious marrow components in mechanically separated poultry may also be minimized in conventional deboners by adjusting deboner settings to reduce yield and monitoring meat color using a colorimeter. The industry must strive for a balance between yield and quality in terms of protein functionality, flavor stability, and color.

References

1. Froning, G. W., Mechanical deboning of poultry and fish, *Adv. Food Res.*, 27, 109, 1981.
2. United States Department of Agriculture, Food Safety Inspection Service, Proposed rules, *Fed. Regist.*, 59, No. 233, December 6, 1994.
3. Federal Register (Title 9 CFR 318 meat inspection; 9 CFR 319 meat inspection and standards of identity; 9 CFR food labeling, poultry and poultry products, standards of identity). Food Safety Inspection Service, U.S. Department of Agriculture, Washington, D.C.

4. D.D.C. Civil Action No. 93–0104.
5. Murphy, E. W., Brewington, C. R., Willis, B. W., and Nelson, M. A., *Health and Safety Aspects of the Use of Mechanically Deboned Poultry, Food Safety and Quality Service*, U.S. Department of Agriculture, Washington, D.C., 1979.
6. Froning, G. W., Mechanical deboning of poultry and fish, *Adv. Food Res.*, 110, 147, 1981.
7. Froning, G. W. and Johnson, F., Improving the quality of mechanically deboned fowl meat by centrifugation, *J. Food Sci.*, 38, 279, 1973.
8. Wallace, M. J. D. and Froning, G. W., Protein quality determination of bone residue from mechanically deboned chicken meat, *Poult. Sci.*, 58, 333, 1979.
9. Young, L. L., Composition and properties of animal protein isolate prepared from bone residue, *J. Food Sci.*, 41, 606, 1976.
10. Schnell, P. G., Vadehra, D. V., Hood, L. R., and Baker, R. C., Ultra-structure of mechanically deboned poultry meat, *Poult. Sci.*, 53, 416, 1974.
11. Froning, G. W., Poultry meat sources and their emulsifying characteristics as related to processing variables, *Poult. Sci.*, 49, 1625, 1970.
12. Grunden, L. P., MacNeil, J. H., and Dimick, P. S., Poultry product quality: Chemical and physical characteristics of mechanically deboned poultry meat, *J. Food Sci.*, 37, 247,1972.
13. Essary, E. O., Moisture, fat, protein and mineral content of mechanically deboned poultry meat, *J. Food Sci.*, 44, 1070, 1979.
14. MacNeil, J. H., Mast, M. G., and Leach, R. M., Protein efficiency ratio and levels of selected nutrients in mechanically deboned poultry meat, *J. Food Sci.*, 43, 864, 1978.
15. Froning, G. W., Arnold, R. G., Mandigo, R. W., Neth, C. E., and Hartung, T. E., Quality and storage stability of frankfurters containing 15% mechanically deboned turkey meat, *J. Food Technol.*, 36, 974, 1971.
16. Babji, A. S., Froning, G. W., and Satterlee, L. D., The protein nutritional quality of mechanically deboned poultry meat as predicted by C-PER assay, *J. Food Sci.*, 45, 441, 1980.
17. Satterlee, L. D., Froning, G. W., and Janky, D. M., Influence of skin content on composition of mechanically deboned poultry meat, *J. Food Sci.*, 36, 979, 1971.
18. Essary, E. O. and Ritchey, S. J., Amino acid composition of meat removed from boned carcasses by use of a commercial boning machine, *Poult. Sci.*, 47, 1953, 1968.
19. Hsu, H. W., Sutton, N. E., Banjo, M. O., Satterlee, L. D., and Kendrick, J. G., The C-PER and T-PER assays for protein quality, *Food Technol.*, 32(12), 69, 1978.
20. Moerck, K. E. and Ball, H. R., Jr., Lipids and fatty acids of chicken bone marrow, *J. Food Sci.*, 38, 978, 1973.
21. Froning, G. W., Characteristics of bone particles from various poultry meat products, *Poult. Sci.*, 58, 1001, 1979.
22. Murphy, E. W., Brewington, C. R., Willus, B. W., and Nelson, M. A., *Health and Safety Aspects of the Use of Mechanically Deboned Poultry, Food Safety and Quality Service*, U.S. Department of Agriculture, Washington, D.C., 1979.
23. McMahon, E. F. and Dawson, L. E., Effects of salt and phosphates on some functional characteristics of hand and mechanically deboned turkey meat, *Poult. Sci.*, 55, 573, 1976.
24. Mayfield, T. L., Hale, K. K., Rao, V. N. M., and Angels-Chacon, I. A., Effects of levels of fat and protein on the stability and viscosity of emulsions prepared from mechanically deboned poultry meat, *J. Food Sci.*, 43, 197, 1978.
25. Orr, H. L. and Wogar, W. G., Emulsifying characteristics and composition of mechanically deboned chicken necks and backs from different sources, *Poult. Sci.*, 58, 577, 1979.
26. Froning, G. W., Satterlee, L. D., and Johnson, F., Effect of skin content prior to deboning on emulsifying and color characteristics of mechanically deboned chicken back meat, *Poult. Sci.*, 52, 923, 1973.
27. Schnell, P. C., Nath, K. R., Darfler, J. M., Vadehra, D. V., and Baker, R. C., Physical and functional properties of mechanically deboned poultry meat as used in the manufacture of frankfurters, *Poult. Sci.*, 52, 1363, 1973.
28. Acton, J. C., Composition and properties of extruded, texturized poultry meat, *J. Food Sci.*, 38, 571, 1973.

29. Lampila, L. E., Froning, G. W., and Acton, J. C., Restructured turkey products from texturized mechanically deboned turkey, *Poult. Sci.*, 64, 653, 1985.
30. Dhillon, A. S. and Maurer, A. J., Stability study of comminuted poultry meats in frozen storage, *Poult. Sci.*, 54, 1407, 1975.
31. Froning, G. W. and Janky, D. M., Effect of pH and salt preblending on emulsifying characteristics of mechanically deboned turkey frame meat, *Poult. Sci.*, 60, 1206, 1971.
32. Froning, G. W., Effect of chilling in the presence of polyphosphates on the characteristics of mechanically deboned fowl meat, *Poult. Sci.*, 53, 920, 1973.
33. Lyon, B. G., Lyon, C. E., and Townsend, W. E., Characteristics of six patty formulas containing different amounts of mechanically deboned broiler meat and hand-deboned fowl meat, *J. Food Sci.*, 43, 1656, 1978.
34. Lyon, C. E., Lyon, B. G., and Townsend, W. E., Quality of patties containing mechanically deboned broiler meat, hand deboned fowl meat and two levels of structured protein fiber, *Poult. Sci.*, 57, 156, 1978.
35. Lyon, C. E., Lyon, B. G., Townsend, W. E., and Wilson, R. L., Effect of level of structured protein fiber on quality of mechanically deboned chicken meat patties, *J. Food Sci.*, 43, 1524, 1978.
36. Janky, D. M., Riley, P. K., Brown, W. L., and Bacus, J. N., Factors affecting the stability of mechanically deboned poultry meat combined with structural soy protein emulsions, *Poult. Sci.*, 56, 902, 1977.
37. Uebersax, K. L., Dawson, L. F., and Uebersax, M. A., Influence of "CO_2-snow" chilling on TBA values in mechanically deboned chicken meat, *Poult. Sci.*, 56, 707, 1977.
38. Uebersax, K. L., Dawson, L. F., and Uebersax, M. A., Storage stability (TBA) and color of MDCM and MDTM processed with CO_2 cooling, *Poult. Sci.*, 57, 670, 1978.
39. Cunningham, F. E. and Mugler, D. J., Deboned fowl meat offers opportunities, *Poult. Meat*, 25, 46, 1974.
40. Mast, M. G., Jurdi, D., and MacNeil, J. H., Effects of CO_2-snow on the quality and acceptance of mechanically deboned poultry meat, *J. Food Sci.*, 44, 364, 1979.
41. Dhillon, A. S. and Maurer, A. J., Quality measurements of chicken and turkey summer sausages, *Poult. Sci.*, 54, 1263, 1975.
42. Dimick, P. S., MacNeil, J. H., and Grunden, L. P., Poultry product quality carbonyl composition and organoleptic evaluation of mechanically deboned poultry meat, *J. Food Sci.*, 37, 544, 1972.
43. Johnson, P. G., Cunningham, F. E., and Bowers, J. A., Quality of mechanically deboned turkey meat: effect of storage time and temperature, *Poult. Sci.*, 53, 732, 1974.
44. Janky, D. M. and Froning, G. W., Factors affecting chemical properties of heme and lipid components in mechanically deboned turkey meat, *Poult. Sci.*, 54, 1378, 1975.
45. MacNeil, J. H., Dimick, P. S., and Mast, M. G., Use of chemcial compounds and rosemary spice extract in quality maintenance of deboned poultry meat, *J. Food Sci.*, 38, 1080, 1973.
46. Moerck, K. E. and Ball, H. R., Lipid oxidation in mechanically deboned chicken meat, *J. Food Sci.*, 39, 876, 1974.
47. Lee, L. M., Surimi process technology, *Food Technol.*, 38(1), 69, 1984.
48. Lanier, T. C., Functional properties of surimi, *Food Technol.*, 40(3), 107, 1986.
49. Ball, H. R., Jr., Surimi processing of MDPM, *Broiler Ind.*, 51(6), 62, 1988.
50. Dawson, P. L., Sheldon, B. W., and Ball, H. R., Jr., Extraction of lipid and pigment components from mechanically deboned chicken meat, *J. Food Sci.*, 53, 1615, 1988.
51. Dawson, P. L., Sheldon, B. W., and Ball, H. R., Jr., A piiot washing procedure to remove fat and color components from mechanically deboned chicken meat, *Poult. Sci.*, 68, 749, 1989.
52. Yang, T. S. and Froning, G. W., Effects of pH and mixing time on protein solubility during the washing of mechanically deboned chicken meat, *J. Muscle Foods*, 3, 15, 1992.
53. Yang, T. S. and Froning, G. W., Selected washing processes affect thermal gelation properties and microstructure of mechanically deboned chicken meat, *J. Food Sci.*, 57, 325, 1992.
54. Yang, T. S. and Froning, G. W., Changes in myofibrillar protein and collagen content of mechanically deboned chicken meat due to washing and screening, *Poult. Sci.*, 71, 1221, 1992.
55. Dawson, P. L., Sheldon, B. W., Larick, D. R., and Ball, H. R., Jr., Changes in the phospholipid and neutral-lipid fractions of mechanically deboned chicken meat due to washing, cooking, and storage, *Poult. Sci.*, 69, 166, 1990.

chapter fifteen

Marination, cooking, and curing of poultry products

Doug P. Smith and James C. Acton

Contents

0-8493-0120-3/01/$0.00+$.50

Introduction

Consumer demand for products with easier and faster preparation, along with the concurrent integration and expansion of the poultry industry, has provided an expanding market for value-added products. Poultry further-processors began marinating meat in the 1950s in the U.S., but the practice became more widespread in the 1970s and then boomed in the next decade in response to the fast food market for chicken products. The overall increase in further-processed items is apparent in Figure 15.1. There are many diverse market outlets supplied by poultry processors (see Table 15.1), and although the categories have changed somewhat over 20 years, there is a massive amount of poultry meat on the market (1997 estimates of 37.5 billion pounds of broiler meat). Finished products are marketed into channels where consumers buy the final product, although many poultry products are modified, prepared, or cooked after distribution but before final presentation to the consumer. Many of these products have been further-processed or prepared (marinated, cured, or smoked) before sale to the consumer, and now a substantial number of products are cooked prior to final sale.

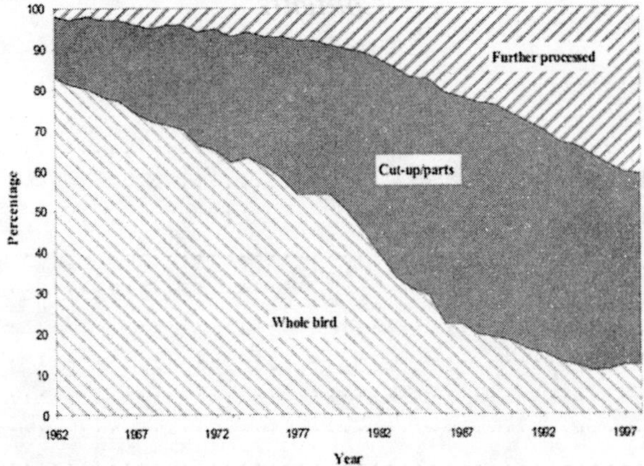

Figure 15.1 U.S. broiler market, percentage of product forms by year, 1962–1998. (From the National Chicken Council, Washington, D.C.)

Table 15.1 **Commercial Market Channels (as Percent of Total Production) for Broiler Products in the U.S.**

Market outlets	1978	1987 (%)	1997
Distributor	48.8	29.7	25.5
Retail grocery	30.4	34.4	23.0
Restaurants	0.6	2.4	5.5
Fast foods	8.2	11.4	5.7
Exports	4.0	5.0	15.7
Further processors	3.6	5.6	7.7
Institution	0.2	0.7	1.5
Government	2.2	1.3	1.6
Other — pet food, renderers, brokers, etc.	2.0	9.5	15.8

Source: (From the National Chicken Council, 1997 Marketing Survey Report and the USDA National Agricultural Statistics Service.)

Marination

Background

Marination, the addition of liquids to meat before cooking, has been practiced in some form for centuries. Soaking in vinegar, oils, or both, in combination with spices improved flavor of the meat and extended shelf-life (or at least masked off-flavors). More recently, marination has been proven to offer additional advantages including improved functionality of product use and improved yield for the processor. Providing improved products to the consumer, as well as improving raw yield at the processing plant has pushed marination as a process into widespread use in the poultry industry. Although this chapter predominately discusses the marinating of broiler chicken, marination is also practiced for other market classes, including spent fowl and Cornish game hens, and for other poultry species, including turkeys and ducks.[1–6]

Market forms of marinated poultry include whole birds, cut-up parts, boneless meat, and chopped and formed items. Many products sold in the raw, unmarinated state will be marinated by the retailer or by the consumer in the home prior to sale or consumption, respectively. The actual amount of marinated poultry products is difficult to determine, but some estimate is available from a 1997 market survey of categories of broiler products sold (Figure 15.2). Approximately 70% of the total edible pounds from the over 8 billion broilers produced in the U.S. were sold domestically; with approximately 15% of the pounds sold as marinated products, and another 29% was in a category that included marinated and non-marinated products. Therefore, somewhere between 15 and 44% of U.S. domestically sold broiler pounds are commercially marinated. When added to the percentage of product sold raw, then marinated at retail fast food outlets or at home by the final consumer, the total percent of broiler meat marinated prior to consumption probably exceeds 50% in the U.S.

Benefits and yields

Meat marination processes, mechanisms and, in particular, imparted product adaptations are available in more detail from technical reviews[7,8] and from articles in trade journals.[9–12] The following applies to poultry products at large and to broilers in particular.

Marination improves poultry meat product quality as perceived by the consumer and by increasing yield for the processor. Both are accomplished in basically the same manner, chemically binding water within the muscle tissue. This can be water that is either already in the meat or that is added to it. The ability of meat to hold water is termed "water-holding capacity." Higher water-holding capacity usually equates to juicier and

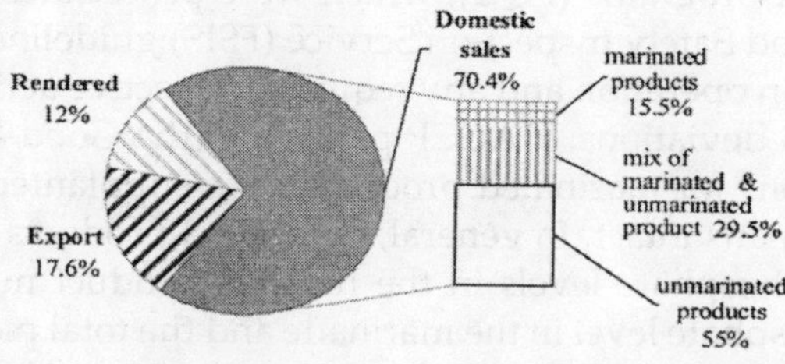

Figure 15.2 U.S. broiler market, percentage of product sold domestically as marinated, unmarinated, or mixed, 1997. (From the National Chicken Council, Washington, D.C.)

more palatable sensory perceptions and an overall improvement in meat quality. From the processor's perspective increased water-holding capacity equals increased yield, or the sale of water at meat prices. Other benefits include flavor enhancement, both by the addition of spices and flavorings, and through reduction in rancidity that develops during storage.[13,14] Physical appearance may be improved by coloring imparted from the marinade or by the presence of spice or seasoning particles in the marinade. Tenderness has been improved through marination by including various components, particularly the basic ingredients of salt and phosphate, plus calcium chloride and papain enzymes.[15–17]

How does marination affect meat structure which in turn improves water-holding capacity?[7–9,18] Poultry meat inherently contains approximately 75% water. The gross structure of poultry meat, with many long thin parallel myofibers, or muscle cells, each surrounded by connective tissue (principally collagen), provides some opportunity for the excess fluid added from marination to be absorbed and held within the tissue. At the protein structure level, some charged muscle proteins (the myofibrillar or contractile proteins) have the ability to attract and bind or "hold" water. The collagen surrounding the myofiber may also have a few charged sites capable of binding water. Salt and phosphates in the marinade act to increase the number of charged sites and in effect, through repulsion, partially unfold or open the space among the protein molecules, allowing more water-binding sites to be available. This occurs due to the ionic characteristics of salt and phosphates. In addition, changes in the aqueous pH surrounding the muscle fibers to more alkaline conditions (via alkaline phosphates) also increases protein-to-protein spatial alignments, allowing increased water binding. There are many types of phosphates available for food use with differing capabilities to promote protein hydration.[9] Most of the phosphates used in the commercial poultry industry is sodium tripolyphosphate (STP).[19] The advantages of STP include availability in bulk powder form (easy to transport and relatively cheap) as well as the enhancement of protein-binding properties.

An example of a typical marination recipe for commercial broiler products is 90% water, 6% salt (sodium chloride), and 4% STP. This recipe is useful for products with a total pickup of 10% of the raw meat weight, but would not be legally acceptable for products with a 15% pickup. This is because the USDA requires no more than 0.5% total phosphate in a finished product. Products with higher pickups are produced with marinade recipes containing less phosphate. As mentioned previously, many types of other ingredients may also be added for flavor, color, physical appearance, and microbial protection (shelf-life or pathogen protection). Normally, marinade preparation begins with adding the phosphate first to cold water, then stirring rapidly for several minutes before other ingredients are added. Otherwise, low solubility of the phosphate may cause it to not dissolve completely if added directly to the salt brine.

Commercial marination of poultry has traditionally been approved and controlled by Partial Quality Control programs (PQC), which were procedures written by a company according to USDA Food Safety Inspection Service (FSIS) guidelines. The USDA inspector monitors the marination operation and any required corrective actions and places product on hold in response to deviations. HACCP programs (the Good Manufacturing Practice sections) with provisions for marinated products have supplanted PQCs, but the overall oversight function remains intact. In general, marinated products must not exceed stated label claims — total phosphate levels in the finished product must not exceed 0.5% as determined by the phosphate level in the marinade and the total pickup of marinade in the batch of product. Marination pickup is defined as the marinated weight minus the initial (green) weight, divided by the initial weight, multiplied by 100.

$$\text{Calculation:} \quad \%\ \text{Marination pickup} = \frac{\text{marinated weight} - \text{initial weight}}{\text{initial weight}} \times 100$$

There are many products produced by the poultry industry that are marinated. Examples include raw whole birds, chickens, or turkeys for retail sale; raw or frozen parts; fresh or frozen whole-muscle fillets, tenders, or nuggets; and fresh or frozen patties and nuggets. Any of these may have been subjected to heat treatment or be fully cooked. Some specific product types include frozen whole turkeys, rotisserie chicken, hot wings, breaded and cooked fillets and nuggets, chicken patties, lunch meat, turkey loaves, frozen dinner entrees, stir fry entrees, turkey or chicken pot pies, and many other products. Much of the raw poultry sold in the U.S. is also marinated by the customer (retail fast food or institutional) before cooking and sale to the consumer. This is usually accomplished by soaking, although some units have small tumblers to prepare bone-in parts or whole-muscle items that will be cooked on site. The largest fast food companies that predominately sell chicken are now purchasing commercially marinated chicken and phasing out their in-store marination operations.

Marination techniques

Soaking or still marination

Several methods are available to marinate meat on a commercial scale. The original and simplest method is soaking, or "still marination." Meat is placed in a container, marinade is added, and the mixture sits, usually for at least 24 h. Containers of product are usually held refrigerated in a cooler (below 4.4°C or 40°F). Benefits include process simplicity and low cost, good adhesion of skin on skin-on products, and the ability to produce small batches and specialty products. Drawbacks include contamination issues from foreign objects dropped into stored containers or microbial growth, somewhat inconsistent marinade uptake, cooler space and container requirements, and extra labor costs to load and unload product. Early work conducted on poultry soaked in marinade consisted of phosphates added to the chill water to improve water retention.[20,21] Soaking postchill carcasses, bone-in parts, and boneless whole muscle in marinade improved meat quality and yield.[22–27] A modification of still soaking has been tried, where agitation of the product in the marinade improved pickup and cooked yield.[28]

Blending

Blending is similar to the tumbling process but generally is used for pieces of whole muscles or coarsely ground meat used to produce chopped and formed products (Figure 15.3). Ribbon-type paddles turn inside the blender to blend the meat and marinade, and sometimes carbon dioxide or other coolants are injected into the mixture during blending. Benefits include a finer control of the process, including coolant injection and more even mixing than tumbling provides. Blenders are expensive, however, and are not used for products such as bone-in parts or whole-muscle parts with skin. Marination of ground meat or chopped and formed products, especially with agitation from the blender, produces products with benefits similar to the soaking method, improving both functionality and yield of finished products.[29–33]

Tumbling

A widespread method of commercial marination is tumbling. A large tank, with a capacity usually ranging from 2000 to 8000 lb, is placed on its side on wheels to allow tank rotation

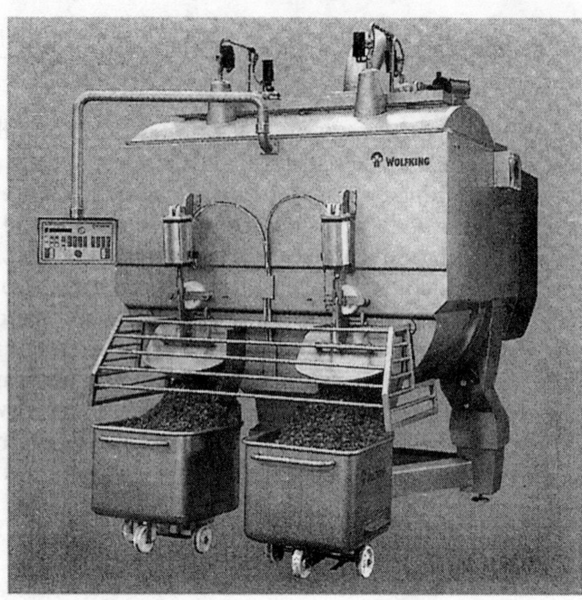

Figure 15.3 Vacuum blender used for mixing marinade and other ingredients with smaller particles of meat. (Courtesy of Wolfking, Inc.)

(Figure 15.4). The speed of rotation is adjustable, and most tumblers have paddles inside to increase agitation of the contents. The walls of the tumblers may be jacketed so refrigerant can be used for cooling the product while tumbling, and some are designed for direct injection of carbon dioxide to the product for temperature control. The tank doors are sealed so that a vacuum can be applied to further increase marinade absorption. Operation of a tumbler is a fairly simple process. Meat and marinade are placed into the tank, with marinade added automatically via a pump or a vacuum tube. Then the tank is closed and settings (vacuum strength, rotation speed, total tumble time, and application of coolant) are chosen for that product. Benefits include large batch size, quick uptake of marinade (20 min), and utility in that a variety of products can be tumbled (skinless, skin-on, bone-in, whole-muscle, and chopped pieces). Tumbling may not be appropriate for some fragile products (chicken tenders) or those with skin loosely attached. The initial equipment cost, including accessory loaders as well as installation can be expensive. In two studies, the agitation from tumbling provided higher raw marinade pickup at a much faster rate than still marination when the two methods were directly compared, but the higher raw pickup was lost during cooking.[34,35] Tumbling products may also show inconsistent individual pickups within a large batch.[36] Overall, however, tumbling provides all of the benefits associated with marination.[37–39] The process is also very cost effective for large, commercial-scale operations.

Mechanical injection

Marinade injection by mechanical means is another method of marination. Turkey processors used an earlier method of single needle injection by hand, where basting fluid was manually pumped into the turkey carcass. New, automated systems consist of conveyor belts that pass meat into a channel where a crosshead assembly of needles is lowered into the product (Figure 15.5). The hollow needles pierce the meat and marinade is pumped in through a small orifice in the side of each needle. Each needle also has independent suspension so bones are not penetrated. The amount of marinade injected generally depends

Figure 15.4 Vacuum tumbler (in unloading position showing internal paddles) used for mixing marinade and other ingredients with larger meat particles or whole muscles. (Courtesy of Wolfking, Inc.)

on the pumping pressure and the speed of the conveyor belt. Injectors are used throughout the meat and poultry industry. This method is especially useful for whole birds and bone-in, skin-on parts. Production line speeds are fast, averaging up to 10,000 lb/h or greater. Injectors cannot adequately marinate chopped and formed products, nor can they inject marinade that contains large solid particles (due to needle clogging). Injection, compared

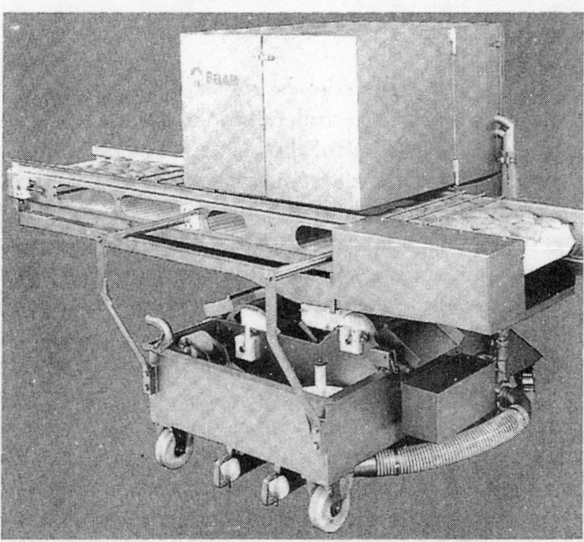

Figure 15.5 Injector showing conveyor belt full of whole-breast muscles passing through injecting cabinet. (Courtesy of Wolfking, Inc.)

to other methods, may also produce products with higher drip loss. Some processors allow the product collected from injectors to soak briefly, absorbing the marinade, before draining and packaging or cooking. The technique in early research mimicked the manual methods, where needles and syringes were employed to inject marinade into the muscle.[40–42] Later, other researchers employed automated injectors available on the commercial market.[43–46] Injection marination produced improved products similar to still marination in studies utilizing both methods.[47,48] One study utilized injection followed by tumbling to maximize marination effects.[49]

Process problems

Overall, marination is an excellent method to improve poultry meat quality. However, there are some general problems that are sometimes associated with this process. Products may exceed the stated label pickup percentage. Inconsistent pickup may occur within the same piece of meat or pieces of meat within the same batch and produce detectable variations in flavor and juiciness. Formulation errors can lead to products exceeding the 0.5% phosphate level in the finished product. Some consumers are sensitive enough to detect and reject product even at lower phosphate levels, complaining of bitterness and dryness in mouthfeel. These problems are usually minimized with proper formulation and step-by-step procedures for ensuring consistent application of marinade, including following written Good Manufacturing Practices and utilizing statistical process control. There are new concerns that marination procedures may introduce surface pathogens into the normally sterile interior of the meat. Proper cooking to appropriate internal temperature alleviates this potential problem but it may remain for products marketed in the raw, marinaded state. The improvements in meat quality recognized by consumers and the higher profits to the processors will insure that marinated products will continue to be produced and marketed.

Cooking

Background

Meat, including poultry, has been cooked or heat processed in some manner since before recorded history. The use of fire to cook or smoke meat and the use of sunlight to dry meat are parts of human survival and cultural practices. The benefits of heat were probably obvious; perhaps consumers were sick less frequently with cooked as compared to raw meat due to destruction of pathogens. In addition, the meat lasted longer before spoilage, and the palatability was changed (although raw meat eaters at first may not have appreciated the new flavor, texture, or color). Cooking can negatively affect meat quality due to potential loss of some nutrients and potential formation of mutagenic compounds.[50,51] Interestingly, marination prior to cooking may decrease mutagen levels of cooked poultry meat.[52]

Traditionally, the commercial poultry processing industry first supplied live product, then raw processed products; the only exceptions were canned items or cooked lunch meat or deli products (usually turkey). Now many items are further processed and cooked, and some retail fast food customers only accept fully cooked poultry into their stores. The variety of different cooking methods for poultry items as presented to customers on restaurant menus in the U.S. is shown in Figure 15.6. A rapid increase in poultry cooking technology has evolved during the past 30 years since consumers expect many different product forms, and restaurants require or expect the processors to provide many of these preheat-treated or precooked items.

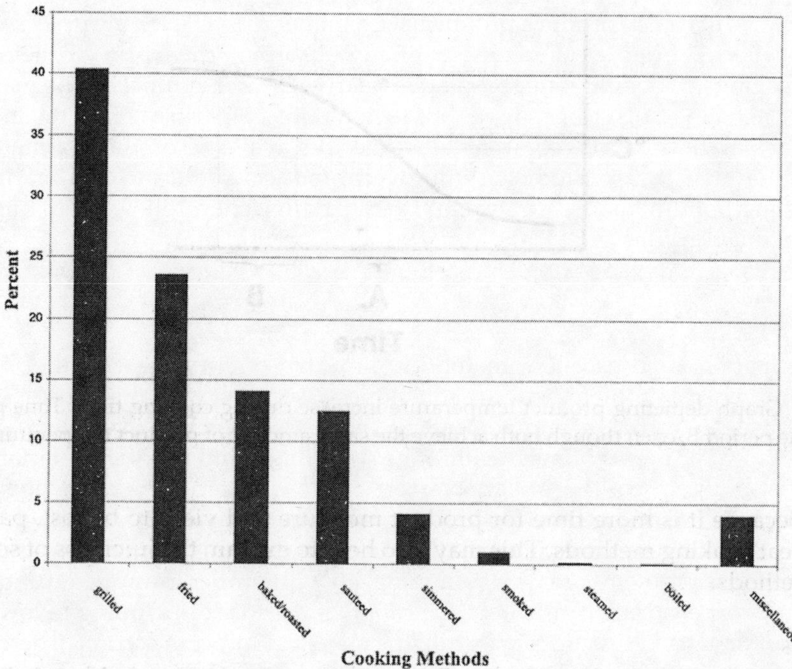

Figure 15.6 Percentage of cooking preparations for chicken items, as listed on U.S. restaurant menus, 1999. (From FlavorTrak database, Foodservice Research Institute, Oak Park, IL.)

Cooking methods

The medium used to apply heat to poultry products (air, steam, oil, or water) is one method of categorizing the different methods of cooking. The commercial processor typically utilizes in-line ovens or fryers whenever possible to maximize throughputs of products adaptable to these methods, including parts, whole-muscle, and chopped-and-formed items. Other products require batch processing under different conditions, such as canned items, kettle-cooked whole birds, and smoked or cured products (see the following section). In general, for commercially processed poultry products, the in-line ovens and fryers are the preferred methods for many of the mass-produced poultry items marketed at retail and to restaurant or food service customers.

Dry-heat cooking methods such as grilling produce drier product with lower yield than moist-heat cooking methods, such as baking; likewise, higher heat rapid methods such as frying or searing can yield moister product by rapidly cooking the surface and sealing in the juice. Marination is a common approach to reduce these cooking losses, thereby increasing yield and quality.

The end point temperature required by law for pasteurization of poultry products labeled as fully cooked are 71°C (USDA standard for commercial cookers) or 74°C (FDA standard for retailer cooking). However, processors generally cook to a slightly higher temperature (75 to 77°C) for a safety margin against process or product variations. Aside from safety, it is important in maintaining yield not to exceed the end point temperature too much. The rate of product temperature increase (°C per min) during cooking gradually decreasesthe closer the product and the cooking (oven, fryer, grill, etc.) temperature become (Figure 15.7). So, the closer the product temperature gets to the cooking temperature, the longer each degree of temperature increase takes to achieve. This longer time

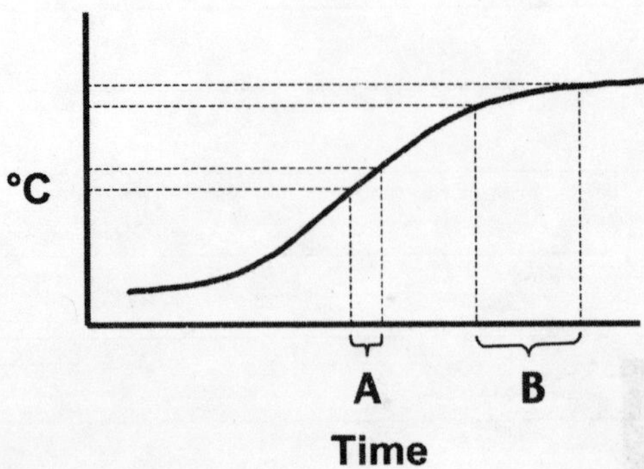

Figure 15.7 Graph depicting product temperature increase during cooking time. Time period A is less than time period B, even though both achieve the same amount of product temperature increase.

is critical because it is more time for product moisture and yield to be lost, particularly with dry-heat cooking methods. This may also help to explain the juiciness of some rapid cooking methods.

Ovens

Ovens are widely used in the industry. Ovens are usually designed either as linear pass through where horizontal space in a plant is not limited; (Figure 15.8) or as spiral pass through where horizontal space is limited but vertical space is available; (Figure 15.9). Product is transported through either oven on a continuous conveyor belt. Heat sources for these ovens are usually direct gas jets heating air inside the oven, gas jets heating air away from the product which is then transferred to the oven, or indirect heat from hot mineral oil heated externally and transferred to the oven to heat the internal air. Indirect heat sources are utilized to prevent the development of pink or red discoloration of some poultry products exposed to gases from the direct heat gas jet. Surface discoloration is a clue that the gas for the oven is incompletely combusted and is producing carbon monoxide,

Figure 15.8 Stein JSO-IV In-line Oven (Courtesy of Stein, Inc.)

Figure 15.9 Stein GCO Spiral Oven (Courtesy of Stein, Inc.)

which readily combines with myoglobin, producing the pink-to-red coloration. Proper adjustment of burner nozzles for oxygen mixture with the gas corrects this problem. In cases of cooking poultry that is basted with a marinade, such as barbequed items, some surface reddening may be desirable.

Ovens may also utilize steam to cook instead of or in addition to the previous methods of heating. Some ovens use impingement, where internal blowers force the heated air onto the product surface at high velocities. All of these types of ovens have controls and sensors to monitor temperature and humidity to effectively and efficiently cook a variety of products. Products typically include grilled, baked, or roasted poultry items, and many fried and fully cooked items are now parfried (partially cooked) in a fryer but then finished in an oven in order to maximize yield that would be typically lost when frying to the fully cooked stage.

Fryers

Commercial fryers are long vats built to contain heated frying oil, where the product passes through the oil on a continuous stainless steel conveyor belt (Figure 15.10). Heat was traditionally applied directly to the internal walls of the fryer through open flame gas jets, but fire safety concerns have prompted manufacturers to produce indirect heat fryers, utilizing hot mineral oil heated in another room then piped into the fryer to heat the frying oil. Oils normally used to cook products are food grade vegetable oils, typically soybean, corn, canola, or blends. Fryers are equipped with filters to remove "fines," which are small particles of product that would otherwise remain in the oil for long periods and burn, reducing both oil quality and product appearance (if adhering to product). Oil quality is critical to good fried product quality as oil problems translate into poor appearance, flavor, and odor of finished product. Fried products are usually battered and breaded items (bone-in parts, whole-muscle items, or nuggets and patties) or unbreaded spicy wings. Small or thin items may be fully cooked in a fryer, but most large items are parfried for a short period to set the breading texture, color, and flavor, and then transported to an oven to finish cooking.

Other methods

Other cooking methods are available but are less widely used. A griddle oven has hot plates above and below product on a conveyor that can heat or cook thin items. Commercial microwave ovens are not widely used due to initial cost, potential for inconsistent

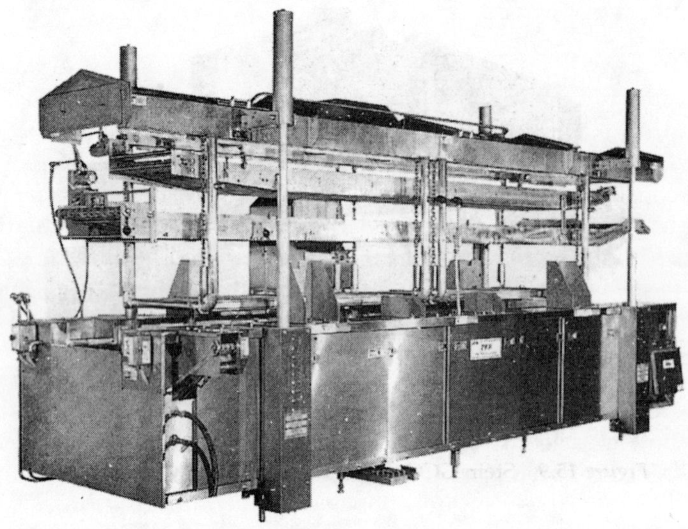

Figure 15.10 Stein TFF-II Fryer, with top raised (Courtesy of Stein, Inc.)

heating, and because of their batch mode of operation. Steam kettles are used for boiling or stewing batches of product and making broth, but again are batch operations. A different method of cooking procedure utilizes an oven for a prepackaged product, in a method similar to canning, but with a flexible package. Examples include some retail pouch products and the military Meal Ready to Eat (MRE) pouches. Cooked poultry is also freeze-dried to produce lightweight and nutritious items with a long shelf-life. Smokehouses are used to cook whole birds and many emulsion or loaf-type items (such as frankfurters, bologna, and deli meat rolls — see next section).

Research has been conducted on many of these cooking methods.[53] A series of experiments was conducted on spent hens, Cornish game hens, and broilers to determine the effects on tenderness from two cooking methods.[1,3,24] Spent fowl dark meat and Cornish breast meat were more tender when smoked than when roasted. No differences in tenderness were observed due to cooking method for spent fowl light meat, Cornish dark meat, or broiler light or dark meat. Bone-in broiler parts steam- or water-cooked were compared and steaming produced better yields and more tender meat.[47] No differences in tenderness or cooked yields were observed for hens either cooked in a microwave or baked in an oven.[17] Duckling breast had higher cooked yields when pan sautéed and broiled than when roasted, although tenderness was not affected by cooking method.[6]

In summary, many types of cooking methods are available to produce ready-to-eat poultry products and a number of different methods are necessary to supply the variety of cooked products sought by consumers. Cooking generally improves the quality of the product although concerns are occasionally expressed over mutagens and loss of vitamins. The use of marination may somewhat alleviate these problems, and continuing research on cooking methods may also provide partial resolution. For example, bone-in parts that were once fully cooked in the fryer (with large yield losses and high heterocyclic amine content) can now be parfried for a few seconds in the fryer and then fully cooked in a humidified oven. This technique lessens both cook losses and mutagen formation. Research has directly contributed to the advances in cooking techniques and in the development of batter and breading formulations, thus improving the poultry processor's yields and quality and consumer safety.

Curing

Introduction

Several developments in the further-processing of poultry contributed to the utilization of "curing" as a means of providing products with a unique flavor, color, and improved safety. As the consumer demanded more cut-up poultry, uses were needed for the fine fibrous, paste-like mechanically deboned meat obtained from residual carcass frames, necks, and backs. Since preparation of highly comminuted meat through extensive chopping of whole tissue was the starting point in frankfurter manufacture, chicken and turkey frankfurters were developed directly with mechanically deboned meat. In addition, mechanically deboned chicken and turkey eventually were permitted as a meat ingredient in combination with pork and beef for manufacture of typical red meat frankfurters.[54] With the improvement of equipment and worker skills for deboning of carcasses; breast, thigh and drums, whole-muscle strips, chunks, and pieces became available to prepare such cured products as poultry hams, pastrami, and cured, smoked breasts. Another category of products made from coarsely ground boneless poultry meat includes various luncheon items such as salamis and fermented summer-style sausages. Today, along with frankfurters made entirely of poultry meat, these other cured products are marketed as intact or portioned chubs or as presliced products and are very popular with consumers.

Background

Meat in early history was primarily preserved with salt that exerted its inhibitory effect on contaminating bacteria while also dehydrating the meat through osmotic effects on muscle cells at the meat surface. Salt was liberally applied as a rub or packed onto meats and the practice was termed "dry curing." This is similar to old style country cured ham manufactured today. Early salt was frequently impure, possessing a brownish-red color, which resulted in reddening of the meat. The contaminant was later found to be sodium nitrate. In the late 1800s, scientists found that nitrate-reducing bacteria yielded the nitrite ion which was the major active compound in curing reactions. Except for approved nitrate use in the manufacturing of dry cured products, most of which are non-poultry except for some turkey jerky, sodium (or potassium) nitrite is the principal permitted curing ingredient today and its use is regulated by the USDA. The initial regulation permitting the use of nitrite as a curing agent was promulgated in 1925[55] and with subsequent revisions as to concentrations permitted, it remains in effect today and applies to all cured poultry products as well as other meats. Both the sodium or potassium form of nitrite may be used as long as the concentration does not exceed a limit depending on the product (156 ppm of nitrite for sausages), calculated as sodium nitrite on an in-going or ingredient additive basis. Such limitations have been instituted to reduce the potential formation of carcinogenic nitrosamine byproduct compounds. In addition to reducing nitrites in the formulations, nitrosamine formation can be limited by assuring an adequate reducing environment with ascorbates or erythorbates to enhance conversion of the nitrites to nitric oxide

In addition to nitrite, several other ingredients are used in poultry meat curing. Nearly all cured products have salt (NaCl) added for flavor as well as antimicrobial and protein functionality purposes. Salt is still considered a primary curing ingredient with nitrite. Examples of other ingredients include various reductants such as ascorbic or erythorbic acid (or their respective sodium salt ascorbate or erythorbate); pH modifiers, such as phosphates, citric acid, or glucono-delta-lactone; and flavorings or flavor enhancers such as sugar, corn syrups, honey, hydrolyzed vegetable protein, autolyzed yeast, spices, and seasonings. Today, the manufacture and safety of cured poultry meat products relies on nitrite addition as the principal preservative agent, but also includes combination with vacuum or modified atmosphere packaging and refrigeration.

Curing as a preservation technique

The primary purpose and benefit of utilizing sodium nitrite for meat curing is its inhibitory property against *Clostridium botulinum* growth and toxin production. *C. botulinum* is an obligately anaerobic, Gram positive, spore-forming rod that is widely distributed in the soil[56] and therefore has the potential to be a contaminant on the feathers, skin, and intestines of poultry. The spores are extremely heat resistant to normal pasteurization or cooking processes for cooked, cured products, excluding those processed by canning. *C. botulinum* is referred to as Type A to G on the basis of the exotoxin produced during vegetative cell growth. Types A and B are more frequently associated with animal foods. The toxins affect the neurological system of the body, paralyzing muscles and eventually can cause death from respiratory failure or cardiac arrest. Antitoxin administered prior to development of severe symptom development increases survival.

As an antimicrobial, nitrite's effectiveness is dependent on factors such as pH, salt concentration, temperature, and level of contamination. While not preventing germination of spores of clostridia, it inhibits the rate of cellular outgrowth. It is also generally inhibitory to other bacteria but this may be due to the combined effect with salt and/or pH, acting together as outgrowth obstacles.[57] The effective compound is thought to be nitrous acid which is rapidly decomposed as the aqueous pH increases. Nitrite's effectiveness generally increases approximately 10-fold as the pH decreases from 7.0 to 6.0 and is most effective in the range of pH 5.5 to 5.0.[58] Nitrite is thought to exert its bacteriostatic effects through several mechanisms such as reaction with iron-sulfur proteins important in energy metabolism and inactivation of catalase and cytochrome by nitric oxide.

Experience with vacuum-packaged, cured poultry products suggests that similar products packaged in modified atmospheres would benefit from the same preservative factors derived from nitrite addition. However, the main weakness for all cured poultry products may prove to be the extensive use of alkaline sodium phosphates to increase the product's functional properties while likely decreasing the bacteriostatic effects of nitrite. These areas will require further research and experience related to safety and shelf-life extension from delayed microbial outgrowth.

Quality characteristics from meat curing reactions

Cured color

The most significant visual result in curing is the development of a characteristic pink to pinkish-red coloration in the cooked end product. The native pigment myoglobin and its oxidized form metmyoglobin (as well as the corresponding pigments of hemoglobin) are converted initially to nitrosylmyoglobin, and with sufficient heating, to nitrosylhemochrome (Figure 15.11). The intensity of color formed depends on the pigment concentration in the raw material. Poultry white meat contains from 0.1 to 0.4 mg of myoglobin per gram tissue whereas the dark meat has a concentration range from 0.6 to 2.0 mg/g tissue. Because mechanically deboned meat will contain some bone marrow, it will have higher pigment levels than the respective hand-deboned tissue[59] and thus yield higher color intensity. Because the USDA requires cured poultry products to reach an internal temperature of at least 68°C (155°F), the final pigment form is nitrosylhemochrome.

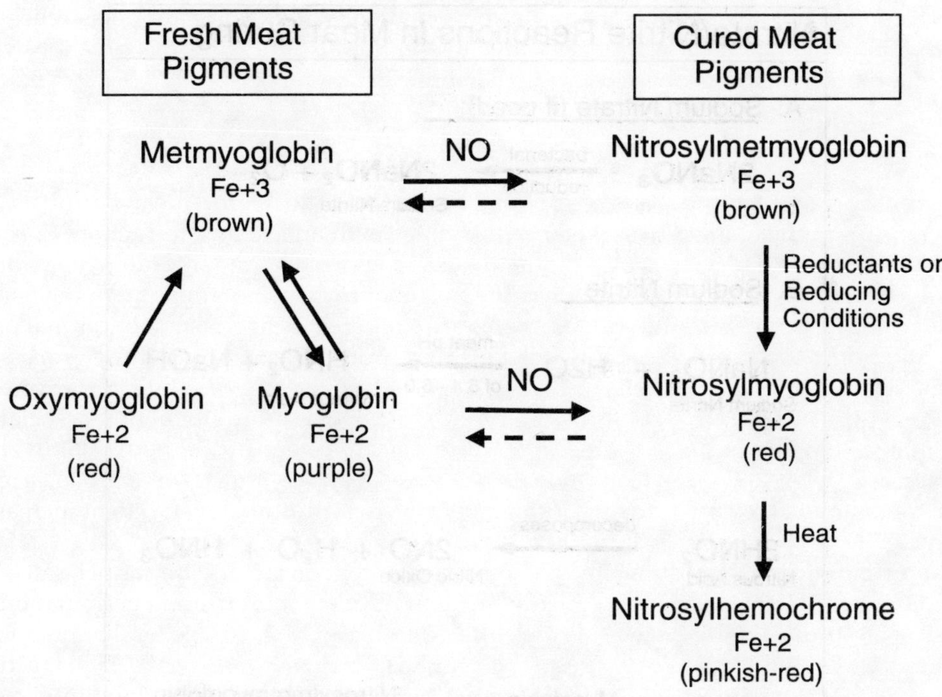

Figure 15.11 Nitric oxide (NO) reaction pathway from fresh meat pigments to form cured meat pigments leading to the final cooked pigment form known as nitrosylhemochrome.

The nitric oxide (NO) which reacts with myoglobin and metmyoglobin is formed from nitrite through a series of reactions (Figure 15.12). Important factors in the rate of NO formation and its ultimate reaction with the globin pigments are meat pH and the presence of reducing conditions. The pH of post-rigor poultry ranges from approximately 5.7 to 5.9 for breast and 6.4 to 6.7 for dark meat.[60] Initially, an equilibrium is established between the concentration of the nitrite ion (NO_2-) and its undissociated conjugate acid, nitrous acid (HNO^2). The slightly acidic environment of the meat keeps the equilibrium shifted toward NO^2- and provides only a low concentration of HNO^2. However, HNO^2 rapidly decomposes yielding NO. The decomposition thereby allows for the continual slow formation of HNO^2 and consequently, more NO formation over time. NO production from nitrite is enhanced by the addition of reductants, such as sodium ascorbate or erythorbate. The interaction pathway of nitrite and ascorbate is significant and complex,[61] involving several intermediates. The compound 2,3-dehydroascorbic acid is believed to be formed and provides for more rapid formation of nitrosylmyoglobin. In addition, the reductants serve to reduce metmyoglobin to myoglobin, which also accelerates the rate of curing.

Nitrosylhemochrome undergoes rapid light-induced dissociation if the product is not properly packaged, resulting in "fading" or surface discoloration to a light gray appearance. Light and oxygen play a central role in a sequence of reactions initially involving NO dissociation from the central iron atom of the hemochrome's porphyrin structure (Figure 15.13). The dissociated NO is then free to recombine with the porphyrin, reforming the pigment and thus maintaining the pink to pinkish-red cured product color. The reestablishment of the pigment is favored by vacuum-packaging with films of low oxygen transmission rate. The dissociated NO can also be oxidized by oxygen and thus not be

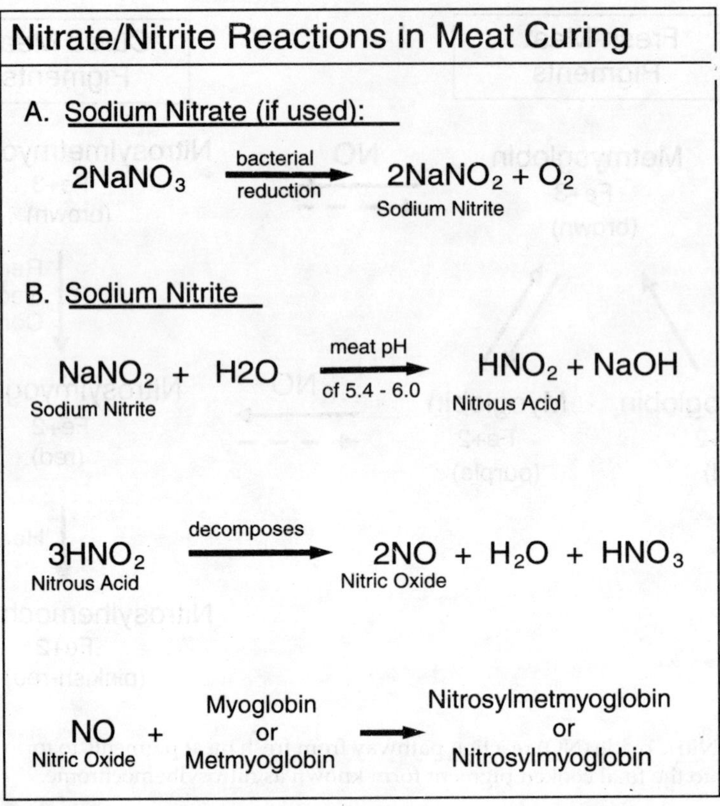

Figure 15.12 Reaction sequence from either sodium nitrate or sodium nitrite leading to the formation of nitric oxide (NO).

available to recombine to form the original pigment.[62] In extensive cases of fading, additional oxygen permeating the film is then free to oxidize the central iron of the hemochrome (Fe^{+2}) resulting in production of a hemichrome (Fe^{+3}). This loss is visually detectable as lightening of the product with a distinct loss of the red hue. Under conditions of extreme oxidation, sites of the hemichrome's structure are oxidized, and the product appears severely bleached. Visual light-induced fading can be minimized by selection of packaging films with oxygen transmission rates generally in the range of 15 to 17 cc/m^2/24h (at 23°C, 0% RH, 1 atm); (Figure 15.14).

In most cured, deli-style poultry products, adequate packaging is utilized for color retention, whether the product is vacuum-packaged or modified atmosphere packaged (MAP) with oxygen exclusion. In some MAP systems, the oxygen absorber is placed in a sachet within the package and fastened underneath the label or at the bottom of the package. With in-going nitrite concentrations of 156 ppm, the average residual nitrite level found in products after a week or so of storage is of the order of 10-fold lower than the initial concentration. The small residual level also provides a potential source for slow NO generation which aids in cured color maintenance.

Cured flavor

There is a significant difference between the flavor of most cured products when compared to their uncured counterparts. Although cured meat flavor development attributable to

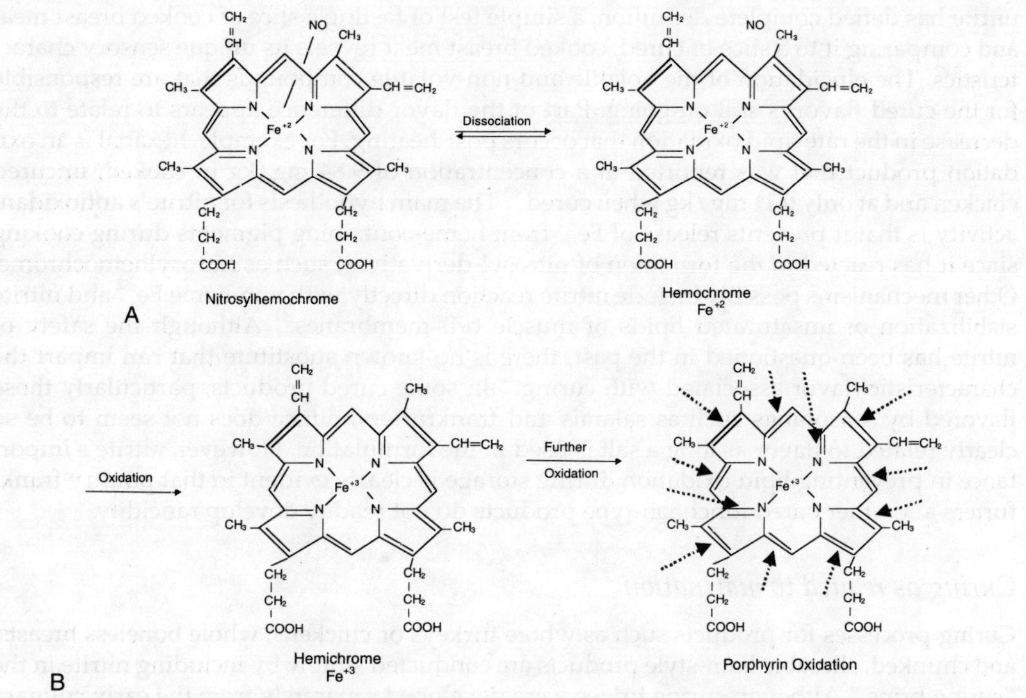

Nitrosylhemochrome

A

Dissociation

Hemochrome Fe^{+2}

Oxidation

Hemichrome Fe^{+3}

B

Further Oxidation

Porphyrin Oxidation

Figure 15.13 The dissociation of nitric oxide from the cured pigment nitrosylhemochrome and its sequential oxidation leading to "fading" of the cured product. (Reprinted from Kartika, S., Dawson, P. L., and Acton, J. C., *Act. Rep. Res. Dev. Assoc.*, 51, 293-299, 1998. With permission of Research & Development Associates for Military Food and Packaging Systems, Inc., San Antonio, TX).

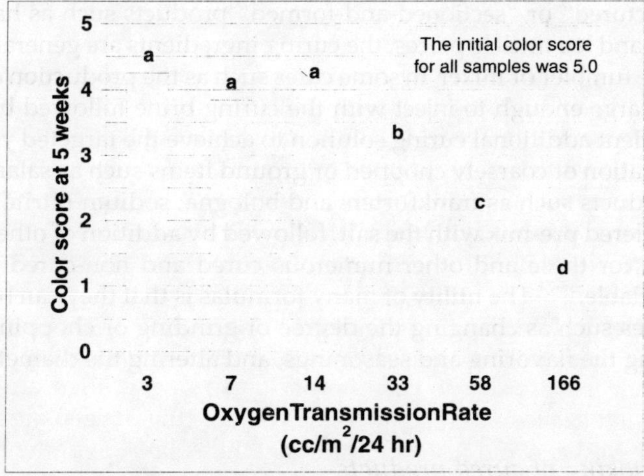

Figure 15.14 Visual color scores at five weeks of lighted display for turkey bologna vacuum-packaged in films differing in oxygen transmission rate. (Bars having a different letter are significantly different at $p < 0.05$). (Reprinted from Kartika, S., Dawson, P. L., and Acton, J. C., *Act. Rep. Res. Dev. Assoc.*, 51, 293-299, 1998. With permission of Research & Development Associates for Military Food and Packaging Systems, Inc., San Antonio, TX).

nitrite has defied complete definition, a simple test of tasting a slice of cooked breast meat and comparing it to a slice of cured, cooked breast meat reveals its unique sensory characteristics. The elucidation of the volatile and non-volatile compounds that are responsible for the cured flavor is still on-going. Part of the flavor difference appears to relate to the decrease in the rate lipid oxidation that occurs post-heating. For example, hexanal is an oxidation product that was reported at a concentration of 9.84 mg/kg in cooked, uncured chicken and at only 0.11 mg/kg when cured.[63] The main hypothesis for nitrite's antioxidant activity is that it prevents release of Fe^{+2} from heme-containing pigments during cooking since it has reacted in the formation of nitrosyl-derivatives, such as nitrosylhemochrome. Other mechanisms possibly include nitrite reaction directly with non-heme Fe^{+2} and nitrite stabilization of unsaturated lipids of muscle cell membranes.[61] Although the safety of nitrite has been questioned in the past, there is no known substitute that can impart the characteristic flavor associated with curing.[64] In some cured products, particularly those flavored by seasonings such as salamis and frankfurters, nitrite does not seem to be so clearly related to flavor as long a salt is used in the formulation. However, nitrite's importance in preventing lipid oxidation during storage is clearly evident in that poultry frankfurters and other cured luncheon-type products do not readily develop rancidity.

Curing as related to marination

Curing processes for products such as whole turkeys or chickens, whole boneless breasts, and chunked, sliceable ham-style products are conducted simply by including nitrite in the "curing brine." Although curing brines were developed separately from the early culinary approaches of marinade development, they share the similarity in that they are principally based on delivering their functions through liquid incorporation into the meat tissue. Whole carcasses or whole boneless meat pieces may be cured by soaking similar to marination, the major distribution for nitrite and other soluble curing ingredients today is also by mechanical injection because it is faster and results in more rapid cure distribution within the tissues. For "restructured" or "sectioned-and-formed" products such as hams and pastrami made from thigh and leg muscle pieces, the curing ingredients are generally included in the fluid added to the tumbler or mixer. In some cases such as the production of turkey ham, the thigh pieces are large enough to inject with the curing brine followed by tumbling in the presence of sufficient additional curing solution to achieve the targeted yield.

In the preparation of coarsely chopped or ground items such as salamis, and for finely comminuted products such as frankfurters and bologna, sodium nitrite is added as a dry granular or powdered pre-mix with the salt, followed by addition of other non-meat ingredients. Formulas for these and other numerous cured and non-cured further-processed products are available.[65–74] The utility of many formulas is that they can be modified to add different attributes such as changing the degree of grinding or chopping for particle size reduction, varying the flavoring and seasonings, and altering the diameter, shape, or form of product.

Cooking and smoking of cured products

The term "smokehouse" is associated with the processing oven utilized for cooking of most cured poultry products. Today's products may or may not be smoked as in the past and the industry now refers to the houses in terms of "processing ovens" rather than smokehouses. Most products that are stuffed into cellulosic or fibrous casings, such as frankfurters and salami or turkey ham, are placed in temperature-, humidity-, and air velocity-controlled ovens that are of batch process (Figure 15.15) or continuous flow-

Figure 15.15 ALKAR batch oven with DDC computer controls for cooking and smoking of poultry products. (Courtesy of ALKAR, A Division of DEC International, Inc., Lodi, WI).

through design (Figure 15.16). In both cases, the heating medium is forced-air and, within limits, regulating the humidity of the circulating air may increase heat transfer between the air and the product's surface. The rate of product heating is primarily determined by the temperature difference between the product surface and its core temperature.[75] Humidity, or more correctly, the wet-bulb temperature setting, is generally increased in the early stages of heating since it will increase the cooler product's surface temperature by condensation and thereby increase the surface-to-core temperature difference. The humidity is then reduced in the latter stages so that the dry-bulb temperature becomes the driving force at the product's surface. Moisture migration and evaporation at the surface result in product "shrink" or weight loss. Although the yield is reduced, the drying effect during the latter stages of cooking is usually beneficial for enhancing the product's appearance. The cured color appears slightly brighter at the dried surface as compared to the more moist, interior of the product.

The final temperature of cooking is specified by the USDA and depends on whether the product is cured or uncured. The regulatory requirement insures that these products are "ready-to-eat" without the necessity of further heating by the consumer. Federal

Figure 15.16 ALKAR inline continuous cook/chill system with flow-through processing for frankfurters or wieners. (Courtesy of ALKAR, A Division of DEC International, Inc., Lodi, WI).

regulations specify an internal temperature of 71°C for uncured and 68°C for cured poultry[76] when heat-processed by any method. However, many processors use an internal temperature of 73 to 74°C for products that are to be sliced, particularly if they are prepared with dark meat. In effect, the cooking temperatures assure pasteurization in the cooking of these products and thus special care is required post-cooking to prevent product contamination prior to packaging. A special case for heating during processing applies to poultry breakfast strips, similar to bacon, which are cured and smoked products that require additional cooking by the consumer prior to consumption. The USDA requires these products to be heated to an internal temperature of 60°C, then cooled to 26°C within an hour and a half, and attaining 4°C or less within 5 h.[76]

Smoking, provided through the controlled oxidation and combustion of wood, contributes to product flavor and aroma, aids in development of surface color, and deposits several bacteriostatic compounds that can aid in extending product shelf-life. With natural smoke generators attached to processing ovens, the generated smoke is metered into the oven as a step in the cooking cycle. In order to avoid streaking while maximizing penetration and deposition, smoke is generally applied to the moist product surface after the early surface condensate on the product has evaporated. It is primarily the phenolic and carboxylic acids of smoke that provide flavor and bacteriostatic effects whereas color enhancement is related to the carbonyl (aldehydes and ketones) content. A noticeable browning effect, in addition to cured color development, results from products of the Maillard reaction between free amino groups of proteins or other nitrogenous compounds and the carbonyls of the smoke. The phenols, as antioxidants, also reduce oxidative rancidity.[77]

Liquid smoke flavorings, applied through sprays or aerosols in the oven or added during mixing or tumbling with other ingredients, are manufactured from entrapment and concentration of smoke volatiles into water or oils. With a more consistent composition, and with removal of polycyclic hydrocarbons having carcinogenic properties, liquid smoke utilization has increased over the past decade.[78] To mimic natural smoking, smoke regenerators are available to heat and volatilize liquid smoke to vapor, which is then transferred to the oven and deposited at the product surface in the same manner as normal smoke vapor.

Innovation in product development has resulted in a wide variety of combinations of curing, smoking, and cooking. These techniques are not limited to their "traditional" uses such as curing and smoking. For example, whole bone-in or boneless products such as smoked, cured turkey or chicken breasts, and cured poultry sausage patties may also be cooked (after smoking if applied) in the continuous conveyor type ovens previously described. Small amounts of various cured poultry products are occasionally thermally retorted in pouches for the military MRE program.

Summary

The development of new poultry products occurring over recent decades has been accompanied by the evolution of marination, cooking, and curing technologies. While marination can enhance eating quality, it also reduced yield losses during subsequent cooking. The many possible cooking techniques and equipment choices not only pasteurize the product and influence its eating quality, they also can affect profitability through the cooking losses they can cause. Curing is a special technique that while imparting many desirable product characteristics, it is also closely regulated because of the possibility for producing potentially harmful byproducts. Marination, cooking, and curing are essential to profitable product development and will undoubtedly advance with future product lines.

References

1. Oblinger, J. L., Janky, D. M., and Koburger, J. A., Effect of brining and cooking procedure on tenderness of spent hens, *J. Food Sci.*, 42, 1347, 1977.
2. Kijowski, J. and Mast, M. G., Tenderization of spent fowl drumsticks by marination in weak organic solutions, *Int. J. Food Sci. Technol.*, 28, 337, 1993.
3. Janky, D. M., Koburger, J. A., Oblinger, J. L., and Riley, P. K., Effect of salt brining and cooking procedure on tenderness and microbiology of smoked Cornish game hens, *Poult. Sci.*, 55, 761, 1976.
4. Maki, A. A. and Froning, G. W., Effect on the quality characteristics of turkey breast muscle of tumbling whole carcasses in the presence of salt and phosphate, *Poult. Sci.*, 66, 1180, 1987.
5. Babji, A. S., Froning, G. W., and Ngoka, D. A., The effect of short-term tumbling and salting on the quality of turkey breast muscle, *Poult. Sci.*, 61, 300, 1982.
6. Smith, D. P., Fletcher, D. L., and Papa, C. M., Evaluation of duckling breast meat subjected to different methods of further processing and cooking, *J. Muscle Foods*, 2, 305, 1991.
7. Hamm, R., Biochemistry of meat hydration, in *Advances in Food Research*, Vol. 10, Chichester, C. O., Mrak, E. M., and Stewart, G. F., Eds., Academic Press, New York, 1960, 355.
8. Gault, N. F. S., Marinaded meat, in *Developments in Meat Science*, Vol. 5, Lawrie, R., Ed., Elsevier Applied Science, London, 1991, chap. 5.
9. Acton, J. C. and Jensen, J. M., Understanding marinade technology, *Poult. Process.*, 9, 18, 1994.
10. Salvage, B., Add value by injecting whole-muscle meats, *Meat Market Technol.*, 7 (4), 34, 1999.
11. Smith, D., Marination: tender to the bottom line, *Broiler Ind.*, 62, 22, 1999.
12. Petrak, L., Liquid solutions, *Nat. Provis.*, 213 (10), 44, 1999.
13. Cassidy, J. P., Phosphates in meat processing, *Food Prod. Dev.*, 11, 74, 1977.
14. Thomson, J. E., Effect of polyphosphates on oxidative deterioration of commercially cooked fryer chickens, *Food Technol.*, 18, 1805, 1964.
15. Young, L. L. and Lyon, B. G., Effect of sodium tripolyphosphate in the presence and absence of calcium chloride and sodium chloride on water retention properties and shear resistance of chicken breast meat, *Poult. Sci.*, 65, 898, 1986.
16. Young, L. L. and Lyon, C. E., Effect of calcium marination on biochemical and textural properties of pre-rigor chicken breast meat, *Poult. Sci.*, 76, 197, 1997.
17. Prusa, K. J., Chambers, E., Bowers, J. A., Cunningham, F., and Dayton, A. D., Thiamin content, texture, and sensory evaluation of postmortem papain-injected chicken, *J. Food Sci.*, 46, 1684, 1981.
18. Dziezak, J. D., Phosphates improve many foods, *Food Technol.*, 44, 79, 1990.
19. Barbut, S., Maurer, A. J., and Lindsay, R. C., Effects of reduced sodium chloride and added phosphates on physical and sensory properties of turkey frankfurters, *J. Food Sci.*, 53, 62, 1988.
20. Schermerhorn, E. P. and Stadelman, W. J., Treating hen carcasses with polyphosphates to control hydration and cooking losses, *Food Technol.*, 18, 101, 1964.
21. May, K. N., Helmer, R. L., and Saffle, R. L., Effect of phosphate treatment on carcass weight changes and organoleptic quality of cut-up chicken, *Poult. Sci.*, 41, 1665, 1962.
22. Froning, G. W., Effect of polyphosphates on the binding properties of chicken meat, *Poult. Sci.*, 44, 1104, 1965.
23. Landes, D. R., The effects of polyphosphates on several organoleptic, physical, and chemical properties of stored precooked frozen chickens, *Poult. Sci.*, 51, 641, 1972.
24. Oblinger, J. L., Janky, D. M., and Koburger, J. A., The effect of water soaking, brining and cooking procedure on tenderness of broilers, *Poult. Sci.*, 55, 1494, 1976.
25. Janky, D. M., Oblinger, J. L., and Koburger, J. A., The effect of salt concentration and brining time on organoleptic characteristics of smoked broiler breeder hens, *Poult. Sci.*, 57, 116, 1978.
26. Post, R. C. and Heath, J. L., Marinating broiler parts: the use of a viscous type marinade, *Poult. Sci.*, 62, 977, 1983.
27. Lemos, A. L. S. C., Nunes, D. R. M., and Viana, A. G., Optimization of the still-marinating process of chicken parts, *Meat Sci.*, 52, 227, 1999.
28. Proctor, V. A. and Cunningham, F. E., Influence of marinating on weight gain and coating characteristics of broiler drumsticks, *J. Food Sci.*, 52, 286, 1987.

29. Froning, G. W., Effect of various additives on the binding properties of chicken meat, *Poult. Sci.*, 45, 185, 1966.
30. Shults, G. W. and Wierbicki, E., Effects of sodium chloride and condensed phosphates on the water-holding capacity, pH and swelling of chicken muscle, *J. Food Sci.*, 38, 991, 1973.
31. Regenstein, J. M. and Stamm, J. R., The effect of sodium polyphosphates and of divalent cations on the water holding capacity of pre- and post-rigor chicken breast muscle, *J. Food Biochem.*, 3, 213, 1979.
32. Young, L. L., Papa, C. M., Lyon, C. E., and Wilson, R. L., Moisture retention and textural properties of ground chicken meat as affected by sodium tripolyphosphate, ionic strength and pH, *J. Food Sci.*, 57, 1291, 1992.
33. Yang, C. C. and Chen, T. C., Effects of refrigerated storage, pH adjustment, and marinade on color of raw and microwave cooked chicken meat, *Poult. Sci.*, 72, 355, 1993.
34. Chen, T. C., Studies on the marinating of chicken parts for deep-fat frying, *J. Food Sci.*, 47, 1016, 1982.
35. Cunningham, F. E., Bowers, J. A., Craig, J., Moore, A. M., and Froning, G. W., Composition and sensory characteristics of hot-boned, tumbled, turkey breast muscle, *J. Food Qual.*, 11, 225, 1988.
36. Heath, J. L. and Owens, S. L., Reducing variation in marinade retained by broiler breasts, *Poult. Sci.*, 70, 160, 1991.
37. Kotula, K. L. and Heath, J. L., Effect of tumbling chill-boned and hot-boned broiler breasts in either acetic acid or sodium chloride solutions on cooked yield, density, and shear values, *Poult. Sci.*, 65, 717, 1986.
38. Lyon, B. G. and Hamm, D., Effects of mechanical tenderization with sodium chloride and polyphosphates on sensory attributes and shear values of hot-stripped broiler breast meat, *Poult. Sci.*, 65, 1702, 1986.
39. Xiong, Y. L. and Kupski, D. R., Time-dependent marinade absorption and retention, cooking yield, and palatability of chicken fillets marinated in various phosphate solutions, *Poult. Sci.*, 78, 1053, 1999.
40. Grey, T. C., Robinson, D., and Jones, J. M., The effects on broiler chicken of polyphosphate injection during commercial processing I: changes in weight and texture, *J. Food Technol.*, 13, 529, 1978.
41. Farr, A. J. and May, K. N., The effect of polyphosphates and sodium chloride on cooking yields and oxidative stability of chicken, *Poult. Sci.*, 49, 268, 1970.
42. Peterson, D. W., Effect of polyphosphates on tenderness of hot cut chicken breast meat, *J. Food Sci.*, 42, 100, 1977.
43. Brotsky, E., Automatic injection of chicken parts with polyphosphate, *Poult. Sci.*, 55, 653, 1976.
44. Hale, K. K., Automated phosphate injection of whole broiler carcasses, *Poult. Sci.*, 56, 859, 1977.
45. Lyon, C. E., Lyon, B. G., and Dickens, J. A., Effects of carcass stimulation, deboning time, and marination on color and texture of breast meat, *J. Appl. Poult. Res.*, 7, 53, 1998.
46. Zheng, M., Toledo, R. T., Carpenter, J. A., and Wicker, L., Yield and sensory evaluation of poultry marinated pre and postrigor, *J. Food Qual.*, 22, 85, 1999.
47. Goodwin, T. L. and Maness, J. B., The influence of marination, weight, and cooking technique on tenderness of broilers, *Poult. Sci.*, 63, 1925, 1984.
48. Hashim, I. B., McWatters, K. H., and Hung, Y.-C., Marination method and honey level affect physical and sensory characteristics of roasted chicken, *J. Food Sci.*, 64, 163, 1999.
49. Froning, G. W. and Sackett, B., Effect of salt and phosphates during tumbling of turkey breast muscle on meat characteristics, *Poult. Sci.*, 64, 1328, 1985.
50. Engler, P. P. and Bowers, J. A., B-vitamin retention in meat during storage and preparation. A review, *J. Am. Diet. Assoc.*, 69, 253, 1976.
51. Chiu, C. P., Yang, D. Y., and Chen, B. H., Formation of heterocyclic amines in cooked chicken legs, *J. Food Prot.*, 61, 712, 1998.
52. Salmon, C. P., Knize, M. G., and Felton, J. S., Effects of marinating on heterocyclic amine carcinogen formation in grilled chicken, *Food Chem. Toxicol.*, 35, 433, 1997.
53. Yingst, L. D., Wyche, R. C., and Goodwin, T. L., Cooking techniques for broiler chickens, *J. Am. Diet. Assoc.*, 59, 582, 1971.

54. Froning, G. W., Armong, R. G., Mandigo, R. W., Neth, C. E., and Hartung, T. E., Quality and storage stability of frankfurters containing 15% mechanically deboned turkey meat, *J. Food Sci.*, 36, 974, 1971.
55. Cassens, R. G., *Nitrite-Cured Meat. A Food Safety Issue in Perspective,* Food & Nutrition, Trumbull, 1990, 21.
56. Hayes, P. R., *Food Microbiology and Hygiene,* 2nd ed., Elsevier Applied Science, New York, 1992, 55.
57. van Laack, R. L. J. M., Spoilage and preservation of meat, in *Muscle Foods,* Kinsman, D. M., Kotula, A. W., and Breidenstein, B. C., Eds., Chapman and Hall, New York, 1994, 378.
58. Kraft, A. A., Meat microbiology, in *Muscle as a Food,* Bechtel, P. J., Ed., Academic Press, Orlando, FL, 1986, 239.
59. Froning, G. W. and Johnson, F., Improving the quality of mechanically deboned fowl meat by centrifugation, *J. Food Sci.*, 38, 279, 1973.
60. Bryan, F. L., poultry and poultry meat products, in *Microbial Ecology of Foods,* Silliker, J. H., Elliot, R. P., Baird-Parker, A. C., Bryan, F. L., Christian, J. H. B., Clark, D. S., Olsen, J. C., Jr., and Roberts, T. A., Eds., Vol. 2, Academic Press, 1980, 410.
61. Skibsted, L. H., Cured meat products and their oxidative stability, in *The Chemistry of Muscle-Based Foods,* Johnston, D. E., Knight, M. K., and Ledward, D. A., Eds., Royal Society of Chemistry, 1992, 266.
62. Kartika, S., Dawson, P. L., and Acton, J. C., Nitrosylhemochrome loss and apparent hemochrome destruction in vacuum-packaged turkey bologna, *Act. Rep. Res. Dev. Assoc.*, 51, 293, 1998.
63. Ramarathnam, N., Flavor of cured meat, in *Flavor of Meat, Meat Products and Seafoods,* 2nd ed., Shahidi, Ed., Blackie Academic and Professional, London, 1998, 290.
64. Gray, J. I., MacDonald, B., Pearson, A. M., and Morton, I. D., Role of nitrite in cured meat flavor: a review, *J. Food Prot.*, 44, 302, 1981.
65. Long, L., Komarik, S. L., and Tressler, D. K., *Food Products Formulary, Vol. 1, Meat, Poultry, Fish, Shellfish,* 2nd ed., AVI Publishing Company, Westport, CT, 1982.
66. Tadelman, W. J., Olson, V. M., Shemwell, G. A., and Pasch, S., Manufactured meat products, in *Egg and Poultry Meat Processing,* VCH Publishers, New York, 1988, 161.
67. Baker, R. C. and Bruce, C. A., Further processing of poultry, in *Processing of poultry,* Mead, G., Ed., Elsevier Applied Science, New York, 1989, 267.
68. Richardson, R. I., Utilization of turkey meat in further-processed products, in *Processing of Poultry,* Mead, G., Ed., Elsevier Applied Science, New York, 1989, 296.
69. Weiner, P. D., Formulations for restructured poultry products, in *Advances in Meat Research,* Pearson, A. M. and Dutson, T. R., Eds., Vol. 3, Van Nostrand Reinhold, New York, 1987, 405.
70. Acton, J. C., Dick, R. L., and Torrence, A. K., Turkey ham properties on processing and cured color formation, *Poult. Sci.*, 58, 843, 1979.
71. Baker, R. C. and Darfler, J. M., The development of a poultry ham product, *Poult. Sci.*, 60, 1429, 1981.
72. Sheldon, B. W., Ball, H. R., and Kimsey, H. R., Jr., A comparison of curing practices and sodium nitrite levels on chemical and sensory properties of smoked turkey, *Poult. Sci.*, 61, 710, 1982.
73. Wisniewski, G. D. and Maurer, A. J., A comparison of five cure procedures for smoked turkeys, *J. Food Sci.*, 44, 130, 1979.
74. Hall, J., *Sausage Made Easy,* 2nd ed., Meat Business Magazine Publishers, St. Louis, MO, 1993.
75. Hanson, R. E., Cooking technology, *Proc. Recip. Meat Conf.*, 44, 109, 1990.
76. Code of Federal Regulations, Title 9, Animals and Animal Products, Vol. 2, Chapter III, Food Safety and Inspection Service, Department of Agriculture, Sec. 381.150 (a), Requirements for the production of poultry breakfast strips, poultry rolls and other poultry products, 1999.
77. Schmidt, G. R., Processing and fabrication, in *Muscle as a Food,* Bechtel, P. J., Ed., Academic Press, Orlando, FL, 1986, 201.
78. Pearson, A. M. and Gillett, T. A., *Processed Meats,* Aspen Publishers, Gaithersburg, MD, 1999, 79.

chapter sixteen

A brief introduction to some of the practical aspects of the kosher and halal laws for the poultry industry

Joe M. Regenstein and Muhammad Chaudry

Contents

0-8493-0120-3/01/$0.00+$.50

Introduction

The kosher dietary laws determine which foods are "fit or proper" for consumption by Jewish consumers who observe these laws. The laws are Biblical in origin, coming mainly from the original five books of the Holy Scriptures. Over the years, the details have been interpreted and extended by the rabbis to protect the Jewish people from violating any of the fundamental laws and to address new issues and technologies. The Jewish laws are referred to as the "halacha."

The Muslim halal dietary laws determine which foods are acceptable to Muslims. These laws are found in the Quran. Again, Muslim leaders have interpreted these laws over the years. Islamic law is referred to as Shari'ah. It is eternal—definite and unalterable—yet it is ever fresh and resilient as applications are adjusted to different times and circumstances. For example, both the Jewish rabbis and Muslim Imans and Mullahs are currently dealing with issues related to biotechnology (see below).

Why do Jews follow the kosher dietary laws? Many explanations have been given. The following by Rabbi Grunfeld is possibly the best written explanation and probably summarizes the most widely held ideas about the subject.[1] Although this explanation is also relevant for halal, it is important to note that, unlike the kosher laws, the health aspects of eating are an important part of the halal laws.[2]

> "And ye shall be men of a holy calling unto Me, and ye shall not eat any meat that is torn in the field" (Exodus XXII:30). Holiness or self-sanctification is a moral term; it is identical with . . . moral freedom or moral autonomy. Its aim is the complete self-mastery of man.
>
> To the superficial observer it seems that men who do not obey the law are freer than law-abiding men, because they can follow their own inclinations. In reality, however, such men are subject to the most cruel bondage; they are slaves of their own instincts, impulses and desires. The first step towards emancipation from the tyranny of animal inclinations in man is, therefore, a voluntary submission to the moral law. The constraint of law is the beginning of human freedom. . . . Thus the fundamental idea of Jewish ethics, holiness, is inseparably connected with the idea of Law; and the dietary laws occupy a central position in that system of moral discipline which is the basis of all Jewish laws.
>
> The three strongest natural instincts in man are the impulses of food, sex, and acquisition. Judaism does not aim at the destruction of these impulses, but at their control and indeed their sanctification. It is the law which spiritualizes these instincts and transfigures them into legitimate joys of life."

The kosher and halal market

The kosher market covers almost 100,000 products in the U.S. In dollar value, about 100 billion dollars worth of products have a kosher marking on them. The actual consumers of kosher food, i.e., those who specifically look for the kosher mark, are estimated to be about 6 to 8 million Americans, and they are purchasing almost 3 billion dollars worth of kosher product. Only about one third of the kosher consumers are Jewish; other consumers include Muslims, Seventh Day Adventists, vegetarians, people with various types of allergy, particularly dairy, grain, and legume, and general consumers who value the quality of kosher products: "We report to a higher authority." AdWeek Magazine has called kosher "the Good Housekeeping Seal for the 90s." By undertaking kosher certification, companies can incrementally expand their market by opening up new markets.

The Muslim market in the U.S. is just emerging. Many urban centers have special halal markets, and most Muslims observe the halal laws. But the real opportunities exist on a worldwide basis—the number of Muslims in the world is around 1 billion people. Many countries of Southeast Asia, the Middle East, and Northern Africa have predominantly Muslim populations. In many countries, halal certification is necessary for products to be permitted to be imported.

Although many Muslims purchase kosher food, these foods, as we will see below, do not always meet the needs of the Muslim consumer. In particular the use of various questionable gelatins in products produced by some kosher supervisions and the use of alcohol (many are permitted in kosher food, if properly prepared) are areas of difference.

Although limited market data is available, the most dramatic data about the impact of kosher has been provided by Coors when they went kosher. According to their market analysis, their share of market in the Philadelphia market went up 18% on going kosher. Somewhat less dramatic increases were observed in other cities in the Northeast.

The kosher dietary laws

The kosher dietary laws predominantly deal with three issues, all in the animal kingdom:

1. Allowed animals
2. Prohibition of blood
3. Prohibition of mixing of milk and meat

However, for the week of Passover (in late March or early April) restrictions on "chometz," the prohibited grains, and the rabbinical extensions of this prohibition leads to a whole new set of regulations, focused in this case on the plant kingdom.

In addition, there is a separate set of laws dealing with grape juice, wine, and alcohol derived from grape products. Basically, these must be handled by sabbath-observing Jews. However, if the juice is pasteurized (heated or "mevushal" in Hebrew), then this juice can be handled as an ordinary kosher ingredient.

Allowed animals and prohibition of blood

Ruminants with split hoofs, the traditional domestic birds, and fish with fins and removable scales are generally permitted. Pigs, wild birds, sharks, dogfish, catfish, monkfish, and similar species along with all crustacean and molluscan shellfish are prohibited. Insects are also prohibited, so carmine and cochineal (natural red pigments) are not used in kosher products.

With specific respect to poultry, the traditional domestic birds, i.e., chicken, turkey, squab, duck, and goose are kosher. Birds in the ratite category (ostrich, emu, and rhea) are definitely not kosher, as the ostrich is specifically mentioned in the Bible.[3] However, it is not clear whether the animal of the Bible is the same animal we know as an ostrich today. Regardless, these and most other birds are prohibited. There have been some attempts to characterize the features that make a kosher bird, but these are not widely accepted, and basically one relies on "tradition." Interestingly, domesticated turkey is considered kosher although wild turkey may not be. Part of the problem is that "hunting" is not permitted under any circumstances.

Furthermore, ruminants and fowl must be slaughtered according to Jewish law by a specially trained religious slaughterman. These animals are also subsequently inspected by the rabbis for various defects. In the U.S., the desire for more stringent meat inspection requirements has led to the development of a kosher meat meeting a stricter inspection requirement, mainly with respect to the lungs, referred to as "glatt (smooth) kosher." This mainly refers to red meats where lung adhesions are a problem and often make an animal not kosher (treife). In general, a glatt kosher animal's lungs have less than three such adhesions. As it is difficult to examine the lungs of poultry, this is not generally done. Yet, to distinguish poultry products as being produced to a stricter standard, some producers will also use the term "glatt." However, we will return to discuss poultry issues in more detail below.

The meat and poultry must be further prepared by properly removing certain veins, arteries, prohibited fats, blood, and the sciatic nerve. In practical terms, this means that only the front quarter cuts of red meat are generally used. Again, a minimal set of rules applies to poultry. To remove the blood, red meat and poultry are soaked and salted within a specified time period. Furthermore, any materials that might be derived from animal sources are generally prohibited because of the difficulty of obtaining them from kosher animals. Thus many products that might be used in the dairy industry, such as emulsifiers, stabilizers, and surfactants, particularly those that are fat-derived, need careful rabbinical supervision to assure that no animal-derived ingredients are used. Almost all such materials are also available in a kosher form derived from plant oils.

Prohibition of mixing of milk and meat

"Thou shalt not seeth the kid in its mother's milk."

This passage appears three times in the Torah (the first five books of the Holy Scriptures) and is thus taken religiously as a very serious admonition. The meat side of the equation has been rabbinically extended to include poultry. The dairy side includes all milk derivatives.

To keep meat and milk separate requires that the processing and handling of all products that are kosher will fall into one of three categories:

1. Meat products
2. Dairy products
3. Pareve (Parve) or neutral products

The latter includes all products that are not classified as meat or dairy. All plant products along with eggs, fish, honey, and lac resin (shellac) are pareve. These pareve foods can be used with either meat products or dairy products, except that fish cannot be mixed directly with meat. Once a pareve product is mixed with either meat or dairy products, they take on the status of meat or dairy, respectively.

Some kosher-observant Jews are concerned with the possible adulteration of kosher milk with the milk of other animals (e.g., mare's milk) and as such require that the milk be watched from the time of milking. This "Cholev Yisroel" milk and products derived from milk are required by some of the stricter kosher supervision agencies for all dairy ingredients, so that dairy products would have to meet these requirements.

In order to assure the complete separation of milk and meat, all equipment, utensils, etc. must be of the proper category. Thus, if plant materials (e.g., a fruit juice) are run through a dairy plant, it would become a dairy product religiously. Some kosher supervision agencies do permit such a product to be listed as "dairy equipment (DE)" rather than "dairy." The DE tells the consumer that it does not contain dairy but was made on dairy equipment (see allergy discussion below). With the DE listing, the consumer can use the product immediately after a meat meal, while a significant wait would be required to use a product with a dairy ingredient. In either case, the dishes would be switched from meat dishes to dairy dishes. A few products with no meat ingredients are made in a meat plant (e.g., a split pea soup), again they may be marked "meat equipment (ME)."

Kosher-observant Jews must wait a fixed time between meat and dairy consumption. Customs vary but generally the wait after meat before consuming dairy is much longer (3 to 6 h) than the wait from dairy to meat (0 to 1 h). However, when a hard cheese (defined as a cheese that has been aged for over 6 months) is eaten, the wait is the same as that for meat. Thus, most companies producing cheese for the kosher market age their cheese for less than 6 months.

If one wants to make the product truly pareve, the plant can usually be made pareve by the process of equipment kosherization (see below).

Passover

During this holiday which occurs in the spring, all products made from the five prohibited grains: wheat, rye, oats, barley, and spelt (Hebrew: chometz) cannot be used except for the specially supervised production of unleavened bread (Hebrew: "matzos"), that are prepared especially for the holiday. Special care is taken to assure that the matzos do not have any time to "rise." In addition, products derived from corn, rice, legumes, mustard seed, buckwheat, and some other plants (Hebrew: kitnyos) are prohibited. Thus, items like corn syrup, corn starch, etc. would be prohibited. Some rabbis, however, permit the oil from kitnyos materials. Some rabbis permit liquid kitnyos products such as corn syrup. The major source of sweeteners and starches generally used for Passover production of "sweet" items is either real sugar or potato-derived products. Some potato syrup is also used. Passover is a time of large family gatherings; however, because of the need for separate Passover dairy dishes, some kosher consumers may not use any dairy products. Overall, 40% of kosher sales for the traditional "kosher" companies occurs during the week of Passover.

Equipment koshering

There are three ways to make equipment kosher and/or to change its status. Which procedure is required depends on the equipment's prior production history. Note that after a plant (or a line) has been used to produce kosher pareve products, it can be switched to either kosher dairy or kosher meat without a special equipment kosherization step.

The simplest equipment kosherization occurs with equipment made of materials that can be koshered that have only been handled cold. These require a good caustic/soap cleaning. However, materials such as ceramics, rubber, earthenware, and porcelain cannot be koshered. If these materials are found in a processing plant, new materials may be required for production and switching between different status conditions will be difficult.

Most food processing equipment is usually operated at cooking temperatures, generally above 48.8°C, which is defined rabbinically as "cooking." However, the exact temperature for "cooking" depends on the rabbi, although an agreement by the major four American kosher certifying agencies has settled on 48.8°C as the temperature at which foods are cooked. To kosher these items which have been used with cooked product, the equipment must be thoroughly cleaned with caustic/soap. The equipment must be left idle for 24 h and then the equipment must be flooded with boiling water (defined between 87.7 and 100°C) in the presence of a kosher supervisor.

In the case of ovens or other equipment that uses "fire," kosherization involves heating the metal until it glows. Again, the rabbi will generally be present while this process is taking place.

The procedures that must be followed for equipment kosherization can be quite extensive, so that the fewer status conversions, the better. Careful formulating of products and good production planning can minimize the inconvenience.

Jewish cooking

Depending on what is being cooked, it may be necessary for the rabbi to "do" the cooking. In practical terms this is often accomplished by having a rabbi light the pilot light, which is then left on continuously.

In the case of cheese making, a similar concept usually requires the rabbi to add the coagulating agent into the vat. However, if the ingredients used during cheese making are all kosher, but a rabbi has not added the coagulant; then the whey derived from such cheese (as long as the curds and whey have not been heated above 48.8°C before the whey is drained off) would be considered kosher. Thus, there is much more kosher whey available than kosher cheese.

Halal dietary laws

The halal dietary laws deal with the following four issues, all except one are in the animal kingdom.

1. Prohibited and permitted animals
2. Method of slaughtering
3. Prohibition of blood
4. Prohibition of intoxicants

The Islamic dietary laws are derived from the Quran, a revealed book; the Hadith, the traditions of Prophet Muhammad; and through extrapolation of and deduction from the Quran and the Hadith, by Muslim jurists.

The Quran (V:3) states:

> "Forbidden unto you (for food) are: carrion and blood and swine flesh, and that which has been dedicated unto any other than Allah, and the strangled, and the dead through beating, and the dead through falling from a height, and that has been killed by the goring of horns, and devoured of wild beasts saving that which you make lawful and that which has been immolated to idols. And that you swear by the divining arrows. This is an abomination. . . . . ."

The Quran (II-172) also states:

> "O you who believe! Eat of the good things wherewith We have provided you, and render thanks to Allah, if it is He whom you worship."

There are 11 generally accepted principles pertaining to halal (permitted) and haram (prohibited) in Islam for providing guidance to Muslims in their customary practices:[4]

1. The basic principle is that all things created by Allah are permitted, with a few exceptions that are prohibited. Those exceptions include, pork, blood, meat of animals that died of causes other than proper slaughtering, food that has been dedicated or immolated to someone other than Allah, alcohol, intoxicants, etc.
2. To make lawful and unlawful is the right of Allah alone. No human being, no matter how pious or powerful, may take it into his hands to change it.
3. Prohibiting what is permitted and permitting what is prohibited is similar to ascribing partners to Allah. This is a sin of the highest degree that makes one fall out of the sphere of Islam.
4. The basic reasons for the prohibition of things are due to impurity and harmfulness. A Muslim is not supposed to question, exactly why or how something is unclean or harmful in what Allah has prohibited. There might be obvious reasons and there might be obscure reasons. To a person of scientific mind, some of the obvious reasons could be as follows:
   - Carrion and dead animals are unfit for human consumption because the decaying process leads to the formation of chemicals harmful to humans.[5]
   - Blood that is drained from an animal contains harmful bacteria, products of metabolism, and toxins.[6]
   - Swine serves as a vector for pathogenic worms to enter the human body. Infections by *Trichinella spiralis* and *Traenia solium* are not uncommon.[7]
   - Fatty acid composition of pork fat has been mentioned as incompatible with human fat and biochemical systems.[7]
   - Intoxicants are considered harmful for the nervous system, affecting the senses and human judgment, leading to social and family problems and in many cases even death.[4,5]
   - Immolating food to someone other than Allah may imply that there is somebody as important as Allah, that there could be two bosses. This would be against the first tenet of Islam: "THERE IS BUT ONE GOD."[8]

   These reasons and explanations, and many more like these, may be acceptable as sound for argumentative purposes, but the underlying principle behind the prohibitions remains the Divine order: "FORBIDDEN UNTO YOU ARE. . . ."

5. What is permitted is sufficient and what is prohibited is then superfluous. Allah prohibited only things that are unnecessary or dispensable while providing better alternatives. People can survive and live better without consuming unhealthful carrion, unhealthful pork, unhealthful blood, and the root of most vices, alcohol. Whatever is conducive to the prohibited is in itself prohibited. If something is prohibited, anything leading to it is also prohibited.
6. Falsely representing unlawful as lawful is prohibited. It is unlawful to make flimsy excuses to consume something which is prohibited, such as drinking alcohol for supposedly medical reasons.
7. Good intentions do not make the unlawful acceptable. Whenever any permissible action of the believer is accompanied by a good intention, his action becomes an act of worship. In the case of haram, it remains haram, no matter how good the intention or how honorable the purpose may be. Islam does not endorse employing a haram means to achieve a praiseworthy end. Islam indeed insists that not only the goal be honorable, but also the means chosen to achieve it be lawful and proper. Islamic laws demand that the right should be secured through just means only.

8. Doubtful things should be avoided. There is a gray area between clearly lawful and clearly unlawful. This is the area of "what is doubtful." Islam considers it an act of piety for the Muslims to avoid doubtful things, in order for them to stay clear of unlawful. Prophet Muhammad said:

   > "The halal is clear and the haram is clear. Between the two there are doubtful matters concerning which people do not know whether they are halal or haram. One who avoids them in order to safeguard his religion and his honor is safe, while if someone engages in a part of them, he may be doing something haram . . ."[4]

9. Unlawful things are prohibited to everyone alike. Islamic laws are universally applicable to all races, creeds, and sexes. There is no favored treatment of privileged class. Actually, in Islam, there are no privileged classes; hence, the question of preferential treatment does not arise. This principle applies not only among Muslims, but between Muslims and non-Muslims, as well.
10. Necessity dictates exceptions. The range of prohibited things in Islam is quite limited, but emphasis on observing the prohibitions is very strong. At the same time, Islam is not oblivious to the exigencies of life, to their magnitude, or to human weakness and capacity to face them. A Muslim is permitted, under the compulsion of necessity, to eat a prohibited food in quantities sufficient to remove the necessity and thereby survive.

Prohibited and permitted animals

Meat of pigs, boars, and swine is strictly prohibited, and so are the carnivorous animals like lions, tigers, cheetahs, dogs, cats, and the like; and birds of prey like eagles, falcons, osprey, kites, vultures, and the like.

Meat of domesticated animals like ruminants with split hoof, e.g., cattle, sheep, goat, lamb, is allowed for food, so is meat from camels and buffaloes. Also permitted is meat from birds that do not use their claws to hold down food, such as chickens, turkeys, ducks, geese, pigeons, doves, partridges, quails, sparrows, emus, ostriches, and the like. Meat of some of the animals and birds is permitted only under special circumstances or with certain conditions. Horsemeat may be allowed to be consumed under some distressing conditions, the discussion of which is beyond the scope of this chapter. The animals fed unclean or filthy feed, that formulated with sewage or protein from tankage, must be quarantined and placed on clean feed for a period of 40 days before slaughter in order to cleanse their systems.

Foods from the sea, namely fish and seafood, is the most controversial among various denominations of Muslims. Certain groups accept only fish with scales as halal, while others consider everything that lives in water all the time or some of the time as halal. Consequently, prawns, lobsters, crabs, and clams are halal but may be detested (Makrooh) by some and hence not consumed.

There is no clear status of insects established in Islam except that the locust is specifically mentioned as halal. Among the byproducts from insects, use of honey was very highly recommended by Prophet Muhammad. Other products like royal jelly, wax, shellac, and carmine are acceptable to be used without restrictions by most, however, some may consider shellac and carmine Makrooh or offensive to their psyche.

Eggs and milk from permitted animals are also permitted for Muslim consumption. Milk from cows, goats, sheep, and buffaloes is halal. Unlike kosher, there is no restriction on mixing meat and milk.

Prohibition of blood

According to the Quranic verses, blood that pours forth is prohibited to be consumed. It includes blood of permitted and non-permitted animals alike. Liquid blood is generally not offered for sale or consumed even by non-Muslims, but products made with and from blood are available. There is general agreement among Muslim scholars that anything made from blood is unacceptable. Products such as blood sausage and ingredients like blood albumin are either haram or questionable at best, and are avoided.

Proper slaughtering of permitted animals

There are special requirements for slaughtering the animal:

- An animal must be of a halal species
- It must be slaughtered by an adult and sane Muslim
- The name of Allah must be pronounced at the time of slaughter
- Slaughter must be done by cutting the throat in a manner that induces rapid and complete bleeding, resulting in the quickest death; the generally accepted method is to cut at least three of the four passages, i.e., carotids, jugulars, trachea, and esophagus

The meat of animals thus slaughtered is called zabiha or dhabiha meat.

Islam places great emphasis on gentle and humane treatment of animals, especially before and during slaughter. Some of the conditions include giving the animal proper rest and water, avoiding conditions that create stress, not sharpening the knife in front of the animals, using a very sharp knife to slit the throat, etc. After the blood is allowed to drain completely from the animal and the animal has become lifeless, only then the dismemberment, cutting off horns, ears, legs, etc. may commence. Unlike kosher, soaking and salting of the carcass is not required for halal. Hence halal meat is treated no differently than commercial meat. Animal-derived food ingredients like emulsifiers, tallow, and enzymes must be made from animals slaughtered by a Muslim to be halal.

Hunting of permitted wild animals like deer and birds like doves, pheasants, and quails is permitted for the purpose of eating but not merely for deriving pleasure out of killing an animal. Hunting during the pilgrimage to Makkah and within the defined boundaries of the holy city of Makkah is strictly prohibited. Hunting by any means, and use of tools like guns, arrows, spears, or trapping is permitted. Trained dogs may also be used for catching or retrieving the hunt. The name of Allah may be pronounced at the time of ejecting the tool rather than catching of the hunt. The animal has to be bled by slitting the throat as soon as the hunt is caught. If the blessing is made at the time of pulling the trigger or shooting an arrow and the hunted animal dies before the hunter reaches it, it would still be halal as long as slaughter is performed and some blood comes out. Fish and seafood may be hunted or caught by any reasonable means available as long as it is done humanely.

The requirements of proper slaughtering and bleeding are applicable to land animals and birds. Fish and other creatures that live in water need not be ritually slaughtered. Similarly there is no special method of killing the locust.

The meat of the animals that die of natural causes, diseases, from being gored by other animals, by being strangled, by falling from a height, through beating, or killed by wild beasts, is unlawful to be eaten unless one saves such animals by slaughtering before they actually become lifeless. Fish that dies of itself, if floating on water or is lying out of water is still halal as long as it does not show any signs of decay or deterioration.

An animal must not be slaughtered in dedication to someone other than Allah or immolated to anybody other than Allah under any circumstances. This is a major sin.

Prohibition of alcohol and intoxicants

Consumption of alcoholic drinks and other intoxicants is prohibited according to the Quran (V:90–91), as follows:

> "O you who believe! Strong drinks and games of chance, and idols and divining arrows are only an infamy of Satan's handiwork. Leave it aside in order that you may prosper. Only would Satan sow hatred and strife among you, by alcohol, and games of chance, and turn you aside from the remembrance of Allah, and from prayer: Will you not, therefore, abstain from them?"

Arabic term used in the Quran is khemr, which means that which has been fermented and implies not only to alcoholic beverages like wine, beer, whiskey, brandy, but to all things that intoxicate or affect one's thought process. Although there is no allowance for added alcohol in any beverage like soft drinks, a small amount of alcohol contributed from food ingredients may by considered an impurity and hence ignored. Synthetic or grain alcohol may be used in food processing for extraction, precipitation, dissolving and other reasons, as long as the amount of alcohol remaining in the final product is very small, generally below 0.5%. However, each importing country may have their own guidelines which must be understood by the exporters and strictly adhered.

Halal cooking, food processing, and sanitation considerations

There are no restrictions about cooking in Islam, as long as the kitchen is free from haram foods and ingredients. There is no need to keep two sets of utensils, one for meat and the other for dairy as in kosher. Alcohol may not be used even in cooking.

In food companies, haram materials should be kept segregated from halal materials. The equipment used for non-halal products has to be thoroughly cleansed using proper techniques of acids, bases, detergents, and hot water. As a general rule, kosher clean-up procedures would be adequate for halal too. If the equipment is used for haram products, it must be properly cleaned, sometimes by using an abrasive material, and then be blessed by an Imam or Mullah by rinsing it with hot water seven times.

Dealing with kosher and halal supervision agencies

Kosher or halal supervision is taken on by a company in order to expand its market opportunities. It is a business investment which, like any other investment, should be examined critically. In the era of total quality management, just-in-time production, strategic suppliers, etc. it is appropriate for companies to look carefully at how they handle their kosher and halal supervision needs.

Price alone may not be the best criterion for selecting a supervision agency. The agency's name recognition may also not be the most important company consideration. Other important considerations should include: (1) how responsive is the agency to the company, both in terms of paperwork handling and in terms of providing rabbis or Muslim inspectors at the plants as needed; (2) how willing are they to work with the company on problem solving; (3) how willing are they to explain their kosher or halal standards and their fee structure; (4) is the "personal" chemistry right, i.e., are you comfortable with them; and finally (5) what are their religious standards, i.e., do they meet the company's needs in the marketplace.

One of the hardest issues for the food industry to deal with in day-to-day kosher activities is the existence of so many different kosher supervision agencies. Unfortunately, though fewer agencies exist, halal also has various agencies with different standards. How does this impact the food companies? How do the Jewish kosher or Muslim halal consumers perceive these different groups? How do groups beyond the immediate community feel about the different agencies? Because there has not been a central authority for many years in either religion, different rabbis and imans/mullals follow different traditions with respect to their dietary standards. Some authorities tend to follow the more lenient standards, while others follow more stringent standards. Given the availability of choices, the trend in the mainstream kosher community today is toward a more stringent standard. The Muslim community also seems to be moving toward tighter standards.

One can generally divide the kosher supervision agencies into three broad categories. First there are the large organizations that dominate the supervision of larger food companies, i.e., the OU, the OK, the Star-K, and the Kof-K. All four of them are nationwide and "mainstream." Two of these, the OU and the Star-K are communal organizations, i.e., they are part of a larger community religious organization. This provides them with a wide base of support, but also means the organizations are potentially subject to the other priorities and needs of the organization. On the other hand, the Kof-K and the OK are private companies. Their only function is to provide kosher supervision. In addition to these national companies, there are smaller private organizations and many local community organizations that provide equivalent religious standards of supervision. As such, products accepted by any of these mainstream organizations will be accepted by all other similar organizations. The local organizations may have a bigger stake in the local community. They may be more accessible and easier to work with, although often having less technical expertise, they may be backed up by one of the national organizations. For a company marketing nationally, a limitation may be whether the consumer elsewhere in the U.S. knows and recognizes their kosher symbol. With the advent of KASHRUS magazine, and its yearly review of symbols, this has become somewhat less of a problem. (KASHRUS magazine does not try to "evaluate" the standards of the various kosher supervision agencies, but simply "reports" of their existence. It is the responsibility of the local congregational rabbi to inform his congregation of his standards. If he does not know enough about the "far-away" organization, he may be uncomfortable recommending it.)

The second category of kosher supervision includes individual rabbis, generally associated with the "Hassidic" communities. These are often affiliated with the ultra-orthodox communities of Williamsburg and Borough Park in Brooklyn, Monsey, N.Y., and Lakewood, N.J. There are special food brands that cater to their needs. Many more products used in these communities require continuous rabbinical supervision rather than the occasional supervision used by the mainstream organizations. The symbols of the kosher supervisory agencies representing these consumers are not as widely recognized as those of the major mainstream agencies in the kosher world beyond these communities. The rabbis will often do special supervisions of products using a facility that is normally under mainstream supervision, often without any changes, but sometimes with special needs for their custom production.

The third level are individual rabbis who are more "lenient" than the mainstream standard. Many of these rabbis are Orthodox; some may be Conservative. Their standards are based on their interpretation of the kosher laws. The more lenient such a rabbi, the more the food processor cuts out the "mainstream" and stricter markets, but that is a retail marketing decision the company needs to make for itself.

The Muslim community has only one mainstream agency at this time, the Islamic Food and Nutrition Council. Other groups are entering the field, but their standards are not as well defined.

However, ingredient companies should try to use a "mainstream" kosher or halal supervision agency. To sell ingredients to most kosher food-producing companies will require such supervision. The ability to sell to as many customers as possible requires a broadly acceptable standard. Unless an ingredient is acceptable to the mainstream, it is almost impossible to gain the benefit of having a kosher ingredient. In a few circumstances, if the company makes a product that would not be acceptable to the mainstream kosher supervision agencies no matter what the company does, then the company might as well use one of the more "lenient" kosher supervision agencies willing to recognize that ingredient.

In the future, companies will have to pay attention to halal standards. In many cases, a few changes will permit kosher products to also serve the halal community, i.e., the true absence of animal products (see below for a few kosher exceptions) and care to assure that any residual alcohol in products is below 0.1%. Again, a standard acceptable in all or most Muslim countries is desirable.

When looking for a religious supervision agency, one must determine the company's priorities and attempt to find an agency that is compatible with these requirements. Like any purchasing decision, time spent in qualifying the vendor before purchasing is usually rewarded.

With respect to interchangeability between kosher supervision agencies, a system of certification letters is used to provide information from the certifying rabbi to others about the products he has approved. The supervising rabbi certifies that a particular plant produces kosher products, or that only products with certain labels or certain codes are kosher under his supervision. Such letters should be renewed every year and should be dated with both a starting and ending date. These letters are the mainstay of how companies establish the kosher status of ingredients as ingredients move in commerce. Consumers may also ask to see such letters. Obviously a kosher supervision agency will only "accept" letters from agencies they consider acceptable.

In addition, the kosher or halal symbol of the certifying agency or individual doing the certification may appear on the packaging. (In some industrial situations, where kosher and non-kosher products are similar, some sort of color coding of products may also be used.) Most of these symbols are "trademarks" that are duly registered. However, in a few cases, the trademark is not registered and more than one rabbi has been known to use the same kosher symbol.

With respect to kosher and halal markings on products, two issues need to be highlighted:

1. It is the responsibility of the food company to show its labels to its certifying agency prior to printing labels to ensure the label is marked correctly. This responsibility includes both the agency symbol and the documentation establishing its kosher status, e.g., dairy or pareve for most dairy plant items. Many agencies currently do not require that "pareve" be marked on products; others do not use the "dairy" marking. The kosher supervision agencies, the food companies, and the consumer would be better served if all kosher products had their status marked. In addition to

providing the proper information, it would challenge everyone to pay more attention to properly marking products, avoiding the many recalls/announcements of mismarked products.

2. The labels for private label products with specific agency symbols on their labels cannot be moved easily between plants. This is why some companies, both private label and others, use the generic "K." Thus, if the kosher supervision agency changes, the label can still be used. The sophisticated kosher consumer, however, is more and more uncomfortable with this symbol and questions will be asked. By paying for a "good" symbol and then only using the "K," a company dilutes the value of its investment in kosher certification. In particular, if a company uses the "K," the customer service and sales departments, and those people representing the company at trade shows need to know who the certifying rabbi is.

Thus far the halal community has not gone to a generic halal marking in this country, although this does seem to be used in some other countries.

Gelatin

Important in many food products, gelatin is probably the most controversial of all modern kosher and halal ingredients. Gelatin can be derived from pork skin, beef bones, or beef skin. In recent years, some fish gelatins have also appeared. The first author is currently involved in research in this area. As a food ingredient, fish gelatin has many similarities to beef and pork gelatin, i.e., it can have a similar range of bloom strengths and viscosities. However, depending on the species from which the fish skins were obtained, its melting point can vary over a much wider range of melting points than beef or pork gelatin. This may offer some unique opportunities to the food industry, especially for ice cream, yogurt, dessert gels, confections, and imitation margarine. These gelatins would be fully kosher and halal, and acceptable to almost all of the mainstream religious supervision organizations.

Currently available gelatins, even if called "kosher," are not acceptable to the mainstream kosher supervision organizations. Many are, in fact, totally unacceptable to halal consumers because they may be pork gelatin-based. However, a recent production of gelatin from the hides of kosher slaughtered cattle has been available in limited supply at great expense, and this has been accepted by the mainstream and even some of the stricter kosher standards.

Among the lenient kosher supervision agencies, one finds a wide range of attitudes toward gelatin. The most liberal view holds that gelatin, being made from bones and skin, is not being made from a food (flesh). Further, the process used to make the product goes through a stage where the product is so "unfit" that it is not edible by man nor dog and as such becomes a new entity. Rabbis holding this view even accept pork gelatin. Most gelatin desserts with a generic "K" follow this ruling.

Other rabbis only permit gelatin from beef bones and hides and not pork. Other rabbis will only accept "India dry bones" as a source of beef gelatin. These bones, found naturally in India (because of the Hindu custom of not using cattle) are aged for over a year and are "dry as wood"; additional religious laws exist for permitting these materials. However, to repeat, none of these products are accepted by the "mainstream" kosher or halal supervisions and thus products with these gelatins are not accepted by a significant part of the kosher and halal community.

Biotechnology

Rabbis, imams, and mullahs currently accept products made by simple genetic engineering, e.g., chymosin (rennin) was accepted by the rabbis about a half year before it was accepted by the FDA. The production conditions in the fermentors must still be kosher or halal, i.e., the ingredients and the fermentor and any subsequent processing must use kosher or halal equipment and ingredients of the appropriate status. A product produced in a dairy medium would be dairy. We believe that the rabbis may soon approve porcine lipase made through biotechnology, if all the other conditions are kosher, but the Muslim community is still considering this issue and a final ruling has not been established. Any product produced by cattle by excretion in the milk would be dairy. The religious leaders of both communities have not yet determined the status of more complex genetic manipulations.

Federal and state regulations

Making a claim of kosher on a product is a legal claim. The Code of U. S. Federal Regulations (21CFR101.29) has a paragraph indicating that such a claim must be appropriate and approximately 20 states, some counties, and some cities have laws specifically regulating the claim of "kosher." Many of these laws refer to "Orthodox Hebrew Practice" or some variant of this term, and their legality in the 1990s is subject to further court interpretation.[9,10]

New York State probably has the most extensive set of kosher laws, including a requirement to register kosher products with the Kosher Enforcement Bureau of the Department of Agriculture and Markets (55 Hanson Pl., Brooklyn, NY 11217). However, the laws in New Jersey—having been written after the state's original laws were declared unconstitutional by the state supreme court—probably have the clearest focus and, it is hoped, no constitutional issue. They focus specifically on "consumer right to know issues" and "truth in labeling." They avoid having the state of New Jersey define kosher. Rather the rabbis providing supervision declare the information that consumers need to make an informed decision. We hope that a similar approach will be adopted by the other states, particularly New York State, and that all of the states will extend the same protection to food products produced with halal certification.

Kosher and allergies

Although it is helpful for many consumers to use the kosher markings as a guideline for determining whether products might meet their special needs, there are also limitations that the particularly sensitive consumer needs to be aware of.

With respect to all kosher products, two important limitations need to be recognized:

1. A process of equipment kosherization is used to convert equipment from one status to another. This is a well-defined religious procedure, but may not lead to 100% removal of previous materials run on the equipment.
2. Kosher law does permit certain ex-post-facto (after the fact) errors to be negated. Thus, trace amounts (less than 1/60 by volume under very specific conditions) can be nullified. Many kosher supervision agencies in deference to the companies desire to minimize negative publicity do NOT announce when they have used this procedure to make a product acceptable.

Products that one might expect to be made in a dairy plant, e.g., pareve substitutes for dairy products and some other liquids like teas and fruit juices may be produced in plants that have been kosherized, but may not meet a very critical allergy standard. Another product that can be problematic is chocolate; many plants make both milk chocolate and pareve chocolate. Getting every last trace of dairy out of the pareve chocolate can be difficult.

Dairy and meat equipment are also problematic. The product was produced on a dairy or meat line, without any equipment kosherization. However, there are no intentionally added dairy or meat ingredients. The product is considered pareve with some use restrictions in a kosher home.

In a few instances where pareve or dairy products contain small amounts of fish (e.g., anchovies in Worcestershire sauce), this ingredient MAY be marked as part of the kosher supervision symbol. Many certifications will not specifically mark this.

For Passover, there is some dispute about "derivatives" of both chometz and kitnyos materials and a few rabbis permit items like corn syrup, soybean oil, peanut oil, and similarly derived materials from these extensions. In general, the "proteinaceous" part of these materials are not used. Thus, people with allergies to these items could purchase these special Passover products from supervision agencies that do NOT permit "kitnyos" derivatives. With respect to "equipment kosherization": supervising rabbis tend to be very strict about the cleanup of the prohibited grains (wheat, rye, oats, barley, and spelt) so these should come closest to meeting potential allergy concerns, but may not be as critical with respect to the extended prohibition.

Consumers should not assume that kosher markings ensure the absence of trace amounts of the ingredient to which they are allergic. How thorough can the dairy line be blocked out? The cleaning probably should go beyond any interlock that exists to lock out the incoming dairy proteins to assure that cross contamination does not occur. Currently what is acceptable for kosher may not meet the needs of allergic consumers. Is the dairy powder dust in the air sufficient to cause problems? A company might want to consider putting a special marking on kosher pareve chocolates produced on lines that also produce dairy products to indicate that these are religiously pareve, but not sufficiently devoid of dairy allergens for very allergic consumers. Furthermore, they may also want to consider checking the chocolate using one of the modern antibody or similar types of tests. For example, regular M&Ms are marked as containing "peanuts" in order to alert people who are very allergic to peanuts, even though the product does not contain peanuts, because common equipment (although cleaned between product runs) is used for both products.

Kosher poultry

We are assuming that the reader has read Chapters 3 and 4, which describe normal poultry processing. This section will simply highlight the major differences when birds are slaughtered for kosher or halal use.

Poultry from kosher slaughter, like regular poultry, may either be raised on contract—so that the slaughtering company controls the supply—or may be purchased on the open market, in which case the company does not control the stock prior to processing.

If the company is in control of the live animals, two issues with poultry are important. The first is the issue of injections. The animals may receive injections but they must be done in such a way as not to be classified rabbinically as a "puncture" that would prevent the animal from surviving for a year. (These standards are rabbinical and should not be thought of in terms of modern scientific discussions.) Of particular concern are injections to the neck region, such as may be used for hormone treatments. Thus, although not religiously required, many of the kosher producers tend not to use hormones or antibiotics requiring shots.

A second issue with the live animal is the feed. Interestingly, the two issues that are of concern are feeds that contain milk and meat, but not those with non-kosher ingredients, and those concerning "baked" chometz. In the latter case, the issue is the processing of poultry during the four (at most) intermediate days of Passover.

Chickens, approximately 60 to 63 days old (slightly older than non-kosher poultry) are used. The birds are kept in the crates until such time as they are removed by a helper at the kill point. Each "shochet" has a helper who holds and positions the bird for the slaughterman. The shochet uses a very sharp knife called a "chalef" to sever the windpipe, the jugulars, and the carotids. The shochet then inspects the birds to check that the cut was made properly. Another helper than hangs the birds. Extra shochtim are responsible for checking the knives and resharpening them. If any of the knives are found to be "nicked," then all the birds slaughtered since the previous knife inspection are declared non-kosher. In order to maintain the normal line speed, less than that for many traditional plants, a team of seven shochets are used to do the slaughtering. The shochets work for one hour and then are off for one hour. Prior to "shechting," the shochet will say a prayer asking for forgiveness. In order to be a shochet, the man must be a pious, observant Jew and must pass a test on both his religious knowledge about the requirements for shechting and on his practical ability to carry out the job correctly. The work of the individual shochets is not monitored, so that there is no pressure on the shochet to keep a specific pace.

The blood from the birds is collected and sawdust is added to return it to the "dust of the earth." Thus, blood collection for rendering is not possible.

The defeathering process uses cold water. The normal procedure of using hot water would "cook" the birds, and this is not permitted. The cold water rinse works best if the water is very cold, so ice is often added. The cold water process makes feather removal more difficult and so the birds are specifically bred to give them a tougher skin so fewer tears will occur during picking. All of the picking machines operate with cold water. Generally many more pickers are required for kosher defeathering than for normal processing. Following defeathering, the birds go through a singer, and then generally are carefully checked for pin feathers. In many cases, more people are assigned to this station that would occur with normal processing.

The main processing sector operates similar to that for non-kosher. However, the level of chlorine used in the water to clean these machines has to be determined by the rabbis and is lower than in the non-kosher plant.

At some point, either immediately before or after USDA inspection, the bodek (internal organ inspectors) will thoroughly examine all parts of the bird. Viscera are inspected to guarantee that each bird was healthy and kosher. The bodek looks particularly closely at the intestines for a particular growth spot that, if present, requires further examination. This is generally done off-line by another bodek. This potential defect is particularly prevalent in the younger birds, i.e., rock cornish hens.

Birds that would ordinarily receive a green tag (requires off-line processing) in a non-kosher plant are simply condemned because such facilities for reprocessing would present a potential kashruth problem as the level of chlorine required by the USDA for reprocessing would be higher than that permitted by the rabbis.

After passing through the inspection cycle, extra cuts are made on the neck and the wing-tips are removed. The knife cuts on the chickens' necks are made with a special three-bladed knife and the birds continue on the line for further blood drainage.

The next step is putting on the "plumbas" (the metal seals/tags) used to indicate that a bird is kosher. These are normally placed on the wings, although with more birds coming as cut-ups, there have been efforts made to have these tags placed on other parts also. One must remember to remove these tags before battering and breading at home or at least check before eating kosher poultry. These tags can be removed and occasionally fall off, making a bird non-kosher. We are also concerned that these tags do not fall into improper hands, as they would make a non-kosher bird appear to be kosher. Companies making these tags need to be reminded that it is inappropriate to give kosher or halal tags as samples. There is still a need for plumbas that would go on more easily, stay on, and be more difficult to remove.

Those birds that have been deemed to be "traife" (non-kosher) but those that have successfully passed the USDA inspection are removed to a special working table, where black plumbas are placed on their wings. These birds are non-kosher and can be sold locally. Control of plumbas, both kosher and non-kosher, are part of the way that the rabbis can assure themselves that birds are properly segregated.

The birds then go into the prechiller for no less than 30 min. The birds then go to the salting station where kosher salt is applied to both the inside and outside of the birds. These birds are then hung on special racks that permit draining of each bird. The time when the rack is fully loaded is marked and one hour must pass before the birds are permitted to be removed. On removal the birds are given a shower followed by three washes with cold, running water before entering the chill tank.

The edible offal, except for liver, is handled the same way, but using smaller equipment. Livers are put into special bins that have good drainage holes following chilling. Livers are then packed into special bags prior to being added back into the bird. The bags are marked to indicate the necessary koshering instructions to the customer. Liver has to be broiled, using special utensils reserved for this purpose, before it is permitted to be used as kosher. Because of its high blood content, liver cannot be koshered using salt.

Following all of this processing, the birds may also be packed in ice in crates that are then also sealed with plumbas. Products for the supermarket are generally put into fully sealed packaging systems, so that the meat will not be touched by "non-kosher" meat, even if placed next to non-kosher meat in a supermarket case.

In order to run a kosher plant, additional kosher supervisers are needed to oversee the complexities of maintaining a kosher plant. In addition, because plants may be isolated, provisions for housing, feeding (strictly kosher), and providing for the religious personnel's religious needs (e.g., prayers three times a day) must be met, generally on site. Scheduling of the plant must be done so as to permit religious personnel to return to their home city in time for the Sabbath (every Friday) and for other religious holidays. Obviously kosher operations will not occur on Jewish religious holidays.

Halal poultry

For commercially processed poultry, the birds are generally acquired from poultry farms that raise the chickens specifically for that purpose or hens and roosters may be acquired from the poultry farms that raise chickens for eggs, when their egg production decreases below a certain level. Chickens of any size, age, and gender may be used for Halal production depending on the end use. Hens and roosters are used for high temperature cooking, such as canning, retorting, or even dehydrating for the purpose of incorporation into soups and other dry blends. In the Middle East, smaller and younger birds than the ones available in the U.S. supermarkets, are preferred because they are used for roasting on the rotisseries. The preferred feed for poultry does not use any animal byproducts or other scrap materials, which is a common practice in the west. Some halal slaughterhouses do use an integrated approach, e.g., where they raise their own chickens on clean feed, but most halal processors do not exert any influence over the feed issue. The Muslim retailers then prefer free range farmed chickens by Amish people, because animal byproducts are not fed to the birds. However, these birds are quite large and may be best used for whole cut-up chicken or for individual parts. Use of hormones in chickens for egg or meat production is discouraged.

Methods of slaughtering

The traditional method of slaughtering in Islam is to slit the throat, cutting at least the carotid arteries, jugular veins, and esophagus, without severing the head. It must be done by a Muslim of sound mind and health while pronouncing the name of God on each bird. In order to carry out the slaughtering process properly, a team of three to seven Muslim slaughtermen may be required at each line for a full day's operation; however, for shorter runs of halal slaughter, we have seen people use only three persons per line for hand slaughter. A common pronouncement is Bismillahi Allahu Akbar, which means, in the name of Allah, Allah is great. Slaughtering by hand is still preferred by all Muslims and quite widely followed in the Muslim countries and other countries where Muslims control the slaughterhouses. Mechanical or machine slaughter of birds, which was initiated in the western countries, is gaining acceptance among Muslims. Almost all countries that import chicken do accept machine killed birds. The method of slaughter by machine devised by the Islamic Food and Nutrition Council of America and approved or accepted by Muslim countries varies from the machine slaughter method used in the industry in several ways as follows:

- A Muslim while pronouncing the name of God switches on the machine.
- One Muslim slaughterman positions himself after the machine to make a cut on the neck, if the machine misses a bird or if the cut is not adequate for proper bleeding. In commercial poultry processing, generally the machine does not properly cut 5 to 10% of the birds. A Muslim then must cut these birds.
- Height of the blade(s) must be adjusted to make a cut on the neck, right below the head, and not across the head or on the chest. The birds should be reasonably close in size to accomplish this requirement.
- A rotary knife should be able to cut at least three of the passages in the neck. It is often difficult to accomplish this requirement with a single knife; hence a double knife set up may be required under such circumstances.
- Any birds that are not properly cut may be tagged by the Muslim slaughterman/inspector, to be used as non-halal.
- Two slaughtermen may be required to accomplish the above requirements, depending on the line speed and efficiency of the operation.
- The machine must be stopped during the breaks and must be restarted using the above procedure.

The birds must be completely lifeless before they enter the hot water bath. The conditions for defeathering, such as water temperature, chlorine level, etc. are the same for halal processing as for regular poultry processing. However, in the poultry processing plants where both halal and non-halal birds are processed, halal birds must be completely segregated during defeathering, chilling, eviscerating, processing, and storing. Containers with halal products should be stamped halal with proper codes and markings, by the authorized halal inspector. A halal certificate issued by the halal inspector in charge of the facility must accompany halal processed items when they are shipped to another facility for further processing. Further processing, like marinating, breading, and application of batters or rubs should also be done under the supervision of a qualified halal inspector, on thoroughly cleaned equipment. Non-meat ingredients, such as spices, seasonings, and breadings must also be halal approved.

Because there is no requirement to salt and soak the birds for halal, the meat is similar to the regular, commercial product. The quality of halal meat may be enhanced because of thorough and complete bleeding and due to the fact that the halal birds are normally calmer and less stressed.

Unlike kosher processing, halal inspectors are not trained to inspect and do not attempt to inspect the internal organs for diseases or any health concerns. This is considered the responsibility of the Department of Agriculture inspectors.

In different countries, metal or plastic tags are generally used around the necks or on the wings. This is becoming difficult, because more and more products are being sold as cut up parts. In the countries where the entire production is halal, tags are generally not used. In certain regions, it is preferable to leave the head on to be removed later by the butcher or the customer. In the countries where Chinese-style slaughtering is also done, the customer can easily differentiate halal birds from non-halal birds. In North America, where heads and necks are removed during processing and cutting up, a practice of leaving the neck attached to the whole chicken is developing, in order to identify halal from regular machine-killed birds. For operating a fully or partially halal plant, several other considerations regarding personnel are important. Provisions for daily prayers and special Friday prayers are highly recommended. It would be prudent to provide a place for offering the prayers in a separate clean room. Muslim workers should also be given time off for the religious holidays.

Summary

Religious and other cultural practices are important considerations in producing poultry for some markets. Although such markets are not common in most of the U.S., there are some areas in which they are significant and some countries in which these practices are law. In addition to the domestic market, companies involved in exporting need to be particularly concerned with the cultural laws and practices in their destination country.

References

1. Grunfeld, I., *The Jewish Dietary Laws*, Soncino Press, London, 1972, 11.
2. Regenstein, J. M., Health aspects of kosher foods, *Act. Rep. Min. Work Groups Sub-Work Groups Res. Dev. Assoc.*, 46(1), 77, 1994.
3. Hertz, J. H., *Pentateuch and Haftorahs*, Soncino Press, London, 1973.
4. Al-Quaradawi, Y., *The Lawful and the Prohibited in Islam*, The Holy Quran Publishing House, Beirut, Lebanon, 1984.
5. Awan, J. A., *Islamic food laws—I. Philosophy of the Prohibition of Unlawful Foods, Science and Technology in the Islamic World*, 1984.
6. Hussanini, M. M. and Sakr, A. H., *Islamic Dietary Laws and Practices*, Islamic Food and Nutrition Council of America, Bedford Park, IL, 1983.
7. Sakr, A. H., *Pork: Possible Reasons for its Prohibition*, Foundation for Islamic Knowledge, Lombard, IL, 1991.
8. Al-Quaderi, Syed J. M., Personal communication, Islamic Food and Nutrition Council of America, Chicago, IL, 1999.
9. Berman, M. A., Kosher fraud statutes and the establishment clause: are they kosher?, *Columbia J. Law and Social Prob.*, 26(1), 1, 1992.
10. Barghout V., Bureau of Kosher Meat and Food Control, U. S. App. LEXIS 27707, 1995.

Selected reading

Chaudry, M. M., Islamic Food Laws: Philosophical Basis and Practical Implications, *Food Technol.*, 46(10), 92, 1992

Chaudry, M. M. and Regenstein, J. M., Implications of Biotechnology and Genetic Engineering for Kosher and Halal Foods, *Trends Food Sci. Technol.*, 5, 165, 1994.

Ratzersdorfer, M., Regenstein, J. M., and Letson, L. M., 1988. Appendix 5: Poultry plant visits, in *A Shopping Guide for the Kosher Consumer*, J.M. Regenstein, C.E. Regenstein, and L.M. Letson (Eds.) for Mario Cuomo, Governor, State of New York.

Regenstein, J. M. and Regenstein, C. E., An Introduction to the Kosher (Dietary) Laws for Food Scientists and Food Processors, *Food Technol.*, 33(1), 89, 1979.

Regenstein, J. M. and Regenstein, C. E., The Kosher Dietary Laws and their Implementation in the Food Industry, *Food Technol.*, 42(6), 86, 1988.

chapter seventeen

Processing water and wastewater

William C. Merka

Contents

Introduction

The purpose of this chapter is not to provide the design and operation of wastewater pretreatment and treatment systems. There are numerous engineering firms and equipment firms that can design and construct wastewater treatment systems that will meet environmental discharge requirements. The purpose of this chapter is, however, to provide essential background information that helps processors significantly increase profits by using the wastewater stream as a source of information to determine the efficiency of processing and further processing plants.

0-8493-0120-3/01/$0.00+$.50

Wastewater analytical measurements

Biochemical oxygen demand (BOD)

Biochemical oxygen demand measures the amount of oxygen consumed by microbes as they digest organics in wastewater. This procedure requires five days to complete. Oxygen is poorly soluble in water so that only approximately 8 mg of oxygen will dissolve in 1 l of water. Wastewater from poultry processing plants contains 300 to 500 times more organic matter than there is available oxygen in the wastewater required for microbial digestion. Therefore, when wastewater enters a water course, microbes rapidly deplete the dissolved oxygen, and aquatic life that extracts oxygen from water dies.

Example calculation:

$$\text{Dissolved oxygen at Day 0} - \text{dissolved oxygen after five days digestion} = \text{BOD}$$

$$DO_0 - DO_5 = BOD$$

$$8 \text{ mg/l} - 5 \text{ mg/l} = 3 \text{ mg/l BOD}$$

$$DO_0 - DO_5 \times \text{dilution factor}$$

$$8 \text{ mg/l} - 5 \text{ mg/l} \times 1{:}500 = 1500 \text{ mg/l BOD}$$

This calculation shows that 1500 mg of oxygen is required by microbes to digest the organic matter in 1 l of the sample. One pound of BOD discharged represents about three pounds of product lost to the waste stream.

Chemical oyxgen demand (COD)

Chemical oxygen demand measures the amount of organic matter in wastewater as determined by conversion from the orange dichromate ion to the green chromium ion during high temperature acid digestion. This procedure requires only two hours, rather than five days for BOD. There is a high correlation of COD to BOD, and COD of poultry-processing wastewater is approximately twice that of BOD.

Total suspended solids (TSS)

Total suspended solids measures the concentration of particulate matter in wastewater and is determined by passing a measured volume of wastewater through a tared standard glass fiber filter. The filter is dried at 103°C and the difference between tared weight and dried weight is used to calculate TSS.

Example calculation:

$$\frac{\text{Dried filter} - \text{tared filter}}{\text{Volume in ml}} \times 1{,}000{,}000 = \text{TSS mg/l}$$

$$\frac{0.3000 \text{ g} - 0.2500 \text{ g}}{100 \text{ ml}} \times 1{,}000{,}000 = 500 \text{ mg/l TSS}$$

Total solids (TS)

Total solids measures the total amount of product, both organic and inorganic, lost to the waste stream. Total solids is determined by pouring a measured volume of wastewater into a tared crucible and drying it to dryness. The increase in weight is the total solids.

Example calculation:

$$\frac{\text{Wt of crucible and dried sample} - \text{dried crucible weight}}{\text{volume in ml}} \times 1{,}000{,}000 = \text{TS mg/l}$$

$$\frac{67.0770\text{ g} - 67.0000\text{ g}}{100\text{ ml}} \times 1{,}000{,}000 = 770\text{ mg/l TS}$$

Fixed solids (FS)

Fixed solids measures the amount of mineral matter in wastewater. Fixed solids is determined by ashing the crucible used to measure total solids. At 550°C, the organic matter burns and leaves only the mineral matter.

Example calculation:

$$\frac{\text{Ashed crucible} - \text{tared crucible}}{\text{Volume of wastewater in ml}} \times 1{,}000{,}000 = \text{FS mg/l}$$

$$\frac{67.0220\text{ g} - 67.000\text{ g}}{100\text{ ml}} \times 1{,}000{,}000 = 220\text{ mg/l FS}$$

Total volatile solids (TVS)

Total volatile solids determines the amount of organic matter in a wastewater sample. It is calculated by subtracting the fixed solids (FS) from the total solids (TS).

Example calculation:

$$\text{TS mg/l} - \text{FS mg/l} = \text{TVS mg/l}$$

$$770 - 220 = 550\text{ mg/l}$$

Fat, oil, and grease (FOG)

Fat, oil, and grease content is determined by extracting the FOG from wastewater with an organic solvent. The organic solvent containing the extracted FOG is separated from the wastewater and delivered into a tared beaker. The solvent FOG mixture is heated to evaporate the solvent so that only the FOG remains in the beaker.

Example calculation:

$$\frac{\text{Beaker containing residual FOG} - \text{tared beaker}}{\text{Volume of wastewater ml}} \times 1{,}000{,}000 = \text{FOG mg/l}$$

$$\frac{45.0250 - 45.000}{1000} \times 1{,}000{,}000 = 250\text{ mg/l FOG}$$

Total Kjeldahl nitrogen (TKN)

Total Kjeldahl nitrogen is determined by converting organic nitrogen in wastewater to ammonia through acid digestion and distillation. The concentration of nitrogen is determined by calculation of ammonia collected in the distillation process. Total Kjeldahl nitrogen is used for the design of biological wastewater treatment facilities. It can also be used to calculate product lost to the waste stream. One pound of TKN comes from 31 lb of meat.

Wastewater treatment

Prior to the Clean Water Act of 1972, little thought was given to either the cost of water received or the environmental effect of wastewater discharged by poultry processing plants. Plants discharged untreated wastewater into municipal sewers, or in some cases directly into streams. Some processors mistakenly felt that the nutrients added to the stream improved fishing. Such assumptions were disproven, as it was determined that the oxygen needed to degrade the discharged biological wastes actually reduced the survival of aquatic life. With enactment of the Clean Water Act of 1972, poultry processors could not continue these practices without suffering severe civil and criminal penalties. Municipalities were likewise required to improve the quality of wastewater discharged into streams. Generally, wastewater discharged into streams was required to have a BOD and a TSS concentration of less than 20 mg/l and a dissolved oxygen (DO) concentration of more than 4.0 mg/l. Poultry processors that discharge directly into streams were likewise required to meet these parameters.

Due to the more stringent discharge requirements, municipalities required that poultry processors reduce the organic concentration of the wastewater to that of domestic sewage (250 mg/l BOD, 200 mg/l TSS, 100 mg/l FOG, and pH in the range of 5 to 10) prior to discharge into a municipal sewer. To achieve these requirements, processors installed various configurations of secondary screening and physical/chemical pretreatment systems. Chemical flocculation-air flotation systems known as dissolved air flotation (DAF) units became a popular method of meeting municipal discharge requirements. Dissolved air flotation is a method of removing suspended material from wastewater. A flocculating agent is added to the water to flocculate or form aggregates of the suspended materials. High pressure air is then injected into the wastewater. The flocculated material adsorbs to the surface of the tiny bubbles and floats to the surface for separation from the water. Although effective in pretreating wastewater, the organics generated by the process became a major industry problem. The material putrefies rapidly due to the concentration of both air and bacteria in the float material. The float material is difficult to render and generally produces a low quality product. Another option available to processors is application of wastewater to spray fields so that vegetation would utilize nutrients and water discharged by the processing plant. Land area required to utilize this zero discharge method of treating wastewater is determined by hydraulic volume and quantities of nitrogen applied to the spray fields.

Figures 17.1, 17.2, and 17.3 show three general wastewater treatment schemes. At any one processing plant, there may be a hybridization of these three schemes. Initially, processors ignored the more stringent regulations until it was apparent that they would face large surcharges and fines in addition to criminal prosecution if they did not comply. To avoid prosecution, poultry processors contracted with engineering firms to construct wastewater treatment facilities to meet municipal discharge requirements, land application requirements, or stream discharge requirements. Average concentrations of contaminants in broiler processing wastewater are given in Table 17.1.

During the past 25 years, environmental engineering firms have become proficient in treating poultry processing wastewater so that processors can meet environmental discharge requirements. To ensure that these requirements are met, processors are required to sample their effluents periodically. Few processors, however, analyze the untreated wastewater discharged by either processing or further-processing facilities to determine the amount of product being lost in the wastewater. By defining those times and processes that discharge excessive water and/or product to the wastewater stream and by correcting and improving those inefficiencies, plant profits can be increased.

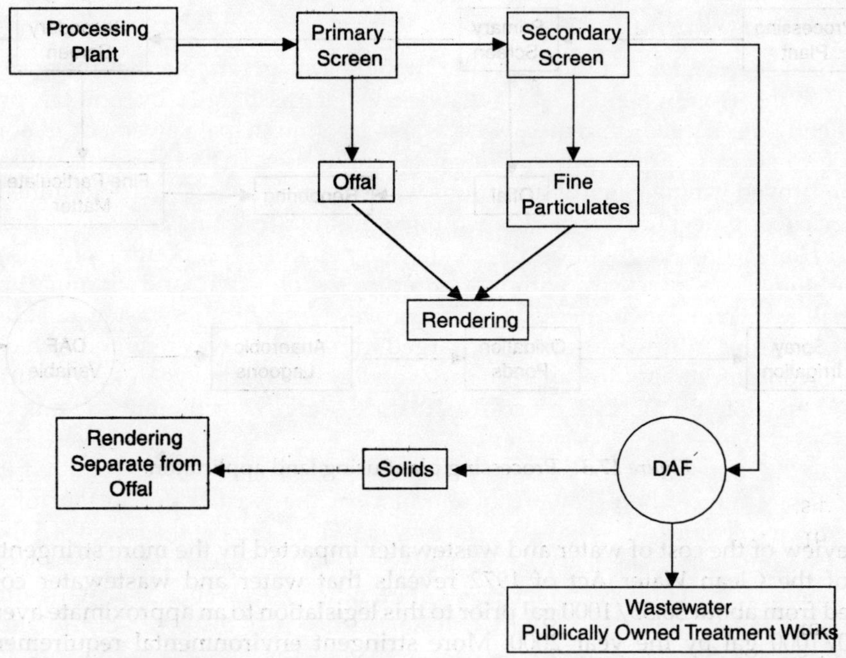

Figure 17.1 Processing plant discharging into a municipal system (POTW, publicly owned treatment works).

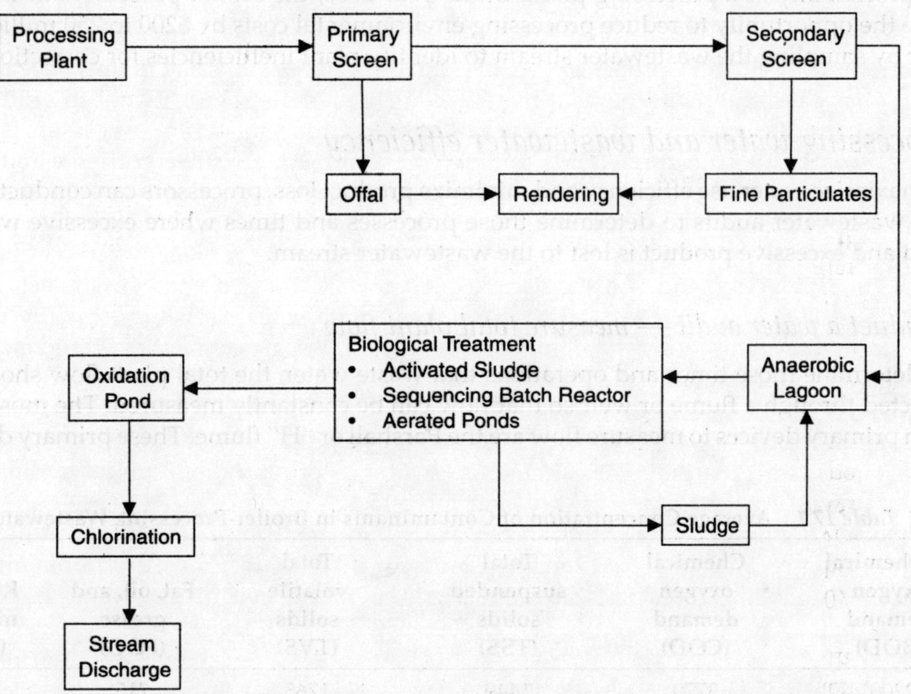

Figure 17.2 Processing plant discharging to a stream.

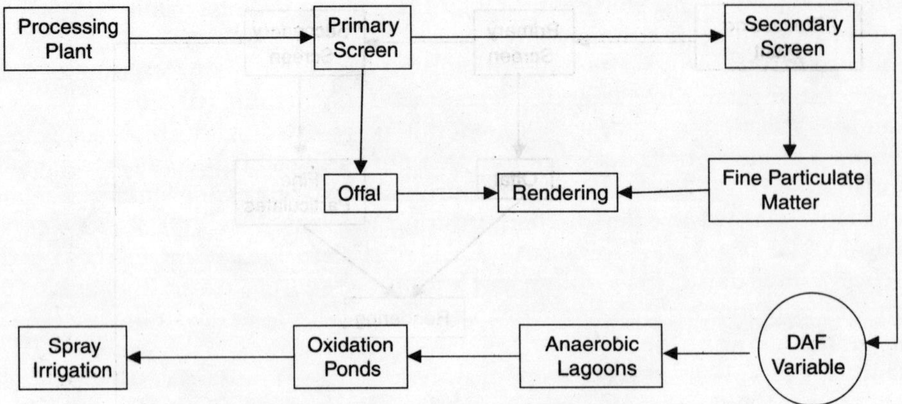

Figure 17.3 Processing plant using land application.

A review of the cost of water and wastewater impacted by the more stringent requirements of the Clean Water Act of 1972 reveals that water and wastewater costs have increased from about $0.33/1000 gal prior to this legislation to an approximate average cost of $5.00/1000 gal by the year 2000. More stringent environmental requirements have caused the cost of water and wastewater to increase rapidly during the last quarter century. At $5.00/1000 gal, each additional gallon per bird increases the cost of a processed bird by 0.5 cent. Therefore, attention to water and wastewater efficiency can significantly increase the profitability of a processing plant. In the year 2000, the broiler processors in the U.S. have the opportunity to reduce processing environmental costs by $200 to 250 million per year by sampling the wastewater stream to identify plant inefficiencies for correction.

Processing water and wastewater efficiency

To maximize water use efficiency and minimize product loss, processors can conduct water and wastewater audits to determine those processes and times where excessive water is used and excessive product is lost to the wastewater stream.

Conduct a water audit — measure total plant flow

To determine those times and operations that waste water, the total plant flow should be directed through a flume or weir so that flow can be constantly measured. The most common primary devices to measure flow are the Parshall or "H" flume. These primary devices

Table 17.1 **Average Concentration of Contaminants in Broiler-Processing Wastewater**

Biochemical oxygen demand (BOD)	Chemical oxygen demand (COD)	Total suspended solids (TSS)	Total volatile solids (TVS)	Fat, oil, and grease (FOG)	Total Kjeldahl nitrogen (TKN)
2200[a]	3770	1440	1765	715	130

[a] mg/h

Source: Adapted from Merka, W. C., *Broiler Ind.*, 52:11, 1989.

allow processors to accurately monitor plant discharge. Commercially available flow height recorders can be programmed to report wastewater flow through these devices in any sequence that the processor would find useful. The value of this measurement is not only to determine total wastewater discharge but also to measure variation in water use.

Example: Variation in water use was determined by dividing diurnal wastewater discharge from a broiler processing plant into first processing shift, second processing shift, and sanitation shift. Wastewater discharge during the two processing shifts was relatively constant. However, wastewater discharge during the sanitation shift would vary from 200,000 to 400,000 gal per day. This variation had no discernable pattern, i.e., need to dump and re-clean chiller, day of week, number of birds processed, etc. An additional 200,000 gal per shift for sanitation cost the company an additional $1000 per day, more than the labor cost for sanitation during the shift.

Example: Continuous measurement of water discharged by a processing plant determined that the average plant discharge during processing was approximately 1000 gal/min; however, the flow would vary from 800 to 1200 gal during any period of time during the processing shift. Efficient DAF pretreatment depends on constant volumes of flow so that flocculent chemical concentrations remain constant. A 50% variation of flow can cause problems in the steady state requirements for successful DAF pretreatment. Large volume users such as slaughter plants can reduce water costs significantly with modest water use reductions.

Example: A processor who processes 1, 250,000 birds per week and pays $5.00/1000 gal will reduce their annual cost by $312,000 by reducing water use by 1 gal per bird. Even companies using small amounts of water can become more profitable by studying water use patterns and developing methods of reducing water use. By conducting water discharge and water use studies, a company that used 10,000 to 12,000 gal of water per day to process and package shell eggs found that they were paying $13,000 per year for water that went on their one-half acre front lawn.

Monitor water use by various processes within the plant

Water meters can be installed at strategic places within the plant to determine the water use and variation in water use by processes and equipment within the plant. Installation of a $250 water meter which leads to a 10 gal/min reduction of water use which has the value of $5.00/1000 gal will pay for itself in about one week.

Example calculation:

$$10 \text{ gal/min @ } 0.5 \text{ cents/gal} = 5 \text{ cents/min}$$

$$5 \text{ cents/min} \times 60 \text{ min/h} \times 16 \text{ h processing per day} = \$48/\text{day}$$

$$\$48/\text{day} \times 5 \text{ days/week} = \$240/\text{week}$$

Measure the time required to collect a measured volume

A five gallon container marked in one quart increments and a stopwatch can be used to calculate the cost of small flows such as handwash stations, spray nozzles, leaks, hose outputs, etc. Simply measure the time required to collect a measured volume and calculate the cost of the flow.

Example calculation:

A hand wash station delivers 6 qt of $5.00/1000 gal water per minute for 16 h/day. What is its annual cost? 1.5 GPM × 60 min/h × 16 h/day × 260 processing days per year = $1872/year.

Emphasize water use efficiency

Water conservation is a highly effective use of labor. A processor hired a person to conserve water in the processing plant. The conservationist reduced water use by 1.75 gal per bird. At \$3.00/1000 gal, the water and wastewater cost was reduced by \$350,000/year. For each hour the conservationist devoted to water conservation, the processing plant received \$100 in cost savings.

Example calculation:

$$\$350{,}000 \text{ per year cost savings} - \$40{,}000 \text{ per year for salary and equipment} = \$310{,}000$$

$$\text{savings } \frac{\$310{,}000/\text{year}}{3000 \text{ h labor per year}} = \$103 \text{ per hour profit}$$

Waste minimization

Waste minimization is the second aspect of reducing environmental costs and increasing profits. The basic premise of waste minimization is that a processor has high environmental costs because too much product is wasted to the drain. The fundamental calculation to determine product loss is the "pounds" equation.

$$\frac{\text{gallons of wastewater}}{1{,}000{,}000} \times 8.34^{*} \times \text{analysis in mg/l} = \text{pounds}$$

Example calculation:

A processor uses 8 gal of water per bird to process 250,000, 5-lb live weight birds per day. A sample of wastewater collected over a 24-h period contained 3000 mg/l of organic matter. What percent of the live bird was discharged in the wastewater?

$$\frac{2{,}000{,}000 \text{ gal/day}}{1{,}000{,}000} \times 8.34 \times 3000 = 50{,}000 \text{ pounds dry weight chicken}$$

Broilers are approximately 70% water. To convert dry weight organics to live weight, divide pounds dry weight by 0.30.

$$\frac{50{,}000}{0.30} = 166{,}800 \text{ lb of live weight in wastewater}$$

$$\frac{160{,}800 \text{ lb lost to wastewater}}{1{,}250{,}000 \text{ lb live weight processed}} = 13.3\% \text{ of live weight}$$

In a slaughter plant, this 50,000 dry weight pounds is from low value product such as blood, fat, viscera, etc. and could be converted to pet food grade poultry meal. However, it must be turned into even lower value DAF skimmings if the processor is discharging into a municipal sewer. These skimmings make a lower quality poultry meal. If the processor is treating wastewater with a biological system, the processor pays excessive operating costs to treat organic matter to an environmentally stable form and does not even recover the value of rendered DAF float material.

*8.34 = weight of one gallon of water in pounds

- 25 tons of pet food grade poultry meal @ \$400 per ton = \$10,000
- 25 tons of DAF quality poultry meal @ \$180 per ton = \$4500
- Lost value = \$10,000 – \$4500 = \$5500 per day

In further-processing plants, analysis of the wastewater can also identify those times and operations that waste product to the drain. The economic impact of product lost during further-processing can be even more significant than in slaughter operations because further-processing uses higher value products such as meat, oil, and flour rather than lower value such as blood, offal, and feathers.

Using the pounds equation, product loss can be determined:

- One pound of organic matter in the waste stream comes from approximately three pounds of meat
- One pound of nitrogen in the wastewater comes from 31 lb of meat.

Example calculation:

A further-processing plant discharges 250,000 gal of wastewater that contains 150 mg/l of nitrogen.

$$\frac{250{,}000 \text{ gal}}{1{,}000{,}000} \times 8.34 \times 150 \text{ mg/l nitrogen} = 313 \text{ lb of nitrogen}$$

Poultry meat brought into a further-processing plant contains protein, fat, water, and perhaps bone. The only significant source of nitrogen is from protein. To convert nitrogen to dry weight protein, pounds of nitrogen is multiplied by 6.25.

$$313 \text{ lb nitrogen} \times 6.25 = 1894 \text{ lb of dry weight protein.}$$

Because poultry meat is approximately 20% protein, the 313 lb of nitrogen in the waste stream came from 9470 lb of poultry meat. If the average value of white and dark meat is \$1.25/lb, this nitrogen loss represents a loss of \$11,800 of product per day.

All processes have loss, but by measuring product lost to the waste stream and identifying loss situations, processes can be improved so that additional product will be recovered for sale and profits will be increased

Example: A further-processor which produces meat, fat, and broth from spent broiler breeders was experiencing difficulty in meeting environmental discharge requirements. Rather than expanding the wastewater treatment facility, the plant took the conservation/minimization approach to improve efficiency of processors so that more product was recovered for sale and less product was lost in the wastewater. Using this approach, most of the environmental discharge requirements were satisfied. There was also an increase in profitability of slightly more than \$1 million/year due to increased product recovery.

Summary

Poultry processors have to operate with a heightened environmental awareness and increasing regulations about water use and wastewater treatment. Although it needs to be cleaned before discharge, wastewater from a processing plant can provide valuable information on water use and product wastage. Such information can increase product yield while reducing water use and the cost of wastewater treatment.

Four steps are necessary to increase profits by wastewater analysis:

1. Commitment by management. Unless this commitment is made, little will happen to improve the efficiency of the processing or further processing facility.
2. Collect data to determine those times and processes that wastewater and product are discharged into the wastewater stream.
3. Based on collected data, commit personnel and labor to correct inefficiencies.
4. Continue this commitment to process efficiency. Without a continuous and long term commitment by management, little will be done to improve efficiency using wastewater stream analysis.

Selected bibliography

Standard Methods for the Examination of Water and Wastewater, 19th ed., Eaton, A. D., Clesceri, L. S., Greenberg, A. E., and Franson, M. A. H., eds., American Public Health Association, American Water Works Association, Water Environment Federation, Washington, DC., 1995.

Biological pretreatment of poultry processing wastewater, Rusten, B., Siljudalen, J. G., Wien, A., Eidem, D., Grabow, W. O. K., Dohmann, M., Haas, C., Hall, E. R., Lesouef, A., Orhon, D., Van der Vlies, A., Watanabe, Y., Milburn, A., Purdon, C.D., and Nagle, P. T., Water Quality International '98, Part 4, *Wastewater: Industrial Wastewater Treatment*, Elsevier Science, Ltd., Oxford, UK, 1998.

Food-processing waste, Walsh, J. L., Ross, C. C., Valentine, G. E., *Water Environment Research*, 65:6, 402, 1993.

Food-processing waste, Borup, M. B., Muchmore, D. R., *Water Environment Research*, 64:4, 413, 1992.

Food-processing waste, Borup, M. B., Ashcroft, C. T., *Journal of the Water Pollution Control Federation*, 63:4, 445, 1991.

Meat, Fish, and Poultry Processing Wastes, McComis, W. T., Litchfield, J. H., *Journal of the Water Pollution Control Federation*, 61:6, 855, 1989.

Meat-, Fish-, and Poultry-Processing Wastes, Litchfield, J. H., *Journal of the Water Pollution Control Federation* (Literature Review issue), 54:6, 688, 1982.

chapter eighteen

Quality assurance and process control

Doug P. Smith

Contents

Introduction

There are many different definitions of quality. Dictionaries, quality experts, and organizations dedicated to promoting quality all provide excellent but slightly different definitions for this word and concept. Another general definition, as may be applied to poultry processing and the resulting product, could be: someone's expectations and perceptions of how acceptable, even desirable, something is at a given price. The quality of a product is usually measured by the number and types of defects, or lack thereof, set against a background scale of price or perceived value. For example, broiler meat emulsions formed into breaded nuggets for a low cost market may appear to be a good value with adequate quality at one dollar per pound, but the perceived value and quality will not be acceptable to most consumers at three dollars per pound, where formed nuggets compete against whole-muscle nuggets and fillets.

Welcome to the last chapter in this book of poultry meat processing, where the subject is probably the most important yet often least appreciated or overlooked by many processors. Quality, as practiced by many companies in the poultry industry, is sometimes given less credit than it deserves in the overall management plan of poultry processing. Yet quality allows processors to retain the most important part of their entire operation: the

0-8493-0120-3/01/$0.00+$.50

customer. Efforts to upgrade quality may be slowed by some factors, including the evident costs of labor to maintain a quality department, the perceived loss of management control due to customer and regulatory requirements, and the inability or unwillingness to document the amount of money saved by the quality department on the company's Profit and Loss Reports. Poultry companies are becoming more receptive of quality departments as their overall importance to customer retention and problem prevention (especially regulatory issues) is recognized.

This chapter will cover, very briefly, organizational structures of quality departments, types of general quality management systems available, and basic functions of the quality department (as defined by the self-developed quality manual, two current quality programs used in the industry, and current quality issues facing the industry). Many excellent resources already exist pertaining to general quality systems, management techniques for quality improvement, and problem resolution. This chapter is designed to give some specific basics on what is currently in use by poultry processors, plus the overall atmosphere in which these quality systems operate in the industry to provide a realistic view of what a new quality employee (whether entry-level or management) should be prepared for regarding basic systems and their application.

Department organization

All poultry processors now have someone, or more often a group of people, organized into a quality department of some type, usually called quality control or quality assurance. Their duties typically include such diverse functions as inspecting incoming products and ingredients, conducting yield studies, performing HACCP duties (recording cooked product temperatures, etc.), and auditing other personnel and the facility itself for quality standards. A very small poultry processor may have one person that serves several roles, including quality, and a large processor and further-processor may employ more than a hundred quality personnel. Every company, and even each plant within a company, will be organized and will operate slightly differently.

A processing plant usually has a quality manager, a quality supervisor for each shift of operation (and, if the plant is diverse enough, a supervisor for each major type of operation), and one or several employees to conduct routine duties throughout the shift for each area (Figure 18.1). The head of the plant quality department may report to a corporate quality department, corporate sales department, or on-site production manager. Each

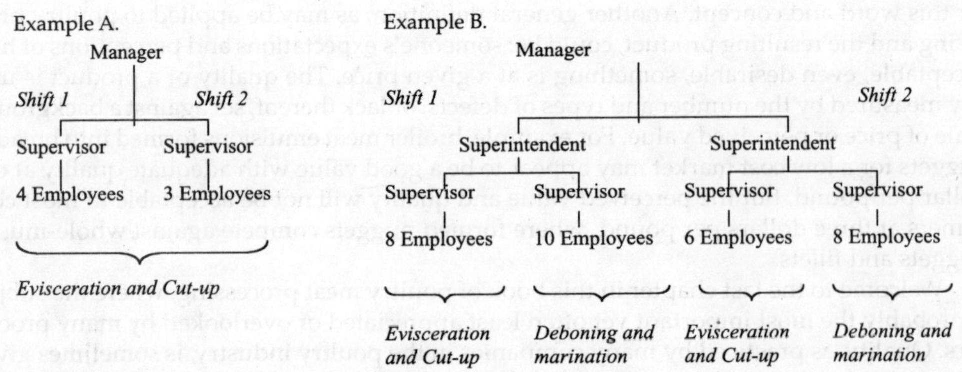

Figure 18.1 Quality department organization typical at a small processing operation (Example A) and at a large processing/further-processing operation (Example B).

reporting structure has advantages and disadvantages. The manager and some supervisors are typically college-educated quality professionals, and some supervisors and most of the employees are either hired from the production lines or are new hires requesting assignment to the quality department. Employees are usually screened using a plant-specific test with basic math and statistics questions. Most companies then train their quality department employees on basic product testing skills and food safety awareness. A few companies have excellent long term training programs, but most use short term on-the-job training. In general, quality department employees receive somewhat higher pay than their production counterparts.

Traditionally an employee is assigned to monitoring one or more production systems or activities deemed important to the processor. The employee normally performs repetitive tasks, manually records data, and reports results to their supervisor as well as production personnel. New quality control computer software combined with handheld computers has allowed a few companies to automate both the data collection and reporting functions. Proper operation of this technology increases the quality and speed of data collection, and can decrease the number of quality employees. Quality of data is important as an inadvertent math mistake on a paper form could result in a large recall the next day, whereas the real-time data gathered by the computer produces fewer mistakes, and quickly alerts plant personnel to occurrences of actual production problems. The total labor force of the quality department is widely variable among plants and companies, but is usually 2% of the total plant employment.

Quality systems

Quality departments at processing companies evolve, usually based on customer and regulatory requirements, into larger, more complex organizations charged with implementing a series of sampling and inspection systems. Each product may have different requirements, but many products plus the overall facility usually operate in a common environment, and many duties standardized into a relatively small number of inspection and auditing functions. For example, a deboning operation will produce many different products, but all product will pass through a standardized inspection process designed to eliminate or minimize bones (to different degrees depending on the specific product specifications), and all production and inspection activities occur in the same room. The deboning plant activities, plus the inspection, sampling and audit functions in the evisceration plant (many of which are required by USDA and the HACCP plan), plus all the other duties performed by the quality department (preoperational facility inspections, incoming product and ingredient inspections, etc.) should be organized and codified into a comprehensive company quality system. Most poultry companies have developed these quality systems on their own over time in response to customer and regulatory requirements.

External quality system templates have also been developed to standardize quality systems within a particular industry. These systems allow quality procedures and practices to be adopted by companies to improve their own self-developed systems and compete nationally and internationally for customers and consumers. Companies that meet certain widely recognized standards have a competitive advantage to market products in other regions or countries, as consumers feel more confident purchasing new products or products from unfamiliar companies if they know the products have met certain standards of quality. The International Organization for Standardization (ISO) is a worldwide group that organizes and standardizes quality requirements. ISO has put together a large number of requirements that companies may voluntarily adopt, then auditors sanctioned by ISO review the companies' efforts and documented performance under ISO standards.

Companies meeting ISO standards are then certified as to the ISO number of the body of standards they have met. Food companies usually attempt the 9000 series for quality systems or the 14000 series for environmental quality. The major benefit to ISO certification is having a comprehensive, objective quality program in place and operational. Externally, it allows companies a better chance to export products with improved consumer acceptance in some countries. The American National Standards Institute (ANSI) does not produce standards, but member groups voluntarily submit standards to ANSI for standardization and compliance. ANSI is the U.S. representative to ISO.

Developed in a different context than the standards for quality systems, general quality management systems provide quality control tools as well as different types of management than the traditional command-and-control/management-by-objective methods. These include Total Quality Management (TQM), zero defect systems, six sigma, and other programs. All of these basically include some statistical process control tools plus prescribed management techniques designed to provide continuous improvement of a plant or process. Of these general systems, only portions of TQM have been implemented to any extent (and with variable success) in the poultry processing industry. In general poultry processors have not adopted TQM systems although they are widely incorporated in other industries. Lack of knowledge is not an issue, as Dr. Fred Benoff published an excellent series of articles in *Broiler Industry* magazine from 1988 to 1992, tailoring specific TQM and process control information to the poultry industry, and many other sources also exist.

The quality manual

Regardless of whether a quality department develops its own system or uses external standards, all of the combined quality department programs, including auditing, inspection, and other functions should be written down and organized into a comprehensive quality manual for use throughout the company. This manual should contain all of the information necessary for the entire quality department's operations, including all areas of the facility, operations, and quality procedures for that site. The manual becomes useful in many other ways, including training for new quality employees. It is usually the first document requested by visiting customers or outside auditors or regulators. It should include forms used by employees, basic product specifications, and instructions on how to conduct audits and inspections for each job assignment. It may be several volumes long and refer to where additional information related to the quality program may be found, such as the location of completed paperwork in a particular filing cabinet in a designated building. Sections that should be included, at a minimum, are pest control program and documentation; liability insurance certificate (usually for minimum of one million dollars); water potability certificate; proof of backflow prevention devices for water lines; letters of guarantee for all ingredients, packaging, and lubricants; calibration program for essential testing equipment (scales, thermometers, etc.); customer complaint file; and, a functioning recall program with documented testing from mock recalls. The HACCP plan and associated good manufacturing practices/sanitation standard operating procedures (GMPs/SSOPs) may be housed in the same general area, but generally are kept separately from the quality manual and associated documentation. Mature and fully comprehensive manuals may also contain additional programs, such as the following: self-inspection reports; incoming goods inspections; waste removal program; air filter cleaning and replacement program; condensation prevention and removal plan; pesticide and chemical storage requirement; metal detection procedures; truck trailer inspection; file of product testing results; grounds maintenance program; and an emergency response plan that includes product testing and

Essential

Comprehensive pest control program	Recall program
Water potability testing certificate	Letters of guarantee
Backflow prevention device verification	Customer/consumer complaints

Recommended

Self-inspection/audit program	Pesticide/chemical storage requirements
Incoming goods inspection program	Metal detection/elimination procedures
Waste removal program	Truck-trailer inspection plan
Air filter cleaning/replacement plan	Product testing results file
Condensation prevention/removal plan	Vendor certification audit reports
Grounds maintenance program	Emergency response plan

Figure 18.2 Components, essential and recommended, to be included in the quality manual.

disposition procedures. Each plant and operation will have different quality manuals and programs (Figure 18.2).

The quality department cannot possibly operate all of these programs at larger facilities, so most departments have evolved into auditing and documenting the programs, plans, and procedures that are conducted in the plant, usually by production, maintenance, and sanitation employees (the most critical checks are still often conducted by quality department employees). The quality department then collects data from its audits and inspections and reports findings to production management. The manual and the quality department itself will be periodically audited and evaluated, as a measure of whether the entire plant is operating adequately to ensure product quality standards on a consistent basis. These auditors include customers that will send their own representatives as well as contracted inspection company inspectors to conduct unannounced audits. Regulators, including USDA compliance teams, and even management within the company may also review and question the quality program practices and procedures. This necessitates an ongoing, flexible, and stringent program review that forces changes in the manual as personnel, production procedures, specification requirements, and other factors change. This process is very similar to the ongoing updates necessary for the HACCP program, but is more extensive due to the much larger compilation of material contained in and addressed by the quality manual.

Inspection systems

The quality department exists for the purpose of monitoring and testing products to ensure that specifications and safety requirements are met. Different tools to perform this function have been developed the past 60 years in response to these needs in industries other than poultry. The two most common forms adopted for use in poultry processing are Acceptance Sampling programs and Process Control techniques. These are very different techniques and approaches for quality control, but both are based on sound science and statistics, and they work very well when used appropriately.

Acceptance sampling

Acceptance Sampling techniques, based on the U.S. Armed Forces Military Standard (Mil Std) 105E inspection program published in 1989, is the predominate form of quality assessment practiced by the poultry processing industry. Many companies still utilize the older version, Military Standard 105D, published in 1963. To save costs the Department of Defense cancelled Mil Std 105E in 1995 and these standards are currently available in the American National Standards Institute (ANSI) document Z 1.4.

This program functions as follows: individual samples, the number and frequency of which are predetermined during program setup, are randomly selected and tested for the appropriate attribute. The samples are counted as pass or fail depending on the specification for that product and specific attribute. The pass/fail criteria is also predetermined during program setup (known as the Acceptable Quality Level for that product, or AQL). The number of fails recorded determines whether that particular batch or lot, in its entirety, will be considered acceptable by the customer or consumer.

For example, if a processor with a deboning operation produces boneless fillets, which are placed in 70-lb capacity plastic totes for delivery to the customer (retail bulk pack or for a further-processor), the customer will require documentation that the product is indeed free of bones or within specifications. A typical Military Standard inspection program may be constructed using the following assumptions: if the average weight of a fillet is 4 oz, a 70-lb tote will contain 280 individuals; if each tote is to be inspected for bones, then each tote becomes a lot. Based on any previous data on customer complaints or in-plant data, the program coordinator must decide if the deboning process is in relative control; if so, then the inspection plan used is normal (tightened plans are for higher numbers of expected defects, and reduced plans for very low numbers of expected defects). Based on this same assumption as well as customer expectations, the Inspection Level may be determined as II (level I is for less discriminatory plans, level III is for highly discriminatory plans). Also, a decision regarding whether a single sampling plan vs. a double sampling will be used depending on labor and space available for inspections. Most processors utilize the single plan to reduce cost and labor. The lot size plus the Inspection Level, and the single plan decision determines the sample size to be taken and tested from each lot. In this example, the lot size of 280 and Inspection Level II equals a sample size of 32 individual samples per lot, based on a single sample plan. That many fillets are to be checked for bone content from each tote. The AQL then must be decided to determine how many fails will be accepted within this sample of 32 to determine if the entire lot fails or passes. An average AQL level used in the industry is 1.0; some strict customers require an AQL of 0.25. At a 1.0 AQL, 0 or 1 bone found per 32 fillets is acceptable and the lot passes, but 2 or more bones found and the lot is rejected (see Figure 18.3). At a 0.25 AQL, 0 bones per 32 fillets is acceptable, but any bones found cause the lot to be rejected, and is essentially a zero defect criteria.

Making honest and valid assumptions when constructing the sampling program and inspection scheme improves the Military Standard program's ability to provide data over

Criteria	Choice
Sampling Plan - Double or Single	SINGLE
Process Situation - Normal, Tightened, Reduced	NORMAL
Batch or Lot Size - numerous choices (2 to 500,000 +)	150-280
Inspection Level - I, II, III	II
Sample Size Code - derived from above choices	G
Sample Size – derived from Code	32
AQL chosen – numerous choices (.065 to 15.0)	1.0
Accept/Reject criteria – derived from AQL	Accept 1, Reject 2

Summary – Based on choices made from program setup, each tote of 280 fillets (approximate) will have 32 fillets removed and inspected for bones; zero or one bone found will pass tote, two or more bones found will fail lot (to be reworked and re-inspected).

Figure 18.3 Military Standard 105E Acceptable Quality Limit (AQL) sampling program setup sequence for a sampling base scheme to inspect totes of deboned fillets for bones and bone fragments.

time regarding the efficacy of the operation, and increases its validity as a means of screening or retesting finished product. Every plan has a risk of error built in, even without human error. The Military Standard program exhibits expected error within each sampling scheme by using Operating Characteristic Curves (OCC), which show, on average, how many acceptable totes will be failed, and how many unacceptable totes will be passed. In this example, if the deboning operation is removing almost all the bones and less than 10% of total totes produced are within limits specifications, the program will be very effective at passing acceptable totes and rejecting unacceptable totes. If more than 10% of the totes are out of specification, the inspection errors increase. This shows that misuse or misunderstanding of this program during setup or failure to understand the OCC will result in products entering the rework cycle at the plant that have no defects (producer risk) or products that are shipped to customers with defects (consumer risk). Producer risks cost money due to space, time, and labor for rework and re-inspection. Consumer risks can be potentially even more expensive (due to market withdrawals, recalls, or loss of customer). Understanding and appropriately applying this program is essential to its success within the capabilities of the program. A number of references regarding the appropriate use of and an understanding of the risks associated with Acceptance Sampling are available and should be consulted before designing a program.[1–3]

Process control

Process control procedures, theories, and applications were developed at approximately the same time as the original Military Standard, but were not then adopted and used by most American industries. W. Edwards Deming incorporated process control into what he developed into Total Quality Management (TQM), which has been used widely in America over the past 20 years. Based on statistical principles, it is applied during the production process in the attempt to provide real time data to allow the process to be corrected before many defects are created. It also uses the discipline that the process will not be corrected unless it is out of control or approaching control limits, which prevents operators from "tweaking" the process based solely on their opinion of how well the process is operating.

There have been many different tools designed to evaluate whether a process is in control. Sampling techniques and analysis tools include X bar charts, R bar charts, histograms, attribute charts, moving average and range charts, process capability charts, Pareto diagrams, and other tools. Using one or more of these enables a process operator the opportunity to take data, analyze it, make a determination of whether the process is in control, then take action, if necessary, to correct the process with little or no defective product produced.

For example, if a further-processor is fully cooking a variable weight and sized split breast product, the end point cook temperature of the product exiting the oven would be considered an important monitoring point, probably even a critical control point in an HACCP plan. The company benefits if it produces all product within specification (greater than 165°F), without any deviation that produces defective (undercooked) product at the maximum production rate. To do this a system can be installed so that an employee checks internal temperatures on five split breasts each 15 min throughout the shift (in reality there would probably be more samples and more frequent checks). Results are recorded and an X bar chart and R bar chart are developed to determine the control limits for this product, based on the oven settings, belt speed, and other production factors to determine if the process is in control and producing acceptable product.

After several sample periods, the average of the readings and the control limits (Upper Control Limit or UCL, and Lower Control Limit or LCL) can be calculated. The average for each subgroup is needed, as well as the overall total sample average (X bar). The range between the highest and lowest number within each subgroup is needed (range), and all of these averaged together equals R bar. Control limits are calculated by the total average (X bar) plus and minus the product of a constant (A2, from Table 18.1) multiplied by R bar. An alternative control limit calculation uses X bar ±3 times the total standard deviation of the samples recorded divided by the square root of the number of samples in the subgroup (UCL sigma and LCL sigma). Either control limit derivation will work and function for process control. After calculations are completed the overall average, control limits, and

Table 18.1 **Process Control X Bar and R Bar Chart Control Limit Constants**

Subgroup size	A2	D3	D4
2	1.88	0	3.27
3	1.02	0	2.57
4	0.73	0	2.28
5	0.58	0	2.00
6	0.48	0	2.00
7	0.42	0.08	1.92
8	0.37	0.14	1.86
9	0.34	0.18	1.82

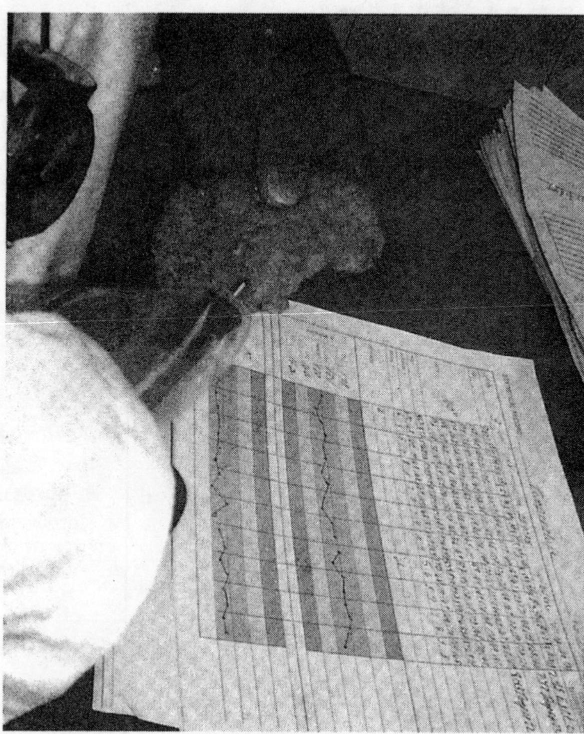

Figure 18.4 A quality department technician checks cooked product temperature and records data on an X bar and R bar "rainbow" control chart.

average of each subgroup are charted. When subgroup averages stay inside the control limits the process is generally considered to be in control. An average outside the control limits, a consecutive series of subgroup averages (6–8) either above or below the average line, or a consecutive series either in incline or decline on the chart is cause for a process correction to return the system to control (according to adjustment guidelines). When all the limit and average lines and subgroup sample readings are graphed on the X and R bar charts, the process is easy to follow and process control is readily determined (Figure 18.4). The control limits, once calculated, can be used for future samplings provided that no major changes have been made to the process. Control limits should be checked and recalculated on a periodic basis (usually weekly or monthly) even without a process change.

In this continuing example of cooked split breasts, internal temperature readings from the first half of the shift, from 8:00 a.m. until 12:00 noon, are shown in Table 18.2, divided into the 15-min intervals with 5 readings each. The subgroup averages are shown, as well as the overall average (182°F). Subgroup ranges are shown and were used to calculate the average range, R bar (10.6). Using a constant from Table 18.1, a subgroup size of 5 equals an A2 value of 0.58. To calculate control limits, 182 + 0.58 (10.6) = 186 (UCL), and 182 − 0.58(10.6) = 176 (LCL). Using an alternative control limit calculation, the overall sample standard deviation is 5.35, and subgroup size is 5, the square root of which is 2.236. Control limits are then 182 + 3(5.35/2.236) = 189 (UCL sigma) and 182 − 3(5.35/2.236) = 175 (LCL sigma).

The X bar, subgroup averages, UCL and LCL are charted on the X bar chart (see Figure 18.5), and the process appears to be in control except at 8:45 a.m. (too low) and at 9:30 a.m. (too high). None of the subgroup averages, nor any individual reading was below the specification limit of 165°F, so no product was held for rework or destroyed. There is room for improvement in the process, however, and a further-processor would probably use this

Table 18.2 **Internal Temperatures of Five Fully Cooked Split Breasts at Oven Exit, Recorded Every 15 Min.**

Time	8:00	8:15	8:30	8:45	9:00	9:15	9:30	9:45	10:00	10:15	10:30	10:45	11:00	11:15	11:30	11:45	12:00
Temp 1	175	190	177	177	178	179	184	187	179	180	181	182	183	175	186	180	185
Temp 2	185	186	187	170	187	188	187	186	189	172	174	181	179	185	183	182	187
Temp 3	180	181	182	172	183	184	189	184	184	185	186	186	188	186	187	181	190
Temp 4	190	170	171	174	172	177	191	183	177	178	179	180	181	175	178	183	186
Temp 5	174	175	176	174	177	178	192	189	182	183	184	185	186	185	179	184	185
Total	904	902	893	867	897	906	943	929	911	898	904	914	917	906	913	910	933
Average	181	180	179	173	179	181	189	186	182	180	181	183	183	181	183	182	187
Range	16	20	16	7	15	11	8	6	12	13	12	6	9	11	9	4	5

Note: $\overline{X} = 181 + 180 + 179 + \ldots + 187/17 = 182$

$\overline{R} = 16 + 20 + 16 + \ldots + 5/17 = 10.6$

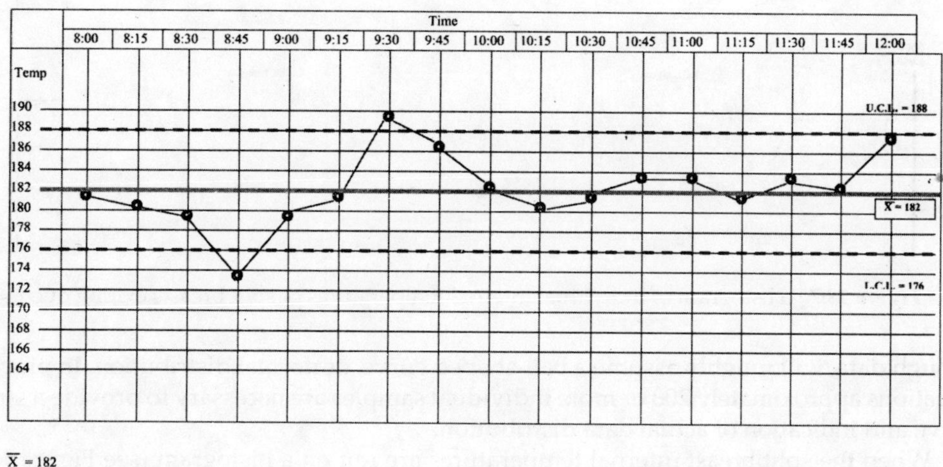

Figure 18.5 X bar chart of internal temperatures of split breasts exiting oven.

data to refine the cooking operation to lower overall product temperatures (to increase yield) yet maintain all product above 165°F to prevent rework or disposal costs, and also to attempt tightening of the deviation between subgroup samples.

For the R bar chart, control limits are determined by R bar multiplied by constants in Table 18.1, D4 for the upper limit, and D3 for the lower limit. The UCL is then 10.6(2.11) = 22; LCL is 10.6(0) = 0. Charting subgroup ranges and the control limits on the R bar chart (Figure 18.6) shows that all ranges are within the control limits. A subgroup range exceeding the control limits would be cause for examination and potential adjustment of the process, similar to the X bar chart adjustment guidelines.

Another graph useful for evaluating data in a process control environment is the histogram. All data collected are assigned to the chart by recording the frequency of occurrence at each level of temperature recording. The subsequent shape, provided there is

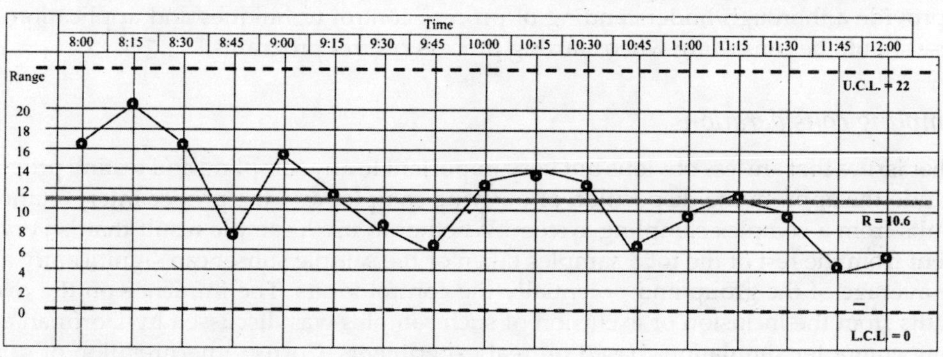

Figure 18.6 R bar chart of internal temperatures of split breasts exiting oven.

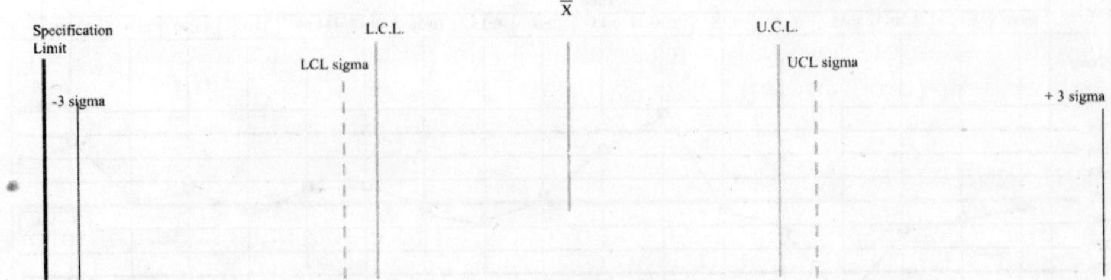

Figure 18.7 Histogram of individual internal temperatures of split breasts exiting oven.

enough data, will roughly assume a bell-shaped curve, or normal distribution. In practical situations approximately 200 or more individual samples are necessary to provide a strong curve and indication of actual data distribution.

When the split breast internal temperatures are put on a histogram (see Figure 18.7), each individual temperature has the number of units with that recorded temperature assigned to it (frequency), as shown by the box above the temperature (one box equals one recorded temperature reading). The average, control limits, the population standard deviation (multiplied by 3, added and subtracted from the average, to provide −3 and +3 sigma lines), and the arbitrary specification limit (165°F) are also on the chart. The distribution of the population can be observed as related to these various criteria. As only 85 observations were charted the actual distribution curve is beginning to take shape but is not yet readily obvious.

Problems with process control as a quality control tool typically occur when a new process is started or when outlying samples are missed when subgroup samples are taken. New processes or existing processes with significant changes usually have more sample variability than existing, stable processes. Some amount of time is needed to set control limits and make necessary adjustments to the process, although special control charts and techniques are available to allow control limits to be set quickly. While out of control, a number of defective individual products will pass through the system. Outlying samples exceeding control and specification limits can occur even in a stable system, and may pass undetected to the customer. At the present time, however, process control is the best system available for producing products of a consistent quality. Several good references exist to provide a thorough understanding of process control techniques and application, and these should be consulting when starting process control programs.[4–7]

Sampling considerations

Other issues that are less obvious but very important to any sampling and testing programs include the appropriate disposition of outlying samples and the proper interpretation of results from a complex sampling system. Whether to use a sample result that is very different from the rest of the total samples taken or the sample subset can significantly affect the average of the group and potentially the control limits. The influence on the overall results from the inclusion or exclusion of such samples was discussed by Dorfman et al.[8] using computer simulations based on real experiments. Correct interpretation of sample results from a program is also important, as Kilsby and Pugh[9] showed that erroneous conclusions are easily made if seemingly obvious factors such as the distribution of defects within a population are not addressed. Flickinger,[10] reporting on a Silliker Laboratory study, similarly stated that certain distribution conditions of defects in products could lead to difficulty in detecting defects and resulting in a potential risk to consumers. Therefore, properly handling outlier samples and awareness of the varied nature of distributions of defects within a sample population are essential to the proper application of any sampling and testing quality program.

Current quality issues

There are many quality problems, issues, and concerns processors must handle on an ongoing basis, plus the addition of new challenges every year. Physical, chemical, and microbiological hazards can occur in processed poultry products, but most quality issues are physical in nature. Problems most commonly encountered are the presence of contaminants or that the physical appearance of the product is not within the customer's specification or expectation. Physical contaminants include bones (in boneless items) and foreign material (plastic, metal, wood, rock, etc.). Physical appearance problems refer to color, flavor/odor, size (length, width, or thickness), weight, or shape (misshapes) of an item. Product color is a crucial quality issue, is extremely subjective, and can be affected by many factors. The inherent color (or discoloration) of the meat itself, ingredients that cause color (browning agents, sugars, and colorings) used in incorrect amounts, processing errors (using old or unfiltered oil for fried products), temperature abuse, and refreezing of fried frozen items (causes severe darkening of the breading after reconstitution) are some color problems. Flavor and odor are closely connected and are chemical problems, but defects usually result from physical mishandling of the product, such as improper meat aging prior to deboning (meat texture and dryness), improper ingredient addition or formulation, improper storage times or temperatures (causing drying, and accelerating the oxidation process or rancidity), and migration of flavors from adjacent products.

Size, weight, and shape issues usually result from improper processing and specification deviations at the plant. Thickness can be a critical problem for certain customers as their operations require a product that can be cooked to 165°F in the minimum time possible. Thicker than expected products may result in undercooked items at the retail level. Larger than normal product portions, for restaurants that buy by weight and serve by the number of pieces per entrée, cause monetary losses for the restaurant. Occasionally product mishandled and broken during shipping causes size and shape problems. Shipping damage is relatively easy to diagnose but more difficult to resolve. Ongoing problems on multiple carriers indicate a need for product, or more likely, a packaging redesign.

Other physical problems that lead to customer/consumer complaints are missing or duplicated giblets in whole fresh carcasses, poor coating adhesion or inclusion of crumbs (for breaded items), inclusion of fat chunks (for raw, frozen items), improper packaging (such as zip lock bags that do not function), and pockets of hot oil contained in fried products after preparation. These and many other problems not listed affect the poultry processing industry, and present ongoing challenges to quality departments.

Certain quality issues affect the entire industry, and although they are well known within the industry, have not been well quantified or categorized. These are important as they form the basis for quality perceptions from customers and consumers nationally and internationally (and often result in stricter customer specifications, tighter government regulations, and tougher regulatory enforcement). Knowledge of these larger issues also benefits quality departments, as they can better focus their efforts on these problems. To determine the most important quality issues facing the industry, data can be collected from many disparate sources over time, such as the information presented in Table 18.3. Recalls for chicken products officiated by the USDA, major U.S. retail chain customer concerns, insurance claims filed against a processor/further-processor, and nationwide consumer complaints against chicken products filed with the USDA illustrate the types of

Table 18.3 **Current Quality Concerns for Poultry Processors Classified by Product Recalls, Customer Quality Issues, Consumer Insurance Claims, and Consumer Complaints During Various Years from 1990 to 1998**

Regulatory — USDA recalls[1]		Customers — quality issues[2]		Consumers — insurance claims[3]		Consumers — complaints[4]	
1	Pathogens	1	Bone	1	Bone	1	Illness
2	Plastic	2	Pathogens	2	Metal	2	Foreign material
3	Underprocessed	3	Breading	3	Illness	3	Bone
4	Metal	4	Portion control	4	Foreign material	4	Metal
5	Bone	5	Foreign material	5	Glass	5	Plastic
6	Spoilage	6	Bird/breast size	6	Insect	6	Insect
7	Undeclared substance	7	Sensory/flavor	7	Wood	7	Glass
8	Chemical/drug	8	Water absorption	8	Rock	8	Allergen
9	Hepatitis A	9	Redness/pinking	9	Plastic		
10	Miscellaneous	10	Process control	10	Gristle		

[1] USDA recalls involving chicken products, 1990 through 1998.
[2] Survey of 11 major fast food and grocery chain poultry retailers, 1998.
[3] Claims filed against chicken processor/further-processor, 1992–1994.
[4] USDA Consumer Hotline chicken product complaints, 1996 to 1998.
Source: Adapted from Smith, D. P., Know your quality, *Broiler Ind.*, 62(7), 22, 1999.

quality problems associated with chicken products on a large public scale.[11] The predominate complaints were bones, microbiological contamination (either fear of, or alleged or actual occurrence of pathogens), and foreign material contamination. Concerns not traditionally associated with chicken that were listed included alleged hepatitis A and allergen contamination. Depending on the source, reported concerns differed among the data sources as may be expected. Retail chain customers included more of the cosmetic quality defect issues pertaining to product appearance, including: breading (color and adhesion), portion control and breast size to bird weight ratio (size and shape), red discoloration or pinkness of cooked product, sensory perception (especially flavors, off-flavors and apparent dryness), and process control. These retail chain customers felt processors should increase process control capabilities throughout production to manufacture products more consistent to all stated specifications. Consumers put more emphasis on obvious physical contaminants and illness from pathogens. Another source of data not listed in Table 18.3 provides a quality department with perhaps its best source information — the customer and consumer complaint file. This information, properly quantified and interpreted, is a valuable tool for focusing attention and effort on tangible quality improvements.

Conclusion

The quality department must be able and equipped to handle ongoing production operations as well as meet new challenges presented by new processing technology, government regulations, and customer requirements. To accomplish this the department must be dynamic, flexible, well-trained, and positively supported by the highest management levels within a company. Accordingly, poultry processors are beginning to realize the importance of quality departments not only to their profit margins but the very existence of the company. Although still not well developed at all companies, the industry is beginning to promote its noteworthy quality accomplishments, which bodes well for the future of quality departments and their employees. This, in turn, will sustain the poultry processors dominance within the meat industry and satisfy the consumer's need for safe and wholesome products.

References

1. Dodge, H. F. and Romig, H. G., *Sampling Inspection Tables,* John Wiley & Sons, New York, 1944, chaps. 1 and 2.
2. Guenther, W. C., *Sampling Inspection in Statistical Quality Control,* MacMillan, New York, 1977.
3. Schilling, E. G., *Acceptance Sampling in Quality Control,* Marcel Dekker, New York, 1982.
4. Hubbard, M. R., *Statistical Quality Control for the Food Industry,* Van Nostrand Reinhold, New York, 1990, chap. 3.
5. Ledolter, J. and Burrill, C. W., *Statistical Quality Control: Strategies and Tools for Continuous Improvement,* Wiley, New York, 1999, chap. 12.
6. Puri, S., Ennis, D., and Mullen, K., *Statistical Quality Control for Food and Agricultural Scientists,* G. K. Hall, Boston, 1979, chap. 6.
7. Derman, C. and Ross, S. M., *Statistical Aspects of Quality Control,* Academic Press, San Diego, 1997, chaps. 5 and 6.
8. Dorfman, J. H., Pesti, G. M., and Fletcher, D. L., Searching for significance: the perils of excluding pseudo-outliers, *Poult. Sci.,* 72, 37, 1993.
9. Kilsby, D. C. and Pugh, M. E., The relevance of the distribution of micro-organisms within batches of food to the control of microbiological hazards from foods, *J. Appl. Bacteriol.,* 51, 345, 1981.

10. Flickinger, B., Quality communication: breaking down barriers with better data, *Food Qual.*, 1(3), 14, 1995.
11. Smith, D. P., Know your quality, *Broiler Ind.*, 62(7), 22, 1999.

Selected bibliography

Books

Juran's Quality Control Handbook, 4th edition, J. M. Juran and F. M. Gryna, Eds., McGraw-Hill, New York, 1988.

Quality Assurance, R. C. Vaughn, Iowa State University Press, Ames, 1990.

Quality Control and Industrial Statistics, A. J. Duncan, Richard D. Irwin, Homewood, 1974.

Quality Control and Statistical Methods, 2nd edition, E. M. Schrock, Reinhold, New York, 1957.

Statistical Methods in Quality Control, D. J. Cowden, Prentice Hall, Englewood Cliffs, 1957.

Statistical Process Control and Quality Improvement, 2nd edition, G. Smith, Prentice Hall, Englewood Cliffs, 1995.

Statistical Quality Control, E. L. Grant and R. S. Leavenworth, McGraw-Hill, New York, 1980.

Total Quality Assurance for the Food Industries, 2nd edition, W. A. Gould and R. W. Gould, CTI Publications, Baltimore, 1993.

Periodicals

Food Quality, 208 Floral Vale Boulevard, Yardley, Pennsylvania, 19067.

Quality Progress, P. O. Box 3005, Milwaukee, Wisconsin, 53201.

Journal of Quality Technology, University of Florida, Department of Statistics, 116A Griffin-Floyd Hall, Gainesville, Florida, 32631.

Quality standards organizations

General

American National Standards Institute, Inc., (ANSI), 11 West 42nd Street, New York, NY, 10036.

American Society for Quality (ASQ), 611 East Wisconsin Avenue, Milwaukee, Wisconsin, 53201.

International Organization for Standardization (ISO), 1, rue de Varembe, case postale 56, CH-1211 Geneve 20, Switzerland.

Food

American Institute of Baking (AIB), P. O. Box 3999, Manhattan, Kansas, 66505.

ASI Food Safety Consultants, Inc., 7625 Page Boulevard, St. Louis, Missouri, 63133.

Index

A

B

C

F

G

H

I

J

O

P

Q

R

S

T

U